# Springer Biographies

More information about this series at http://www.springer.com/series/13617

L. J. Reinders

# The Life, Science and Times of Lev Vasilevich Shubnikov

Pioneer of Soviet Cryogenics

 Springer

L. J. Reinders
Leiden University
Leiden
The Netherlands

and

Panningen
The Netherlands

ISSN 2365-0613          ISSN 2365-0621    (electronic)
Springer Biographies
ISBN 978-3-030-10156-5          ISBN 978-3-319-72098-2    (eBook)
https://doi.org/10.1007/978-3-319-72098-2

Printed on acid-free paper

This Springer imprint is published by Springer Nature
The registered company is Springer International Publishing AG
The registered company address is: Gewerbestrasse 11, 6330 Cham, Switzerland

*To Jury Ranjuk who died at age 82 on 9 February 2017 and whose efforts have done much to preserve the memory of the UFTI scientists who fell victim in Stalin's purges.*

# Preface

The current book has a rather unorthodox history. When following the Master's programme in Russian and Eurasian Studies at Leiden University as a mature student from 2012, I was searching for a topic for my thesis whereby my earlier career as a physicist might be useful. In this search, I stumbled on Lev Vasilevich Shubnikov, a physicist from Russia, who in the late 1920s had spent 4 years at the Physics Laboratory of Leiden University, at that time the most advanced laboratory for low-temperature physics in the world. At Leiden he became the co-discoverer of the Shubnikov-De Haas effect, the first observation of quantum-mechanical oscillations of a physical quantity, in this case the resistance of bismuth, at low temperatures and high magnetic fields. I had heard about the effect, was aware of its importance, but that was about all. Shubnikov turned out to be an excellent topic for my Master's thesis and I used him as the guiding figure in a description of the development of fundamental physics in the Soviet Union in the 1930s. The research for this thesis set me on a course through mainly Russian language literature about Soviet fundamental physics and about Shubnikov's life and his physics. It gradually dawned on me that Shubnikov's work in physics, not only at Leiden, but especially at the Ukrainian Physico-Technical Institute (UFTI) in Kharkov, where he founded the first low-temperature laboratory in the Soviet Union, was of such high level that he deserved a proper (scientific) biography. Shubnikov's efforts and those of the Soviet Union's greatest theoretician Lev Landau, who was one of Shubnikov's best friends and from 1932 also worked at UFTI, in less than 5 years turned UFTI into the foremost physics institute in the Soviet Union, and one of the most important low-temperature laboratories in the world, where a number of important discoveries were made, among which most notably the discovery of type-II superconductivity.

In 1937 disaster struck Shubnikov and many of his colleagues at UFTI when the NKVD, a forerunner of the KGB, decided to launch an assault on UFTI, virtually destroying the laboratory. Shubnikov was one of the scientists to be executed, and his tragic end at the hands of Stalin's ruthless security services resulted in his name and work almost being forgotten in the wider physics community until it was revived in 1957 by Alexei Abrikosov (1928–2017), who mentioned Shubnikov's 1937 paper in his seminal work on the theory of type-II superconductivity. But the details of his life, the events that led to his arrest and execution, and the astonishing range of research topics he was working on have remained virtually unknown to physicists in Western countries. I hope that this book will be able to fill that void.

Shubnikov was an experimental physicist who contributed fully to the many exciting developments following the formulation of quantum mechanics in the 1920s. The physics in the book is kept to a minimum, no formulas or calculations of any kind, just (simplified) descriptions of Shubnikov's most important work. The story of this book should be accessible to everybody with an interest in the history of Soviet science and especially the development of Soviet physics in the 1930s, but also to anybody interested in Soviet history, in the writings of which science only plays a meagre role.

Jury Ranjuk (1935–2017) (picture taken by the author in June 2016 at the UFTI building)

As regards the dramatic events in Kharkov in the late 1930s, the book is unique in the sense that it relies both on detailed knowledge of what took place there at that time which was the reason for the NKVD to clamp down on the institute, and on extensive documents from the KGB archives. The latter became available in the early 1990s, mainly through the efforts of the Ukrainian nuclear physicist Jury Ranjuk, who unfortunately passed away early in 2017 and to whom this book is dedicated. Professor Ranjuk worked all his life at UFTI, and without his efforts the details of the reasons why Shubnikov and other outstanding UFTI physicists were purged in the 1930s would probably still be largely unknown. When I wrote to him about my plans for a biography of Shubnikov he almost immediately invited me to come to Kharkov and I was fortunate to be able to visit him and the institute in June 2016, about half a year before his death. I hope that this book will contribute to restoring Shubnikov and his colleagues to their proper place in the history of physics.

Leiden\Panningen, The Netherlands                                           L. J. Reinders

# Acknowledgements

I am very grateful to Prof. Van Lunteren for offering me a guest research position at Leiden University, which provided me with access to the library and other university facilities, which were indispensable for the writing of this book. I also thank the members of the Russian department at Leiden University who tolerated me in their midst during both the Bachelor and the Master's programme, in particular the then head of department Dr. Otto Boele and my thesis advisor, for both the Bachelor and Master's thesis, Dr. Andries van Helden.

For the research of this book I contacted various archives in Germany, Finland and the United States. The employees at these archives were all very helpful and I am especially grateful to Mrs. Raija Ylönen-Peltonen of the Finnish National archives who managed to dig up a large amount of material pertaining to Shubnikov's sailing trip across Lake Ladoga in 1921. This material has been vital in clearing up this obscure event in Shubnikov's life.

I am also grateful to Dr. Sergei Lebedev and Dr. Mikhail Volkov of the Ioffe Physics Institute in St. Petersburg, who were very helpful in obtaining information on Shubnikov's connection with the Ioffe Institute. I thank Dr. Misha Gryaznevich for using his contacts at the Ioffe Institute to my advantage. Special thanks go to the Ukrainian Physico-Technical Institute, and in particular to the late Prof. Ranjuk and to Prof. Sokolenko, who received me most cordially during my visit in June 2016.

Finally, I have to thank my wife, Mayke Visser, for tearing apart a first, long-winded version of this book which set me on course for the version that is now in front of you, and Kate Bellamy for a thorough proof-reading of the manuscript.

# Contents

# Abbreviations

| | |
|---|---|
| CNIGRI | Central Research Institute of Geological Prospecting (*Центральный научно-исследовательский геологоразведочный институт*) |
| Glavnauka | Main Scientific Administration of Narkompros (*Главное управление научными, научно-художественными и музейными учреждениями*) |
| Gosplan | Central State Planning Commission (*Государственный комитет по планированию*) |
| LFTI | Leningrad Physico-Technical Institute |
| Narkompros | People's Commissariat for Education (*Народный комиссариат просвещения*) |
| Narkomtjazhprom | People's Commissariat of Heavy Industry (*Народный комиссариат тяжёлой промышленности*) |
| NEP | New Economic Policy, 1921–1929, mixed state-market economic system, introduced by Lenin |
| NKVD | People's Commissariat of Internal Affairs (*Народный комиссариат внутренних дел*) |
| OGPU | Joint State Political Directorate (*Объединённое государственное политическое управление при СНК СССР*) |
| OSGO | Deep-cooling research station (*Опытная станция глубокого охлаждения*) |
| Phys. Z. Sow. | Physikalische Zeitschrift der Sowjetunion |
| Sovnarkom | Council of People's Commissars (*Совет народных коммиссаров*) |
| UFTI | Ukrainian Physico-Technical Institute |
| VAK | Supreme Certifying Commission (*Высшая Аттестационная Комиссия*) |

| | |
|---|---|
| Vesenkha | Supreme Soviet of the National Economy; All-Union Economic Council (*Высший Совет Народного Хозяйства, ВСНХ*) |
| VIET | Voprosy Istorii Estestvoznanija i Techniki (*Вопросы истории естествознания и техники*) |
| Z. Physik | Zeitschrift für Physik |
| ZhETP | Zhurnal Eksperimental'noj i teoreticheskoj fiziki (*Журнал Экспериментальной и Теоретической физки*) |

# Chapter 1
# Introduction

The Soviet low-temperature physicist Lev Vasilevich Shubnikov lived from 1901 to 1937. The year of his death, 1937, already indicates that he was a victim of the infamous Stalinist purges of the late 1930s. Nevertheless, in this short span of just 36 years during a very tumultuous and cruel time he became a pioneer of low-temperature physics in the Soviet Union, founded the first cryogenic laboratory in that country, made significant contributions to this branch of physics and lived a rather adventurous life with a very sad and dramatic end. Shubnikov was a talented physicist and organizer of physics, and his fate, so unnecessary and wasteful, is illustrative of the callousness of a regime that did not tolerate the slightest independent thinking among its citizens and ruthlessly destroyed anyone who showed, or threatened to show, even the smallest deviation from the path thought out by Stalin and the party. Apart from Shubnikov, many more physicists were killed, imprisoned or sent to labour camps in those years, but among them Shubnikov was without doubt the most prominent and promising.

Attention to the oppression of natural scientists by the Soviet regime is dominated today by the case of Trofim Lysenko's pseudoscience in genetics [1], and it is generally thought that the other natural sciences escaped relatively unscathed, with regard to both physical and ideological oppression. It is indeed true that Marxist ideology had relatively little impact on physics, chemistry and mathematics, although the 1930s saw a rather vicious debate between Marxist pseudo-philosophers[1] and physicists about dialectical materialism on the one hand and the 'idealistic deviations' of quantum mechanics and relativity theory on the other. This debate also played a role in the story that will unfold in this book. At the end of 1936, it led to a failed attempt to dismiss Lev Landau, Shubnikov's friend and colleague at the Ukrainian Physico-Technical Institute (UFTI), as professor at Kharkov University on the accusation of 'idealism'.

---

[1]Most philosophers of any significance that had survived the revolution and civil war had been sent packing by Lenin in the early twenties (see L. Chamberlain, *Lenin's Private War: the Voyage of the Philosophy Steamer and the Exile of the Intelligentsia* (St Martin Press, New York, 2006)) and those that were left were definitely of a lesser calibre and faithful followers of the party line.

© Springer International Publishing AG 2018
L. J. Reinders, *The Life, Science and Times of Lev Vasilevich Shubnikov*,
Springer Biographies, https://doi.org/10.1007/978-3-319-72098-2_1

It caused other UFTI scientists, including Shubnikov, who occupied positions at Kharkov University, to hand in their resignation. This action which was dubbed an 'anti-Soviet strike' and played a minor role in the cases against Shubnikov and other physicists constructed by the NKVD, the People's Commissariat of Internal Affairs, the precursor of the KGB.

The philosophical debate on the compatibility of quantum mechanics and relativity theory with dialectical materialism died a rather sudden death when, after World War II, Soviet physics was mobilised to build the atomic bomb, for which dialectical materialism offered little practical assistance. But that was later. The late 1930s actually saw a massive onslaught on the physical and other sciences. Astronomy was greatly affected with the virtual destruction of the Pulkovo observatory, which also had a considerable impact on physics in Leningrad. Other Leningrad physicists were swept away in purges taking place among geologists, or were targeted for personal reasons. But, as will be set out in this book, the most systematic and savage attack on physics had to be endured by the Ukrainian Physico-Technical Institute. When Stalin's henchmen were finished with it, its status as a world-class physics laboratory and the leading physics establishment in the Soviet Union, built up in less than a decade with Shubnikov and Landau as its most prominent members, was virtually destroyed.

Apart from his scientific papers published in national and international physics journals, very little has survived written by Shubnikov himself; no letters, diaries or other documents, just a few pages from laboratory notebooks. Some correspondence from his Leiden days and with scientists in Leiden (Netherlands) after he had already moved to Kharkov in Ukraine was confiscated by the secret police upon his arrest in 1937 and never returned to his family. These papers remain hidden in the currently inaccessible archives of the KGB/FSB. One of the consequences is that Shubnikov's biography must be compiled almost exclusively from external sources, from official documents that became available in the early nineties through the efforts of some Ukrainian scientists, in particular Jury Ranjuk, and from the reminiscences of his wife Olga Nikolaevna Trapeznikova and of friends and colleagues. Fortunately there is quite a lot of such material, in the form of articles and books of which the following deserve special mention: a memorial volume published in 1990 by the Physico-Technical Institute of the Ukrainian Academy of Sciences in Kharkov [2] and the book by Ju.V. Pavlenko, Ju.N. Ranjuk and Ju.A. Khramov [3] on the repression at UFTI, published in 1998. The latter book includes a large number of documents from the KGB archives, which were declassified in the early 1990s. Both books are in Russian and large parts, especially documents of the NKVD cases against Shubnikov and other physicists, have been translated by the author and reproduced in the appendices to the current volume. They form fascinating reading and provide a unique insight into the workings of the NKVD apparatus, about the way in which they let their victims construct their own cases along the lines of a seemingly predetermined script, and were meant to lend a semblance of justice to the whole proceedings.

Shubnikov's scientific activity divides itself naturally into three periods: a Petrograd/Leningrad period from his enrolment at the university in Petrograd in

September 1918 until his departure for Leiden in November 1926, a Leiden period at the Leiden Physics Laboratory from 1926 to 1930, and a Kharkov period at UFTI from 1930 until his death. The Leningrad period was interrupted by an extended stay abroad from 1921 to 1923 after a sailing trip that ended up in Finland. His scientific work after returning from this trip in 1923 was devoted to growing single crystals and investigating deformations in crystals. In the Leiden period this experience was used for preparing (for that time) extremely pure single crystals and applying these in investigating the magnetic resistance of bismuth at low temperatures. This led to the discovery of the Shubnikov-De Haas effect, the first observation of quantum mechanical magnetic oscillations of a physical quantity in metals at low temperatures. In Kharkov he built up and headed the first low-temperature laboratory in the Soviet Union with a world-class scientific programme comprising a large number of themes: superconductivity, liquid helium, low-temperature magnetism, solidification of gases, nuclear physics, and applied technical research. It was the high point of his career and of the Ukrainian Physico-Technical Institute where he worked.

In Kharkov, Shubnikov and co-workers discovered what is now known as type-II superconductivity or the Shubnikov phase. Type-I superconductivity, discovered in 1911 in Leiden by the Dutch physicist Heike Kamerlingh Onnes, is characterised by the fact that the superconductivity is abruptly destroyed when the strength of an applied magnetic field rises above a critical value. Type-II superconductors are generally characterised by two critical fields, whereby an increase of the applied magnetic field above the first critical field leads to a mixed state in which an increasing amount of magnetic flux penetrates the material, while the electric resistance remains zero as long as the current is not too large. The superconductivity is only destroyed after increasing the applied magnetic field above a second critical value.

In addition, Shubnikov and his team are credited with the discovery of nuclear paramagnetism and, independently of and almost simultaneously with Meissner and Ochsenfeld, they observed the complete diamagnetism of superconductors (the Meissner effect).

The range of subjects explored at his laboratory at UFTI was unusually broad. In fact in the first years of its existence a whole series of very important lines of research was developed, almost fully comprising what is now called low-temperature physics, so by 1934 the UFTI cryogenic laboratory had become one of the leading cryogenic centres in the world. At the end of this book a complete list of Shubnikov's scientific papers, 62 in total, published in just over 10 years, has been included. They are witness to the astonishing diversity of his scientific interests. The last six papers of 1938 and 1939 do not have Shubnikov as a co-author as they were published after he had been declared an 'enemy of the people' and his name could no longer be mentioned in the Soviet Union.

In 1937 Shubnikov was arrested and, following an inquiry by Stalin's repressive organs that in no way resembled a fair process, although some pretence of legality was maintained, he was executed for no other reason than that he had shown evidence of independent thought. This murder, along with that of some of his colleagues at UFTI, delayed the appreciation of Shubnikov's work until the mid-1950s when Alexei Abrikosov quoted his work in his seminal paper on the theory of type-II superconductivity. The subject of type-II superconductivity had not been taken up anywhere else after Shubnikov's discovery in 1937 due the fact that pure alloys, in which it typically occurs, were difficult to prepare. After Abrikosov's theory it still took five years to prepare sufficiently pure alloys to test the theory; alloys that had however already been available to Shubnikov and his co-workers in 1937. As Kurt Mendelssohn states: "*of the laboratories engaged in low temperature research in the thirties Shubnikov's group in Kharkov had evidently the best metallurgical know-how*" (Ref. [4], p. 209), and this several decades before others managed to get similar know-how.

This book is first and foremost about Shubnikov, who deserves to be honoured with a proper biography, so that future books about Soviet physics will not repeat the glaring absurdity that "*Landau's low-temperature laboratory was the first in the Soviet Union to work with liquid helium*" (Ref. [5], p. 254). It was Shubnikov who headed this laboratory and liquified helium in Kharkov at the end of 1932. At that time Landau had only just arrived in Kharkov and had no involvement whatsoever in this achievement. But it is also about Soviet physics in general, about the upsurge in physics research activity in Russia after the Russian revolution, which almost coincided with the revolution in physics exemplified by quantum mechanics and relativity theory. And it is in particular also about the repression which struck physics rather viciously at the end of the 1930s.

After the devastation of World War I, the revolutions of 1917 and the ensuing civil war, the 1920s had become an exhilarating time for Russian physics as it was for the whole of physics. Quantum mechanics and relativity theory had brought a completely new understanding of the natural world and opened up exciting possibilities for research in atomic and nuclear physics. With help from the West, many young Russian physicists obtained plenty of opportunities to take part in these developments, by spending time abroad in Göttingen, Copenhagen, Leiden, Cambridge or elsewhere in Western Europe. A number of them became real specialists in the new theories and made many significant contributions.

Near the end of the twenties the climate in the Soviet Union began to change. The 'gay' twenties were over, Lenin's New Economic Policy (NEP) had ended, Stalin's cultural revolution was in full swing to rid the country of the last vestiges of bourgeois influence, in particular also of 'bourgeois' science, and the country was labouring to try to fulfil the first five-year plan. Backward Russia needed to be modernized at lightning speed and accomplish in a decade what the West had done in centuries. Travel to and from the West too virtually came to a standstill. Everything and everybody had to be mobilised for the Great Break (*Veliky Perelom*). Some physicists and other scientists were aware of the sinister turn events were taking and preferred to leave the country while this was still possible. Examples are the important chemist Vladimir Nikolaevich Ipatiev (1867–1952), who because of his background as a typical 'bourgeois' scientist had reason to fear that arrest would be imminent and did not return home after visiting a conference in Munich in 1930, and the Leningrad physicist George Gamow (1904–1986) who in

1933 after several failed attempts succeeded in getting exit visas for himself and his wife to attend the Solvay conference in Brussels, from which he likewise did not return. But some were still abroad at the end of the twenties like Lev Vasilevich Shubnikov in Leiden in the Netherlands and Pëtr Leonidovich Kapitsa (1894–1984) in Cambridge (England), who both returned to Russia in the 1930s, Shubnikov voluntarily after a stay of four years in Leiden, Kapitsa involuntarily after he was prevented from returning to Cambridge when making one of his regular visits to the Soviet Union in 1935.

In many respects Shubnikov's career reflects the capricious nature of developments in the Soviet Union from the uncertainties and suffering of the revolutionary years and the ensuing civil war, the years of relative freedom in the twenties with Lenin's New Economic Policy and their tumultuous ending with Stalin's cultural revolution, the drive for industrialization and the first five-year plan, followed by the stifling, often deadly oppression at the end of the thirties.

Shubnikov was one of the first to enrol at university after the 1917 October Revolution and personally lived through the hardship in the early years of the Bolshevik experiment. He then disappeared from the Russian stage for a few years, returned to Petrograd to finish his studies and in 1926 embarked on an extended *komandirovka* (official business trip) to Leiden, returning in 1930 to go to Kharkov in Ukraine, where he set up the first cryogenic laboratory in the Soviet Union. He and Lev Davidovich Landau (1908–1968), arguably the greatest Russian physicist that ever lived and a close friend of Shubnikov, were jointly responsible for the great prestige, both nationally and internationally, that the Ukrainian Physico-Technical Institute in Kharkov built up in just a few years from the early 1930s onwards. The foundation of UFTI, its history and Shubnikov's work will be discussed in detail in this book.

The remainder of the book is devoted to the onslaught on UFTI by Stalin's repressive machine, as well as to the why and how of this repression. It struck UFTI particularly heavily with all department heads, save one, being arrested and several of them executed. Shubnikov's laboratory and Landau's theory group, the two most prestigious departments of the institute, were devastated. Landau fled to Moscow where the NKVD caught up with him a year later, while Shubnikov was arrested and a few months later shot dead.

The question that will take prime place in this last part of the book is why and how this happened. Was the repression that devastated the physics community in the thirties random, as is often argued, something that could happen to anybody as a bolt from the blue, or can at least some of that repression, although unnecessarily brutal and completely out of proportion, be viewed as a more or less logical step in the stage of development Soviet society found itself, where the regime demanded absolute loyalty of its subjects? One of the main conclusions will be that it was not random, but that the NKVD had carefully planned the destruction of in particular UFTI as it was perceived to be a hotbed of anti-Soviet activity, i.e. of independent thinking. It will be argued that in the paranoid atmosphere of the 1930s it did indeed have reason to hold that view as the physicists at UFTI failed to show the absolute loyalty the regime demanded. It is in no way intended to give the impression that the accused were guilty of any 'crime' or that it was their own fault, but is an attempt at giving an explanation why precisely these people were selected by the NKVD and not others. Their arrests were not random occurrences, but they were carefully chosen, as the NKVD thought to have reason to suspect their loyalty to the regime. And it must be said that they did this skilfully. Afterwards neither UFTI, nor any other physics laboratory, ever again gave problems.

There were two conflicts that arose in a rather short time. The first one occurred at Kharkov University where many of the UFTI scientists, including Landau and Shubnikov, taught. It had chiefly to do with Landau's teaching which largely went over the heads of the students, who in many cases lacked the proper training. It also had an ideological aspect as Landau and Shubnikov were accused of allowing attacks on dialectical materialism. However, as is clear from the documents of Shubnikov's case, in the actual NKVD inquiry ideology hardly played a role. The second conflict, which unfolded at UFTI and led to the charge of sabotage, was much more decisive for the fate of the UFTI scientists. The conflict arose over the direction research at the institute was supposed to take, whether the institute would engage in applied and defence-related research or be purely devoted to fundamental physics. It was blown out of proportion by the intransigent attitudes of various actors in the drama, leading to deadly consequences when in the middle of 1937 the NKVD decided to go over to action.

A large number of the best scientists were imprisoned, sent to camps or special research institutes, where they were essentially forced labourers, or even executed. Those who remained at liberty were subjugated, their institutes not run by scientists, but by Communist party administrators. Did science suffer? It undoubtedly did, but the regime did not care. They saw science as just another tool in reaching their objectives, whereby it was assumed that nobody is irreplaceable. A certain individual might be more useful than others for reaching the aims the party and the government had set, but if he or she caused trouble or threatened to cause trouble the vast mass of Soviet workers would always yield another one who would be able to do what was expected of him or her. And it seems that they were not far wrong. The scientific achievements of Soviet scientists in certain fields were first class. In this connection it is often argued that science is 'democratic' and that the free exchange of ideas between scientists all over the world is essential or in any case the best guarantee for science to prosper. That may indeed be so, but there is no reason why such exchange should necessarily take place at a global level. Once a country is large enough a proper development may be perfectly possible within a single country, certainly when that country has access to scientific developments that take place elsewhere, in the rest of the world, without revealing its own achievements. Like 'socialism in one country', 'science in one country' is also easily imaginable. The scientific enterprise may be democratic in the sense that every scientist is listened to and that one's reputation does not imply immunity from scrutiny, but once the 'truth' (or a partial truth) has been established it acts as a dictator sweeping away everything that is considered contrary to that truth. The important point in this respect is of course that such truth must be established by the scientists themselves on the basis of rational criteria and/or experimental facts, within the community of scientists, and is not imposed on them. Contrary to genetics where the pseudoscience of Lysenkoism was imposed on the scientific community and did great harm to Soviet genetics causing it to drop far behind international developments, this never happened in physics. Soviet physicists have always been able to determine the 'truth' within their subject, whereby they were always aware of developments in other countries, even though they kept, or were forced to keep, their own results mainly to themselves. The only requirement was that they gave priority to researching the topics that party and government thought relevant to the development of the country. After the repression of the late 1930s, and further sweetened by ample funding they were very happy to do so as the history of the Soviet nuclear programme and other scientific and technological developments in the Soviet Union have shown.

The book does not have the ambition to give sweeping general answers to such deep questions or to questions of why and how the repression took place, but will mainly provide an illustration on the basis of Shubnikov's experience at UFTI and will, where necessary, also pay attention to events at the Leningrad Physico-Technical Institute (LFTI) and in the Moscow physics community, which both had close connections with UFTI.

## References

1. See e.g. Vadim J. Birstein, *The Perversion of Knowledge* (Westview Press, 2001); Ethan Pollock, *Stalin and the Soviet Science Wars* (Princeton University Press, 2006) and a multitude of other books.
2. B.I. Verkin, V.G. Manzhely, Ju.A. Khramov, O.N. Trapeznikova, S.A. Gredeskul, L.A. Pastur, Ju.A. Frejman, V.G. Gavrilko, and L.K. Snigireva (eds.), *L.V. Shubnikov–Izbrannye Trudy. Vospominanija* (L.V. Shubnikov–Selected Works. Reminiscences) (Ukrainian Academy of Sciences, 1990), also to be referred to as B.I. Verkin et al. (1990).
3. Ju.V. Pavlenko, Ju.N. Ranjuk and Ju.A. Khramov, *"Delo" UFTI 1935–1938* (The "UFTI" Case 1935–1938) (Feniks, Kiev, 1998), also to be referred to as Ju.V. Pavlenko et al. (1998).
4. K. Mendelssohn, *The Quest for Absolute Zero* (New York, 1966).
5. Simon Ings, *Stalin and the Scientists–A History of Triumph and Tragedy 1905–1953* (Faber&Faber, London, 2016).

# Chapter 2
# Shubnikov's Early Years in St. Petersburg/Petrograd/Leningrad

Lev Shubnikov was born on 29 September 1901 in St. Petersburg into the family of the accountant or bookkeeper Vasily Vasilevich Shubnikov (b. 1882). His grandfather, Vasily Mikhajlovich Shubnikov (1845–1889), had also been a bookkeeper working in Moscow for the firm 'Danilovskaja Manufaktura', one of the largest companies in Tsarist Russia. In 1889 he hurt his leg in an accident, developed gangrene and died at the early age of 44, leaving a wife and six children, five boys and one girl,[1] the youngest only a few months old, in rather poor conditions. Lev's grandmother Anna Ivanovna (1853–1924) was forced to take a job as a seamstress in order to be able to clothe and feed her rather large family. Lev's father Vasily was the eldest of the children, and just seven years old when his father died. His father's old firm 'Danilovskaja Manufaktura' paid the school fees for him so that he could attend Moscow Commercial College. Lev's uncles Aleksej and Leonid also attended this school. Aleksej finished in 1906, after which he went to Moscow University, where he studied with Georgy Viktorovich Wulff (1863–1925). He became a well-known crystallographer, an Academician and the founder and first director of the Institute for Crystallography of the Academy of Sciences, which after his death was named after him. He was the leader of this branch of science in the Soviet Union for almost 50 years. These and other biographical details of Lev's forebears are taken from Aleksej's biography (Refs. [1, 2]). His brother Vasily, Shubnikov's father, hardly features in this biography, and his nephew Lev Vasilevich does not at all. Perhaps the relations of the eldest brother with the rest of the family were not very good. In this respect it should also be noted that Lev's father was only 19 when Lev was born and already living in St. Petersburg. He must have moved there a short time before as at that age he would have just finished his studies at Moscow Commercial College. But in 1901 his mother still had five other children on her hands and a supplement to the family income would have been more than welcome. In Aleksej's biography it is said that after finishing at

---

[1]Vasily, Elizaveta, Nikolaj, Boris, Aleksej and Leonid, all born in a timespan of just seven years.

© Springer International Publishing AG 2018
L. J. Reinders, *The Life, Science and Times of Lev Vasilevich Shubnikov*,
Springer Biographies, https://doi.org/10.1007/978-3-319-72098-2_2

**Fig. 2.1** Lev's parents
Vasily Vasilevich and Ljubov
Sergeevna Shubnikov (*from*
B.I. Verkin et al. (1990))

Moscow Commercial College in 1906 he and his brother Leonid were looking for ways to help the family (Ref. [1], p. 14).

Apart from his uncle's biography, some information about Shubnikov's direct family, albeit very little, is available from the reminiscences of Lev's wife Olga Nikolaevna Trapeznikova (1901–1997) (Ref. [3]). Their families had dachas next to each other in Finland and often spent the summers together, so Olga and Lev knew each other from childhood. As a child Lev apparently was rather unruly. His parents (Fig. 2.1) complained that he was an impish boy who could not be left alone for one moment and was always looking for someone to fight with. Lev had a younger sister Ljudmila (b. 1904) and a younger brother Kirill (b. 1915). In that time neither Lev nor his sister Ljudmila made a favourable impression on Olga.

In 1911 Lev's parents decided to send him to one of the best secondary schools in St. Petersburg: the gymnasium named after L.D. Lentovskaja (*gimnazija im. L.D. Lentovskoj*) founded in 1906 in the wake of the revolution of 1905, and considered a progressive school. In the autumn of 1918 the 8th class at the school was cancelled and Shubnikov, still only 16 years old, enrolled in the mathematics department of the Faculty of Physics and Mathematics at Petrograd[2] University (in 1918 there was not yet a separate physics department at the university). Why he chose physics and mathematics is not clear. Times were extremely unfavourable, not only for harbouring thoughts about an academic career, but also simply for living, since Petrograd as a city had virtually ceased to exist. Many people left. Between 1918 and 1920 the population dropped from 3.2 million to just 720,000 (Ref. [4], p. 452) and there was a great lack of food and fuel. Pitirim Sorokin, who after having spent a few months in the country returned to Petrograd at the end of 1918, writes in his *Leaves from a Russian Diary* (Ref. [5], p. 208): "*What I saw from the Nicolaevsky Station was the abomination of desolation. It was as though a devastating plague had swept the town. No trams, no droshkies. No shops open. Broken and dirty*

---

[2]The German sounding name St. Petersburg was renamed Petrograd in 1914 at the start of World War I, and after 1924 it was named Leningrad after Vladimir Lenin.

*windows revealed dark emptiness. All sign boards had been torn down. The streets were indescribably dirty, and in many places the pavements had been torn up. (...) And the people!–in rags, with emaciated and pallid faces, the few pedestrians plodded along as though crushed by poverty and grief".* The winter of 1918/1919 was particularly bad and conditions remained very tough until well into 1922, when Lenin's New Economic Policy gathered steam. In 1920 Emma Goldman returned to Petrograd, to the city she had known as a teenager in the 1880s and 1890s, and witnessed a similar devastation as Sorokin had done before (Ref. [6], pp. 8–9): *"I found Petrograd of 1920 quite a different place. It was almost in ruins, as if a hurricane had swept over it. The houses looked like broken old tombs upon neglected and forgotten cemeteries. The streets were dirty and deserted; all life had gone from them. The population of Petrograd before the war was almost two million; in 1920 it had dwindled to five hundred thousand. The people walked about like living corpses; the shortage of food and fuel was slowly sapping the city; grim death was clutching at its heart. Emaciated and frostbitten men, women, and children were being whipped by the common lash: the search for a piece of bread or a stick of wood. It was a heart-rending sight by day, an oppressive weight at night".*

Why on earth would a boy of not yet seventeen years old want to start studying at a university in such an atmosphere? And why physics? In view of his family background another career would have been more logical. Perhaps his uncle Aleksej, whose interest in physics and particularly in crystallography was aroused following a lecture at his secondary school in Moscow by his future teacher Georgy Wulff, had something to do with it, but there is no evidence that they had any extensive contact during Lev's secondary school years. Lev spent these years in Petrograd, while Aleksej was living in Moscow and from 1914–1918 was drafted into the army. In 1918 he returned to Moscow to continue his work with Wulff (Ref. [2], pp 648–649). From 1925 Aleksej worked in Leningrad, during which time Lev and his uncle had actually frequent contacts (Ref. [3], p. 260).

Lev's life in Petrograd was further complicated by the fact that soon after he enrolled at the university his family, like many others, had moved to Moscow, leaving him alone in an empty apartment (Ref. [3], p. 257). It is hardly imaginable, however, considering the housing situation in Petrograd at the time, that he (a 16 year old boy) would have had a whole apartment to himself. Shubnikov's family returned to Petrograd/Leningrad a couple of years later, since, as Trapeznikova (Ref. [3], p. 260) recalls, after her marriage to Shubnikov in 1925 they lived for some time with Lev's parents. It is not known why the parents had left for Moscow, but probably because life in Petrograd had become virtually impossible, certainly with small children like Lev's 2-year old brother Kirill, and they had family living in Moscow. Moscow had suffered much less from the revolution, and many people were actually fleeing to Moscow. Sorokin (Ref. [5], p. 134) recalls that where Petrograd impressed him as a dying capital, Moscow reminded him of a disturbed anthill.

Was it youthful stubbornness why Lev did not accompany them to Moscow and instead enrolled at Petrograd University (at the age of 16!) or were there other reasons? In the winter the apartment was cold and since it was almost impossible to get any fuel for heating, making a piece of firewood the most valuable present one could give or receive (Ref. [5], p. 218), Lev moved to the Physics Institute and spent the nights among the equipment in a laboratory room (Ref. [7], p. 50). Actually most of the university was "*dark, except the physics building where, by some miracle, Professor Khvolson managed to get enough kerosene to keep his laboratory windows alight*" (Ref. [5], p. 223). Apparently Khvolson[3] also managed to keep it heated as Frish (Ref. [7], p. 51) recalls that "*during the day there were many people at the institute; as before, it remained an oasis in the general desert that prevailed in the other buildings of the university. The main building was not heated at all and the lecture halls were closed. Mathematics, biology, chemistry— they all had to make use of the physics institute. But in the evening the institute became empty. In the quiet building Lev and I spent long hours, discussing the current physics news.*"

Lev was the only student of his year, a reflection of the fact that student numbers had dropped by 80–90% compared to the numbers before World War I. He therefore attended lectures, which had continued since the October revolution with some interruptions when part of the Petrograd faculty had been evacuated, together with the students of the previous year. These included Vladimir Aleksandrovich Fock (1898–1974), who became a well-known theoretical physicist, and Sergej Èduardovich Frish (1899–1977), who became a notable specialist in spectral physics and was one of Shubnikov's best friends.[4] Later they were also joined by students of the following year, among whom Lev's future wife Olga Trapeznikova (Fig. 2.2).

From 1919, after just one year at university with lectures often interrupted because of the uncertain political and military situation in Petrograd, Shubnikov started to work as a laboratory assistant (like many other students) at the State Optical Institute of Dmitry Rozhdestvensky.[5] One of the main reasons for these appointments may have been the fact that such jobs entitled their holders to extra rations (so-called *atomnyj paek*, a special ration for scientists working on the structure of the atom). Such rations were a bitter necessity as is witnessed by the death from hunger and other hardship of the historian Aleksandr Sergeevich Lappo-Danelevsky (1863–1919) and several other university professors and lecturers (Ref. [5], p. 229 ff). A second advantage was that the collaborators of the Optical Institute were exempt from military service, which was especially relevant when in 1919 a general mobilization was proclaimed to the offensive of the tsarist general Nikolaj Nikolaevich Judenich (1862–1933) (Ref. [7], p. 43).

In 1919 Judenich started a campaign against Petrograd with some (rather reluctant) support from Western allied forces, but without support from the Finns

---

[3]Orest Danilovich Khvolson (1852–1934), professor of physics at Petrograd University.

[4]In 1929 Frish was on *komandirovka* in Germany for a few months; in that period he also visited the Shubnikovs in Leiden. From the autumn of 1930 he spent almost a year in Groningen in the Netherlands where he worked with Dirk Coster (Frish 2009, p. 83ff).

[5]Dmitry Sergeevich Rozhdestvensky (1876–1940).

**Fig. 2.2** Frish, Shubnikov, Timoreva and Trapeznikova at Petrograd University (*from* B.I. Verkin et al. (1990))

and the Estonians as he refused to recognise Estonian and especially Finnish independence. The Finns had a proper, well-equipped army under the command of general Mannerheim and had defeated the Red Army in the spring of 1918. Lacking their support Judenich's campaign, which came threateningly close to Petrograd in October 1919, failed at the end of that year and, after the armistice which the Bolsheviks signed with Estonia (recognising its independence, something which Judenich and other White Army generals were never prepared to do), Petrograd was firmly in the hands of the Bolsheviks.[6]

It is unknown how Shubnikov reacted to these military developments, but Frish tells in his memoirs that, in spite of being exempt from military service as a collaborator of the Optical Institute, he volunteered to dig trenches and barricade buildings when Judenich was threatening to march into the city (Ref. [7], p. 44). It is likely that other collaborators at the Optical Institute, including Shubnikov, did the same. Whether this work was indeed voluntary as Frish wants us to believe is doubtful. The Bolsheviks seldom relied on unreliable volunteers and were used to issuing orders. Sorokin (Ref. [5], p. 224) recalls that such work (the transportation of timbers) was compulsory and that "persons whose names begin with A to M must be at the Admiralty Quay" at a certain time the next morning and others at the

---

[6]See Ref. [8] (p. 671ff) on this campaign.

same place on the following day. "Those who disobey this order will be arrested as counter-revolutionaries."

## Sailing trip to Finland

Following the revolution the old sailing clubs had been closed and the boats requisitioned by the new rulers. In the summer of 1920 life at the imperial yacht club on Krestovsky Island started to pick up, with the former owners of some boats hiring themselves out as instructors (Ref. [7], p. 51).

Both Olga Trapeznikova and Sergej Frish say that among the students and lecturers of the physics faculty there were quite a few, including Shubnikov, who were keen on sailing and enthusiastic participants of sailing trips in the Gulf of Finland and on Lake Ladoga, which at the time was shared with Finland.[7] For this purpose they used to rent a yacht at the former imperial yacht club.

In view of the military activity in and around the city until the end of the civil war in 1922 it must have been quite hazardous to make sailing trips on Lake Ladoga and certainly in the Gulf of Finland. For a ship to enter the Gulf of Finland from Petrograd or Petergof it had to pass Kronstadt on Kotlin Island, which housed a naval base guarding the approaches to Petrograd. Moreover, British ships blocked the harbour of Petrograd and at one time even attacked Kronstadt, so it must have been virtually impossible to enter the Gulf of Finland, which is confirmed by Frish (Ref. [7], p. 58).

In August 1921, independent of the activity of the physics faculty, a sailing trip on Lake Ladoga was organised, with the apparent purpose to escape from the country. The organisers were looking for crew members and also asked Shubnikov, without revealing to him the real purpose of the trip. He readily agreed to come along, after all it was still the summer holiday period. Various versions of the event exist, one by Trapeznikova (Ref. [3]), another by Frish (Ref. [7], pp. 58–59 and Ref. [9], p. 293), who says that he himself was also asked to be part of the crew, but happened to be out of town when they left, and by Shubnikov himself as told to the NKVD investigators[8] after his arrest in 1937. All these versions disagree in some of the details. Trapeznikova states that chemists from the university were involved and she suggests that the trip was in the Gulf of Finland, which is unlikely as explained

---

[7]After the Winter War of 1939–1940 it became an internal basin of the Soviet Union, under the terms of the Moscow Peace Treaty of March 1940. It is the largest lake in Europe with an average surface area of 17,891 km$^2$. Its north-to-south length is 219 km and its average width is 83 km.

[8]The documents of his case, which will be discussed in detail in later chapters, have been made available on the Internet http://www.ihst.ru/projects/sohist/document/ufti/ufti.htm (the website of the Institute of the History of Science and Technology (*Institut istorii estestvoznanija i tekhniki im. S.I. Vavilov RAN, IIET*)) as part of the UFTI affair (*Delo UFTI*) and have been published in Ref. [10]. Parts of this book have also been published in *Universitates*, the journal of the V.N. Karazin Kharkiv National University.

above. According to Frish several young people were recruited as crew members without being aware of the actual purpose of the trip. He does not mention any chemists, but states that the former manager of the yacht club had set up this trip in order to escape from Russia. The version as told by Trapeznikova is slightly different from the one by Frish. According to Frish Shubnikov left on this trip at the end of August 1921 and stayed in Germany for almost two years, which is hardly possible if he returned in 1922 as both Trapeznikova and Frish seem to agree upon. This date of 1922 is incorrect, as Shubnikov himself stated in 1937 that he returned in the autumn of 1923 and documents from the Ioffe Institute confirm that Shubnikov started work there in the autumn of 1923.

Enquiries with the Finnish National Archives[9] have now established some basic facts about this sailing trip. According to the documents in the files of the Finnish State Police (*Etsivä Keskuspoliisi*) Lev Vasilevich Shubnikov was arrested together with eight other people by border guards in Salmi Mantsinsaari on the (then Finnish) northeast coast of Lake Ladoga on 30 August 1921. The trip had been organized by Boris Leonidovich Nagornov, born in 1898, jointly with Boris Konstantinovich Merezhkovsky, born in Yalta in 1883 and a professor of chemistry at Petrograd University.[10] This name agrees with the one given by Shubnikov to the NKVD, while Nagornov comes close to Nagorny, whom Shubnikov names as the captain of the yachting club. According to Nagornov's statement to the Finnish police he had enrolled at Petrograd Polytechnic in 1917, but was conscripted into the army that same year and served in an artillery battery. After being discharged in February 1918 he became a translator at the Commissariat for Road and Waterway Construction, translating foreign professional literature into Russian. At the end of 1919 he resigned from this job and set up the so-called Railway Workers' Sailing Club, where he became a Senior Instructor, paid for by the Soviet Government. Only railway workers were accepted into the club, where they received free sailing instruction. During the early days of August 1921 he had received written orders from the administration of the 1st Territorial Brigade to prepare the club's yacht 'Lorna-Doon' for departure. He was to assume captaincy of this vessel and to find an adequate crew for it. Later he was officially informed that the Petrograd division of the Russian Hunting Association intended to conduct research on the southern (Russian) coast of Lake Ladoga. For this purpose the 'Lorna-Doon' had been provided, and the chemist Boris Merezhkovsky and Johan Kuusik, an official at the Estonian embassy in

---

[9]I thank Mrs. Raija Ylönen-Peltonen for searching both the Finnish National Archives and the files of the Vyborg police, which had survived the 1939 Winter War, for traces of Lev Vasilevich Shubnikov and for sending me copies of documents from the files.

[10]Boris Konstantinovich Merezhkovsky, indeed a chemist who worked in Petrograd with the future Academician and inventor of the first commercially viable synthetic rubber, Sergej Vasilevich Lebedev (1874–1934) (Sergej Vasilevich Lebedev, *Zhizn' i Trudy* (Leningrad, ONTI, 1938)). Merezhkovsky was the only son of the prominent biologist and botanist Konstantin Sergeevich Merezhkovsky (1855–1921), who had emigrated to Switzerland in 1917 and had just died by suicide in Geneva in January 1921. He also was a nephew of the well-known Russian writer Dmitry Sergeevich Merezhkovsky (1865–1941), who had emigrated in 1920.

Petrograd, had volunteered to carry out this task on behalf of the Association. Later in his statement Nagornov admitted, however, that the expedition was actually a joint initiative instigated by himself and Merezhkovsky. The latter was the chairman of the Petrograd division of the Hunting Association and had for some time wanted to go to Finland. He had asked Nagornov to arrange a sailing trip on Lake Ladoga in order to take Merezhkovsky, his wife Nadezhda Merezhkovskaja and their belongings to Sviritsa,[11] from where they would seek access into Finland under the escort of fishermen. But Nagornov himself also had a secret agenda, as his father, Leonid Nagornov, who had occupied a high position in the imperial Tsarist government as a state counsellor, had fled to Finland after the 1917 revolution and now ran a transport company in Helsinki. The son wanted to join his father in Helsinki, although he claimed in his statement that on this trip he did not want to escape from Russia, but intended to make a second attempt at a later date.

That they chose 1921 as the year to escape is not surprising. The food situation in Petrograd had been bad since 1917, but became even worse from mid-1920 when a drought devastated the harvest and the requisition policy of the Bolsheviks in the countryside had resulted in the peasants drastically reducing their crop production. Until the local Petrograd government issued a decree forbidding any kind of commercial transaction, a few shops were still open and illegal markets, which had sprung up in all towns, had been semi-tolerated, but they now also had to close. The government was, however, in no position to supply the town with food. Official rations were distributed irregularly and even in smaller amounts than stipulated. Workers would receive, at irregular intervals, the equivalent of 700–1,000 calories a day. The famine of 1921, which was the result of the combination of a poor harvest and inadequate government policies, has gone down in history as one of the most severe famines the country had ever experienced, and lasted through 1922. It was also the year though that convinced Lenin to change his policy and decree the New Economic Policy on 15 March 1921.

As the crew of the vessel, Nagornov had selected Ilja Aleksandrovich Dessler,[12] Lev Shubnikov, the mechanic Nikolai Mikhailovich Kortin, an Estonian, and the railway official Mikhail Shchilkin,[13] as well as Kira Podlesskaja, Nagornov's fiancé

---

[11]Just a few hundred kilometres from St. Petersburg, on the mouth of the river Svir which flows westward from Lake Onega into Lake Ladoga.

[12]A student at the Mining Institute in Petrograd, twenty-seven years old, who, according to Shubnikov in his statement to the NKVD, had a brother living in Helsinki.

[13]A great surprise among the crew members is the name of Michael Schilkin (1900–1962) (transliteration of his name in Russian: Mikhail Nikolaevich Shchilkin) who according to his current English Wikipedia page (consulted on 18 April 2017): "In 1921, while sailing with a local sailing club on Lake Ladoga, (…) accidentally crossed the Finnish border. The team was arrested by the border guards and subsequently released". How accidental this crossing indeed was remains to be seen. Whether he was aware of the intention of some of his companions on the boat to escape from war-torn Russia is not completely clear. But after having arrived in Finland he stayed there and went into the employment of Leonid Nagornov, Boris Nagornov's father. He later studied at the Aalto University School of Arts, Design and Architecture and became one of Finland's best known ceramicists making his career with the Finnish ceramic firm Arabia.

and a theatre school student, as a guest. Merezhkovsky's luggage and other cargo were loaded onto the yacht at the club. He had obtained permission for three or four boxes of technical literature, but Nagornov knew that Merezhkovsky had only two boxes of such literature, the third box containing paintings by various painters. Nagornov pointed out that this was unknown to the Soviet officials, and if they had known they would have forbidden the paintings from being loaded. On 15 August 1921 the party, that is: Merezhkovsky and his wife, Kuusik, Shubnikov, Shchilkin, Kortin, Nagornov and Podlesskaja, gathered at the club and set out up the River Neva in the tow of a tugboat. When they arrived at Shlisselburg, at the head of the River Neva on Lake Ladoga 35 kilometres east of Petrograd, it turned out that Nagornov was the only one on the boat who could navigate the yacht by compass. While the others waited, Merezhkovsky returned to Petrograd to fetch Ilja Dessler, who was an accomplished sailor, and had signed up for the trip, but failed to show up. On 20 August the party finally set out from Shlisselburg to Lake Ladoga, but after only a few kilometres the main sail became torn, and they had to return to Shlisselburg to have it mended, after which the trip started again on the evening of 25 August. So before any sailing had been done, ten days had already passed, and one would think that all crew members would by now have been familiar with the real purpose of the trip. Had Shubnikov wanted he still could have gone back at this stage, but he didn't.

The trip continued to be beset by problems. While sailing during the first night the fork of the boom gaff became damaged, and the party was forced by the strong wind to sail into the harbour of Sortanlahti (present-day Vladimirovka on Lake Ladoga). There they arrived on 26 August and were greeted by a customs boat, which ordered them to stay on the yacht until further orders. The next day a doctor and several customs officials boarded the yacht and carried out inspections, without finding anything to complain about in terms of the party's health or the goods on board (the box containing the artwork was not opened and the customs officials were left with the impression that it contained literature like the other two boxes). On 29 August in the evening the boat was towed out of the harbour and ordered to sail back across the border. They now set course for Sviritsa as originally intended and sailed straight until the morning of 30 August. The strong headwind prevented them from approaching the shore and after several failed attempts to land, the decision was made to sail into cover at Mantsinsaari, an island in the north-east Finnish part of Lake Ladoga. The yacht ran into shallows, after which troops sailed out to meet them from a nearby anchored vessel. Later a military patrol arrived at the yacht from the shore. Nagornov was taken ashore to be interrogated by an artillery officer, who gave orders to transport the whole party ashore and be guarded by soldiers. From their accommodation they walked to the shops in the company of soldiers and bought various items, such as clothes and food, all under the impression, according to Nagornov, that they could freely sail back to Russia, with the exception of Merezhkovsky and his wife, who earlier had expressed the desire to stay in Finland.

A couple of days later, on 5 September, they were interrogated by the border guard and transported to Sortavala where an office of the State Police was situated.

From there all were taken to the prison in Vyborg, where Merezhkovsky and his wife, Shubnikov, Shchilkin, Dessler and Podlesskaja were held until 14 September. Kortin and Kuusik had returned to Estonia (a newly independent country that had friendly relations with Finland), while Nagornov had been released after having obtained a valid permit to enter the country on 8 September (probably through the intervention of his father) and somewhat later a residence permit to stay in Mikkeli (St. Michel).

The Finnish State Archives also contain an interrogation report dated 17 September 1921 of the six persons mentioned above in which Shubnikov is reported to have stated that *"he has no monetary possessions, but has the intention to settle down in France together with Professor Merezhkovsky. He does not want to go to Russia under any circumstances."*

Merezhkovsky, who had considerable foreign funds in his possession as well as a number of paintings and other valuables, declared that *"the purpose of the trip was to make it to France, where his uncle the author Merezhkovsky is living"*. He also declared to be prepared to finance the journey to France for Shubnikov. After the interrogation the apartment in which they were all staying was searched. In Merezhkovsky's room a bundle of silver wire and metallic platinum in sheets, bars, pieces and wire with a total weight of about 5 kilos was found. This confirms Trapeznikova's statement that the chemists organizing the sailing trip had stolen precious metals from the university laboratory (Ref. [3], p. 259). These items were seized by the police and stored. Whether they were returned to Merezhkovsky is not clear.

Shubnikov and Nagornov's fiancé Podlesskaja remained under investigation while their application for a residence permit was rejected due to a lack of *"information concerning their trustworthiness in matters of state"*, as the Archive document states. Finally on 18 October 1921 Shubnikov was issued with a residence permit valid until 1 April 1922, giving him permission to stay in the province of Mikkeli. In this period he was also allowed to make a two-week trip to Helsinki. It is not known for what purpose.

In 1937 Shubnikov declared (Ref. [10], p. 246) to the NKVD that Merezhkovsky wanted to set up his own chemical laboratory in Helsinki and proposed that Shubnikov come and work for him for bed and board. When after 5–6 months Merezhkovsky had failed to get the laboratory organised, he decided to move to Berlin taking Shubnikov with him. This does not agree with Merezhkovsky's statement to the Finnish police that he wanted to go to France, but it agrees more or less with a document from the Finnish Archives dated 31 January 1922 listing all foreigners staying in Mikkeli during January 1922. It states that Shubnikov and Dessler had left Finland and had travelled to Germany, namely well before the expiry of their residence permit, which suggests that their departure was well planned and had a purpose, for else it would have been more logical for them to stay in Finland at least until the expiry of their permits. The Merezhkovskys are not on the list, so they must have already left Mikkeli at an earlier date. Shubnikov had supposedly been waiting for a couple of months in Mikkeli for Merezhkovsky to get things organised in Helsinki, and when that did not work out he left the country in January 1922.

In Germany Shubnikov joined the thousands of Russian refugees who in the years since 1917 had fled Russia after the revolution and civil war in order to escape the turmoil and chaos. During 1922 and 1923, which was the high point of the influx of Russian refugees, in Berlin alone there were more than 300,000 Russian refugees (at a total population for Berlin of close to 4 million) with a total of about 600,000 for the whole of Germany (Ref. [11], p. 262). In addition, the Russian diaspora[14] also included a large number, probably around 400,000 Russian prisoners of war, a legacy of World War I. An entire Russian culture had sprung up with its own newspapers, publishing houses, theatres, restaurants etc. There were hundreds of organisations and associations, most of them concentrated in the Charlottenburg quarter, which by some for this reason was called Charlottengrad. For instance Ilja Ehrenburg, who lived in Berlin from 1921–1924, testified about this: "*I don't know how many Russians there were in Berlin then. Probably very many—one heard Russian being spoken all over. Dozens of Russian restaurants were opened—with balalaika, with salmon, with gypsies, with bliny, with shashlik and naturally, with the obligatory excitement. There was cabaret. Three daily papers and five weeklies were on offer. Within a year, seventeen Russian publishers began to operate*" (Ref. [13], p. 47). So if Shubnikov had wanted to and had the means to go out and about, he could very much have felt at home in Berlin. Moreover, as Sammartino writes (Ref. [14], p. 187), the Russian refugees were seen as victims of Bolshevism and tolerated by Germans from across the political spectrum with a remarkable lack of resentment, a climate that started to change somewhat after the murder, in March 1922, of Vladimir Dmitrievich Nabokov (1870–1922), the father of the Russian-American author Vladimir Nabokov. It was felt that by this murder the Russian assassins and, by extension, the émigré community as a whole had violated the *Gastrecht*, the formal and informal rules foreign guests were expected to adhere to. Another important turning point in the political relation between Germany and the Soviet Union was the Treaty of Rapallo signed between the two countries on 16 April 1922, whereby they renounced all territorial and financial claims against each other following World War I and decided to restore full diplomatic relations. This led to the (re)opening of the former Russian embassy in a vast and luxurious Rococo Palace at Unter den Linden 7. It was a former property of Tsar Nicolas I and had functioned as the Russian embassy since 1840/41. Now it was destined to become the centre of the German-Soviet scene in Berlin until 1941.

A second factor that had a great impact on the refugee situation in Berlin and in Germany in general was the inflation which started a few months before

---

[14]See Ref. [12] for information about the Russian diaspora in Berlin, especially the chapter *Sankt Petersburg am Wittenbergplatz: Eine Hauptstadt im Jahrhundert der Flüchtlinge* (pp. 78–110).

Shubnikov's arrival in Berlin. In September 1921 one dollar cost 101 Marks, at the end of October the price had already risen to 4,475 Marks and in November 1923 it had reached the colossal amount of 4.2 billion Marks. For some it represented a great chance: small dollar amounts paid for the rent of houses or for accommodation in expensive hotels. A few dollars could last for months (Ref. [12], pp 96–97). But for most Russian refugees who, like Shubnikov, were destitute and dependent on charity, the living conditions became very tough. Many decided to return home or move on to other countries, like France or the USA.

Shubnikov was also affected by the Decree of the All-Russian Central Executive Committee and the Council of People's Commissars of 15 December 1921 which stipulated, among other things, that: "*persons who had left Russia after 7 November 1917 without the authorization of the Soviet authorities*" were automatically deprived of their Soviet citizenship.[15] This obviously applied to him and implied that he, like many of his compatriots, was stateless while in Berlin. As he had no money and no means of existence he must have lived, probably illegally, with Merezhkovsky whose address can be found in the Berlin address book of 1924.[16] He is not in the 1923 address book, which is understandable as it takes time to process new entries, but he is not in the 1925 address book either, suggesting that by 1925 he had already left Berlin.

In the interrogation protocol of the NKVD, Shubnikov further stated that in Berlin he worked for Merezhkovsky on the development of daylight developer for photographic plates, a type of research that was carried out at the time by many people and in the course of the next few years several patents were taken out. When Shubnikov failed to get any results, he was fired in the autumn of 1922 and, almost dying from hunger and with a severe flu, he went for help[17] to Dessler.[18] Dessler was also interested in Shubnikov's work on the developer and decided to give him money to live on and carry out experiments. After four months of work the results were still poor, and the work on the developer was stopped. Dessler now demanded another invention and Shubnikov decided to continue the work on obtaining transparent quartz from sand which he had also worked on at the Optical Institute in Leningrad. From this too, no invention resulted after several months of work.

---

[15]Dekret VTsIK i SNK ot 15 dekabrja 1921 g. "O lishenija prav grazhdanstva nekotorykh lits, nakhodjashchikhsja za granitsej" (*On the forfeiture of citizenship rights of certain persons who are abroad*) SU RSTSR 1921, no. 72, p. 578.

[16]Mereschkowsky, Boris, Prof., Mariendorf, Königstraße 40.

[17]Even though at the time there were a large number of very active Russian charity organisations that could have helped him both financially and as regards a possible return to Russia. The official policy in Germany was that Russians who had fled Russia for political reasons were not sent back. (See Aleksandr Ushakov, *Die russische Hilfsorganisationen in Deutschland zu Beginn der 20er Jahre*, in Karl Schlögel (ed.), *Russische Emigration in Deutschland 1918 bis 1941* (Akademie Verlag, Berlin, 1995), pp. 131–137.)

[18]For that matter, no Ilja Dessler can be found in the Berlin address books of 1923–1925, so he either did not live in Berlin or had a shady status.

Shubnikov further states that in Berlin he met two staff members of Rozhdestvensky's Optical Institute, Arkhangelsky[19] and Grebenshchikov,[20] who advised him to apply for a restoration of his Soviet citizenship.[21] As he had worked at the Optical Institute before his departure for Finland, Shubnikov probably knew them quite well. He followed their advice, but a restoration of citizenship is unlikely to have been simple. The early 1920s were a busy time for the repatriation of Russian refugees from Western countries, especially also prisoners of war who, after diplomatic relations had been restored, flocked from the camps in Germany to the hopelessly overloaded Soviet representation agencies in order to be repatriated (Ref. [12], p. 122). Perhaps Shubnikov managed to slip through rather easily because of this pressure on the embassy, but other sources say that returning refugees were first sent to a detention centre in the town of Velikiye Luki in West Russia, not far from the border with Latvia, in order to be cleared for entrance into Russia as the Soviets did not want to let saboteurs and spies seep into the country.[22]

In the meantime Dessler had obtained a job for him at a bank, a type of currency-exchange shop, called the Berlin Credit and Trade Bank (*Berliner Credit-und Handelsbank*).[23] Before Shubnikov's departure to Petrograd in the autumn of

---

[19]Probably Aleksandr Alekseevich Arkhangelsky (b. 1890), who was in Berlin several times around this time. Trips by Arkhangelsky to the West are mentioned by several people (e.g. by Ehrenfest in his letter to Abram Ioffe of 27 September 1920, and by Aleksej Krylov in his memoirs), but always in combination with Vladimir Mikhajlovich Chulanovsky (1889–1969); I have not seen Arkhangelsky mentioned in connection with any journey together with Grebenshchikov. According to Frish (Ref. [7], p. 76) later in the twenties Arkhangelsky was a member of the Soviet trade mission in Berlin for some time.

[20]Ilja Vasilevich Grebenshchikov (1887–1953), a chemist and one of the organisers of the State Optical Institute in 1918, Academician from 1932. In 1922 he was sent on *komandirovka* to Germany for half a year to purchase equipment and reagents for the Optical Institute, so Shubnikov may indeed have met him there early in 1923.

[21]This does not agree with Trapeznikova's version, who mentions as Shubnikov's rescuer M.M. Glagolev, a lecturer at Petrograd University, who later became the head of the roentgenographic analysis laboratory of the Technical Department of the Leningrad Physico-Technical Institute and was also on a business trip in Germany.

[22]Emil Draitser, *Stalin's Romeo Spy* (London, 2011) (p. 46) tells about the future KGB spy Bystrolyotov, who in June 1922 had expressed a desire to return to Russia to the Soviet naval attaché in Berlin, being first sent to Velikiye Luki. On the repatriation of Russian refugees from Western countries after the end of the Russian civil war, see also Martyn Housden, *The League of Nations and the organisation of peace* (London, 2012), p. 62 ff. The repatriation described by Housden took place via the Bulgarian Black Sea port of Varna and also involved security checks by Soviet officials before letting people back into the country.

[23]This bank is indeed mentioned in the 1921–1925 Berlin address books with its office at Alvenslebenstrasse 12a, but no information about it can be found in the archives. Until 1924 a 'Berliner Credit- und Handelsgesellschaft' (which originated from the 'Kugel und Kugellagergesellschaft' in Goldap (now in Poland, at the time East Prussia)) was registered at the same address, but no bank. To add to the confusion at the time there was also a 'Credit- und Handelsbank' (without Berliner in its name) with registered office in the Dorotheenstrasse in Berlin.

1923[24] Dessler allegedly recruited him for espionage work, forcing him to provide information about applied physics work done in Petrograd. Apart from this espionage business, which could very well be an NKVD fabrication, there is nothing in Shubnikov's story that sounds improbable.

For some reason, which is not further explained, this escapade to Finland and Germany had no further direct consequences for Shubnikov when he arrived back in Petrograd in 1923. Upon his return he must however have been interviewed by the NKVD and it can be easily imagined that he could have been charged with aiding and abetting in an illegal escape from the country, with the theft of a yacht and of the precious materials taken from the university laboratory.

# References

1. N.V. Belov and I.I. Shafranovsky (eds.), *Aleksej Vasilevich Shubnikov* (Nauka, Leningrad, 1984).
2. A.V. Shubnikov, Autobiographical data and personal reminiscences, in: P.P. Ewald (ed.), *50 Years of X-ray Diffraction* (Oosthoek, Utrecht, 1962), p. 647–653; http://www.iucr.org/publ/50yearsofxraydiffraction.
3. O.N. Trapeznikova in B.I. Verkin et al. (1990), p. 256–291.
4. A. George and E. George, *St Petersburg. The First Three Centuries* (Sutton Publishing, 2004).
5. Pitirim A. Sorokin, *Leaves from a Russian Diary – and Thirty years After* (The Beacon Press, Boston, 1950).
6. E. Goldman, *My Disillusionment in Russia* (Dover Publications, 2003).
7. S.E. Frish, *Skvoz' prizmu vremeni* (Through the prism of time) (Solo, St Petersburg, 2009).
8. Orlando Figes, *A People's Tragedy: The Russian Revolution 1891–1924* (Random House, London, 1996).
9. S.E. Frish in B.I. Verkin et al. (1990), p. 292–297.
10. Ju.V. Pavlenko, Ju.N. Ranjuk and Ju.A. Khramov, *"Delo" UFTI 1935–1938* (The "UFTI" Case 1935-1938) (Feniks, Kiev, 1998).
11. Jochen Oltmer, *Migration und Politik in der Weimarer Republik* (Göttingen, 2005).
12. Karl Schlögel, *Berlin Ostbahnhof Europas: Russen und Deutsche in ihrem Jahrhundert* (Siedler Verlag, Berlin, 1998).
13. Ilja Ehrenburg, "Zwei Jahre lebte ich hier in Angst und Hoffnung", in *Berliner Begegnungen: Ausländische Künstler in Berlin, 1918–1933, Aufsätze, Bilder, Dokumente*, ed. Klaus Kändler (Berlin, Dietz, 1987).
14. A.H. Sammartino, *The Impossible Border: Germany and the East, 1914–1922* (Cornell University Press, 2010).

---

[24]This does not agree with Trapeznikova and Frish who both state that Shubnikov returned in the autumn of 1922. A document in Shubnikov's personnel file at the Ioffe Institute confirms, however, that he was employed from September 1923 and not from autumn 1922. The first article by Shubnikov and Obreimov, which appeared in volume 25 of *Zeitschrift für Physik* in 1924, was received by the journal on 15 April 1924. This means that he had barely six months to settle down in Leningrad, learn some physics, set up an experiment, do measurements and write a paper. Probably quite a lot of preparatory work had already been done by Obreimov, but it shows, as would also become apparent later when Shubnikov worked in Kharkov, that he could work at incredible speed.

# Chapter 3
# Shubnikov's Scientific Work in Leningrad; Papers with Obreimov

Back in Russia Shubnikov did not return to the university, but started to study at the Petrograd Polytechnic Institute,[1] combining this with work at the State Physico-Technical Radiological Institute, directed by Abram Ioffe,[2] in the laboratory of Ivan Obreimov.[3] According to his personnel file at the Ioffe Institute he started work in September 1923. The reasons for the switch from the university to the Polytechnic Institute are not very clear, nor who gave him advice in this matter. According to Trapeznikova (Ref. [1], p. 259) it surprised everybody, and the result was that it delayed his studies by another year. The point was, however, that the situation regarding physics in Petrograd had changed considerably since Shubnikov left in 1921. The most active and talented physicists at the university, including Pëtr Lukirsky,[4] Viktor Bursian,[5] the first head of the theory department of Ioffe's institute, Ivan Obreimov and Jakov Frenkel,[6] were now working either part-time or full-time at Ioffe's Physico-Technical Institute and teaching at the Polytechnic Institute. Some of these people had been Shubnikov's teachers before 1921 and their presence at the Polytechnic Institute may well have been the reason for him to switch, or for being advised to switch. Another consequence of this switch was that Shubnikov hardly got to know Lev Landau who arrived, 16 years of age, at the university in 1924. In the end Shubnikov finished his studies, at the level of a current master's degree, in 1926. The results of his work with Obreimov were published in two papers in *Zeitschrift für Physik* co-authored with Obreimov. Both papers, but especially the first one on a new method for growing single crystals of

---

[1]Founded in 1899 as the most advanced engineering school in Russia, it opened its doors to students in 1902. Now it is called the *Sankt-Peterburgsky Politekhnichesky Universitet Petra Velikogo*. In the past it had various names. From 1923–1924 it was called the Petrograd Polytechnic Institute and from 1924–1930 Leningrad Polytechnic Institute.
[2]Abram Fëdorovich Ioffe (1880–1960), who would become the godfather of Soviet physics.
[3]Ivan Vasilevich Obreimov (1894–1981).
[4]Pëtr Ivanovich Lukirsky (1896–1954).
[5]Viktor Robertovich Bursian (1886–1945).
[6]Jakov Ilich Frenkel (1894–1951).

© Springer International Publishing AG 2018
L. J. Reinders, *The Life, Science and Times of Lev Vasilevich Shubnikov*,
Springer Biographies, https://doi.org/10.1007/978-3-319-72098-2_3

metals, were of interest for the work he was going to do in Leiden. The second paper gave him experience in working with crystals.

Shubnikov's supervisor Ivan Obreimov was born in Annecy (France) in 1894. In 1871 his father, Vasily Ivanovich Obreimov (1843–1910), had been dismissed from his position as a gymnasium teacher in mathematics for anti-religious propaganda, for proclaiming the equality of men and women and for distributing socialist writings among students. He had been living in exile in France since the early 1890s. In 1896 the family returned to Russia, where for quite a few years Obreimov's father lived under a false name. After having been pardoned, he and his family settled in St. Petersburg where in 1904 he started to teach mathematics at the newly founded commercial college in Lesnoy, part of the Vyborgsky District in St. Petersburg. His son Ivan also attended this school, where from fifth grade he was taught physics by Abram Ioffe.[7] In 1910 Ivan enrolled at St. Petersburg University, where Jakov Frenkel, also born in 1894, was one of his contemporaries. From 1912 Obreimov took part in the famous seminar held by Paul Ehrenfest,[8] later taken over by Ioffe, and a year later, when still a student, he went to Göttingen where he spent half a year working with Gustav Tammann (1861–1938), at the time a leading authority in crystallography. As for many others, this first trip abroad on a scientific mission had a profound influence on Obreimov and on the research topics, molecular physics and in particular crystals, he would choose to work on later in his career. He asked Rozhdestvensky to supervise the work for his master's degree (*diplomnaja rabota*) and after finishing university started work as Rozhdestvensky's assistant at the production facility of optical glass, making glass for lenses, prisms, mirrors and suchlike for use in microscopes and other optical instruments, which had just been set up at the Imperial Porcelain factory.[9] Here he developed, among other things, methods for improving the measurement of the refractive indices of pieces of glass. After the revolution, the production facility was mothballed, and it finally took the Soviet Union until 1926 before it could produce its

---

[7]That Abram Ioffe who from 1906 was back in Russia ever taught at this school is not mentioned by Horst Kant in his biography of Ioffe [2]. The information here comes from [3]. From 1906 Ioffe had a position as laboratory assistant at the Polytechnic Institute, which did not come under the Ministry of Education, but under the Ministry of Trade and Industry. The commercial college had close connections with the Polytechnic Institute and with the Forestry Institute. The latter two were both close to the Polytechnic Institute which explains Ioffe's involvement as a teacher. According to Kant, Ioffe taught at the faculties of electromechanics and metallurgy at the Polytechnic Institute, at the Mining Institute and at Lesgaft's school of physiology (Ref. [2], p. 22–24), so not much time can have been left for teaching physics at another school.

[8]Ehrenfest's wife was a Russian mathematician and he had spent the years from 1907–1912 at the university in St. Petersburg where he played a large and active role in introducing the new physics to Russian physicists. His influence was large, and after the revolution he again was instrumental in getting support for Russian physics and physicists from the West. In 1912 he became the successor of H.A. Lorentz as professor of theoretical physics at Leiden University.

[9]Founded in 1744, it became the State Porcelain Factory after the revolution and from 1925 the Leningrad Lomonosov Porcelain Factory. In 2005 it returned to its pre-Soviet name of Imperial Porcelain Factory.

own optical glass and become independent of imports. Obreimov is credited with having made large contributions to this success (Ref. [3], p. 287).

When in 1918 the State Institute for Roentgenology and Radiology was set up, Obreimov became one of the first employees of Rozhdestvensky's optical division, and later of the State Optical Institute. In the early twenties he became interested in the study of electronic properties of matter at low temperatures, and in 1922 started to work half-time at Ioffe's Physico-Technical Institute, where he eventually became director of the laboratory for molecular physics and deputy director, under Abram Ioffe, of the institute (Ref. [2], p. 53). So Obreimov was involved from the very beginning in the organisation of physics after the Bolshevik takeover of power.

The work he was going to do with Shubnikov was very much in line with the general interest at Ioffe's institute. Ioffe had studied with Röntgen in Munich and had since been interested in the electrical and mechanical properties of crystals. Röntgen had proposed that he investigate the connection between the elastic after-effects due to deformation and the piezoelectric effect[10] in quartz.[11] Heterogeneities in the crystal structure and impurities made it difficult to study the process of piezoelectricity. There were unexplained phenomena which were later attributed to such heterogeneities and impurities. These anomalies, as they were called, had Röntgen's special interest. From this followed Ioffe's interest in the mechanism of the deformation of crystals, which he studied from 1914–1916 with the help of X-rays with M.V. Kirpicheva, a chemist by training, and one of the first women at the Polytechnic Institute to grow and purify crystals. She was Ioffe's constant assistant-demonstrator during lectures and became his closest collaborator in the research on the electro-conductivity of crystals. Kirpicheva died very young in 1923 after an unsuccessful operation. In 1919 she and Ioffe carried out some classic work on the elastic deformation of crystals. Especially the second paper by Obreimov and Shubnikov was a continuation of this, but also the first work on crystal growth fitted in perfectly with Ioffe's research program. So, it is clear that the initiative and the underlying ideas for this work were coming from Obreimov, and not from Shubnikov, very much in line with his status as a fourth-year undergraduate student. He had some experience, however, in working with quartz crystals from his time in Germany and before that at the Optical Institute, but was just arriving from abroad and plunged right into this research from September 1923.

Shubnikov's first work with Obreimov involved the growing of single crystals. A lot of the preparatory work must have been done by Obreimov before Shubnikov was back in Petrograd, as it would have been impossible for a young inexperienced physicist as Shubnikov was in 1923 to partly invent and set up (for the time) fairly complicated apparatus, do experiments and write a paper within a little more than half

---

[10]The piezoelectric effect is the phenomenon that crystals of certain materials under pressure, for instance by bending, produce an electric current and the other way round that they deform when electric voltage is applied to them. The word *piezo* has been derived from the Greek word *piezein*, which means to press.

[11]Ioffe's doctoral thesis, submitted in 1905, was entitled *Elastische Nachwirkung im krystallinischen Quarz*. The thesis was published in book form in 1906, with an extract in *Annalen der Physik* (20 (1906) 919–980).

a year. This first paper was entitled *"Eine Methode zur Herstellung einkristalliger Metalle"* (A method for preparing single-crystal[12] metals) and was published in the leading physics journal *Zeitschrift für Physik* [4]. Crystal growth is typically achieved by forming a solid from another state of matter, for example, a liquid (melted) metal is cooled to form a solid. The procedure predominantly used at that time came from the Polish physicist Jan Czochralski (1885–1953),[13] but, as mentioned by Obreimov and Shubnikov, it has a number of drawbacks. The most important of these are that it is rather expensive, no regular forms of crystals can be achieved and impurities (e.g. oxygen) are difficult to avoid. Moreover deeply penetrating permanent deformations occur when the crystals obtained are worked mechanically. In the late twenties and early thirties it was superseded by what is now called the Bridgman-Stockbarger method [5, 6]. The first part of the long and very detailed paper by Bridgman, which is mainly concerned with measuring the properties of the various crystals obtained, describes the crystallization method. It was received by the journal on 6 November 1924, while Shubnikov and Obreimov's paper was received by *Zeitschrift für Physik* on 15 April 1924, more than half a year earlier.

The procedure developed by Obreimov and Shubnikov for metals with a low melting point, which very much resembles the Bridgman-Stockbarger method or at least the earlier version used by Bridgman, seems to have been largely forgotten.[14] It is essentially a perfection of the method developed earlier by Tammann for bismuth and extended to other metals. Tammann was the first to prepare single-crystal metals, and from his stay at Tammann's institute ten years earlier Obreimov was probably familiar with this work. Obreimov and Shubnikov mention Tammann in their paper, but without giving a specific reference.[15] Tammann's method failed for other metals than bismuth, probably as it generated many centres from which the crystallization could start. The point according to Obreimov and Shubnikov is to make sure to start from a single crystallization centre, for instance by inserting a seed crystal, and let that grow. Their procedure is still very rarely, mainly in Russian or Soviet publications, referred to as the Obreimov-Shubnikov method.

---

[12]In a single crystal the atoms are ordered in an almost perfect periodic arrangement, therefore the crystal lattice of the entire sample is continuous and unbroken to the edges of the sample, to be distinguished from a polycrystal consisting of many (microscopic) crystals (called crystallites or grains) and an amorphous solid in which there is no periodic arrangement at all.

[13]J. Czochralski, Ein neues Verfahren zur Messung der Kristallisationsgeschwindigkeit der Metalle [A new method for the measurement of the crystallization rate of metals], *Zeitschrift für Physikalische Chemie* 92 (1918) 219–221. Czochralski is the most frequently cited Polish scholar. See also Anna Pajaczkowska, Jan Czochralski: Brief sketch of his life and achievements, *Journal of Crystal Growth* 401 (2014) 5–6.

[14]Several others employed similar methods around this time, but it is clear that Obreimov and Shubnikov were the first. See Brian R. Pamplin ed., *Crystal Growth* (Pergamon Press, Oxford, 1980), p. 7. It would be fairer to call this method the Tammann-Obreimov-Shubnikov method.

[15]In a later paper written in 1930 in Leiden on the preparation of bismuth crystals (L. Schubnikov, Über die Herstellung von Wismuteinkristallen, *Proc. Roy. Acad. Amsterdam* 33 (1930) 327–331) Shubnikov refers to Tammann's 1923 book *Lehrbuch der Metallographie*. In this paper he also mentions the similarity with Bridgman's method.

The method works as follows: a metal is melted in a vacuum glass or quartz capillary tube, which tapers off to a point at the bottom, in a specially designed two-zone oven (the upper zone and lower zone could be heated separately and were divided by an asbestos sheet). The tube was placed in the oven such that the bottom part (1–2 mm) finished (through the asbestos sheet) in the lower zone, after which both zones were closed off, the tube drawn vacuum, and the melting could start. The two zones of the oven were very evenly heated. It was crucial for the melted metal to penetrate down into the very tip of the tube (for this to happen the pressure in the tube was increased). Then the cooling started by switching off the heating for the lower zone, which caused the crystallization to start in the bottom tip of the tube. The crystal grew while the tube is slowly lowered from the hot zone of the furnace into the cool zone. The cooling was carefully controlled. If in the cooling process the growth rate of the crystal is large, the resulting crystal tends to have more defects. It is therefore paramount to keep the growth rate under control by cooling the melted metal slowly. Whether single crystals were obtained depended crucially on the cross section of the tube used; for zinc the cross section had to be less than 1.5 mm$^2$ to guarantee that there was only a single crystallization centre; for antimony a tube with a cross section of up to 3 mm$^2$ could be used. In this way Obreimov and Shubnikov managed to grow single crystals for bismuth, antimony, tin, zinc, magnesium, aluminium and copper with a length of up to 30 cm. This work already shows the meticulousness which would be characteristic for Shubnikov in substantiating and checking the results of his work. Three methods were used to test the mono-crystallization of the specimen grown: the nature of cleavage, the form of the cleavage plane under plastic deformation and X-ray analysis. (Ref. [7], p. 21.)

To illustrate the striking similarity with Bridgman's method I quote here from his paper: "*The general method is that of slow solidification from the melt. A tubular electric furnace, in a vertical position, is maintained at a temperature above the melting point of the metal in question. The metal in the molten condition in a suitable mold of glass or quartz tubing is slowly lowered through the bottom of the furnace into the air of the room or into a cooling bath of oil. Solidification thus starts at the bottom of the tube and proceeds slowly along its axis, keeping pace with the lowering. If the lowering is at a speed less than the velocity of crystallization and also slow enough so that the latent heat of solidification may be dissipated by conduction, then the metal will usually crystallize as one grain, provided that only one nucleus started in the tube at the bottom.*" (Ref. [5], p. 307.)

Just like Obreimov and Shubnikov Bridgman also states that "*the most difficult and important is to secure the formation of only one grain at the bottom end,*" namely to make sure that there is just one centre of crystallization.

The second paper, which was sent to *Zeitschrift für Physik* after Shubnikov had already left Leningrad for Leiden, was entitled "*Über eine optische Methode der Untersuchung von plastischen Deformationen in Steinsalz*" (On an optical method for investigating plastic deformations in rock salt) [8].[16]

---

[16]A Russian version of the paper appeared in the *Journal of the Russian Physico-Chemical Society* (*Zhurnal Russkogo fiziko-khimicheskogo obshchestva* LVIII (1927) 817–828).

Deformations of a body result from stresses in the body caused either by applied forces or due to temperature changes in the body. Two types of deformation can be distinguished: elastic deformations which are reversible, meaning that the body returns to its original state when the stress has been removed, and plastic deformations which are irreversible, namely the body remains deformed after removal of the stress. Such deformations occur due to slip or dislocations at the atomic level when the stresses have exceeded a certain threshold value (the *elastic limit* or yield stress). When the deforming forces are further increased the body "hardens", implying that both the plastic and elastic deformations increase. Until this point, observations of the mechanical properties of the crystal were mostly made by measuring the dependence between the applied force and the resulting stress. These methods only gave information about surface deformation. What was happening inside the crystal could only be deduced indirectly. In 1924 Ioffe and his collaborators Kirpicheva and Levitskaja had used X-rays to investigate plastic deformation in crystals. The optical method had also been used for investigating the stress distribution in deformed bodies, but has as major disadvantage that it can only be used for transparent bodies. Therefore Obreimov and Shubnikov used rock salt, also called *halite*, in their experiment. Rock salt is a mineral form of sodium chloride ($NaCl$) and forms isometric (or cubic) crystals. It is typically colourless or white, but may also have various other colours depending on the amount and type of impurities. Moreover it is transparent, can easily be obtained in large blocks, and is optically isotropic (uniform in all directions). It was one of the most favourite research objects not just at Ioffe's institute, but all over Europe. For the period 1920 to 1935, the journal *Zeitschrift für Physik* alone published 289 papers which in one way or another discussed properties of rock salt.

Obreimov and Shubnikov made careful observations of refraction patterns caused by dislocations in rock salt during plastic deformation, i.e. caused by the force applied on the material. For a non-crystalline (amorphous) material the refractive index is proportional to the applied deforming force. When a tensile force is applied to such material it will become uniaxially double refractive, whereby the double refraction is proportional to the applied force. A light ray falling on the material will be split into two rays travelling in different directions. One ray (called the extraordinary ray) is bent, or refracted, at an angle as it travels through the medium; the other ray (called the ordinary ray) passes through the medium unchanged. A crystal will also become double refractive under application of a tensile force, but now the value of the double refraction will depend not only on the applied force, but also on the crystallographic direction. An optically isotropic material like rock salt becomes anisotropic by the application of a force, and will then show double refraction of light falling on the material, whereby the value of the refractive index is different for different crystallographic directions. Such phenomena had already been studied and measured in the final decades of the 19th century.

Obreimov and Shubnikov placed their rock salt sample between the crossed Nicol prisms[17] (for producing a polarized beam of light from an unpolarised beam) of a polarisation microscope and immersed the crystal in a liquid of the same refractive index (*nitrobenzene*), and in this way were able to observe dislocations in the crystal when a force was applied. They took photographs of the double refraction patterns and could optically determine the deformation of the crystal and the elastic limit, as well as the increase of the deformations with the increase of the applied force.

Double refraction made it possible to determine the residual pressure in the region of plastic deformation (near the line of cleavage) in transparent crystals. For the first time it was possible to clearly see what happens inside a crystal after reaching the elastic limit, which for transparent crystals supplemented the picture shown by X-ray analysis.

One of the major challenges of this type of research was the preparation of very pure samples of crystals. Obreimov and Shubnikov chose the most faultless natural rock salt they could obtain and split it into small $5 \times 5$ mm prisms (60–80 mm in length). The sides of the prisms were polished and the whole crystal heated in an oven until the residual stresses, which are always present in natural rock salt and were still increased by the polishing, had completely disappeared. Such crystal preparation very much resembled the work Shubnikov was going to do later in Leiden with crystals of bismuth, and it was a skill he was very good at. All in all they prepared a total of 25 crystals, of which only five were suitable for the experiment. The other ones contained too many impurities. The work Shubnikov was doing here was, probably not accidentally, very similar to that of his uncle Aleksej Vasilevich. Aleksej had worked in Sverdlovsk at the Mining Institute from 1920–1925 where he *"was personally in charge of the preparation of the thin plates of natural crystals for the courses in crystal optics. [He] developed an interest in the problems of cutting, grinding, and polishing of crystals and stone."* (Ref. [9], p. 649.) In May 1925 he came to Leningrad to a position at the Mineralogical Museum of the Academy of Sciences and stayed there until 1934 when the Academy moved to Moscow. Trapeznikova (Ref. [1], p. 260) says that Lev often met his uncle during 1925 and 1926 both at home and at the Academy.[18] Although no more details are known about this, it is only natural to assume that his uncle gave him good advice on how to prepare his crystals. Aleksej was sent on a *komandirovka* (business trip) to Germany and Norway in 1927 and again in 1929, during the time that Lev was in Leiden. They do not seem to have met during any of these trips.

---

[17]A Nicol prism is a type of polarizer, an optical device used to produce a polarized beam of light. It is made in such a way that it eliminates one of the rays by total internal reflection, i.e. the ordinary ray is eliminated and only the extraordinary ray is transmitted through the prism (*Wikipedia*).

[18]In A.V. Shubnikov's biography by Belov and Shafranovsky nothing is said about such meetings.

The suitable crystals were placed in an experimental set-up such that pressure could be applied on them through a lever press to observe pressure deformations and in a second set of experiments to observe deformations under tensile forces. In Abram Ioffe's book *The Physics of Crystals*, a write-up of his lectures at the University of California in 1927, Ioffe describes the method and I quote from his book: "*The construction of the apparatus used for the loading was carefully worked out in order to produce a uniform tension through the whole cross-section of the crystal. This was tested by the optical method itself. Gradually loading the crystal and observing the picture in polarized light, we* (i.e. Obreimov and Shubnikov) *observed the first appearance of a permanent set at stresses about one-tenth of those causing the destruction as observed by means of X-rays. A sharp bright line on a general dark background appeared and remained after removal of the load. At a constant load exceeding by a little the elastic limit, a second line parallel to the first one appeared after some time; then a third one and so on. All these lines represented projections of planes ( ... ). More often they suddenly cross the whole crystal, sometimes beginning at one side and extending rapidly through the crystal to the opposite side. ( ... ) The phenomenon of a plastic deformation of this sort consists of a series of slips parallel to some crystallographic plane and is produced by the shearing stress in this plane.*" (Ref. [10], pp. 47–49.)

This second piece of work became the subject of Shubnikov's diploma work, entitled "Optical method for studying elastic and residual deformations in crystals", which he defended on 7 July 1926. Although his work was of considerable interest, the show at the defence ceremony was stolen by Lev Sergeevich Termen[19] who demonstrated the new musical instrument invented by him, the first-ever electronic musical instrument, the termenbox, which played and sang.

The research on deformations in crystals was continued by Obreimov and co-workers at the Ukrainian Physico-Technical Institute in Kharkov.

During the time that Shubnikov was working with Obreimov, his future wife Olga Nikolaevna Trapeznikova finished her studies at the university. For her master's degree she also worked on rock salt, in her case the electrolysis of rock salt, i.e. the separation of *NaCl* into sodium (*Na*) and chloride (*Cl*) by means of an electric current, work she did under the supervision of Pëtr Lukirsky. The work was reported by Abram Ioffe at the Solvay conference in Brussels in 1924 and published in the *Journal of the Russian Physico-Chemical Society* [11], as well as in *Zeitschrift für Physik* [12], after which her fellow-students jokingly called her the "Queen of Salt". Her co-authors were her supervisor Lukirsky and the chemist

---

[19]Lev Sergeevich Termen (1896–1993), better known as Leon Theremin, the inventor of the first mass-produced electronic musical instrument, the theremin. In 1927 he embarked on a lengthy tour in the West, staying in the United States until 1938 when he hurriedly returned to the Soviet Union and was put to work until 1947 in a *sharazhka* together with the aircraft engineer Andrej Nikolaevich Tupolev and the rocket designer Sergej Pavlovich Korolev. A *sharazhka* (also spelled: *sharashka* or *sharaga*) is a prison in the form of a secret research and development institution or design bureau where imprisoned scientists were set to work. In his novel *The First Circle* Solzhenitsyn gives an account of life in such a *sharazhka*.

Sergej Aleksandrovich Shchukarev (1893–1984) who did the chemistry while Trapeznikova did the experiment (Ref. [1], p. 259). Shchukarev later became a professor at Leningrad State University.[20]

After she finished university, at least two years ahead of Shubnikov, Trapeznikova became a graduate student with Lukirsky. Shubnikov's friend Sergej Frish had already finished in 1921 and was now working at the university, combining this with a job at the State Optical Institute. Frish's wife and Trapeznikova's friend, Aleksandra Vasilevna Timoreva,[21] finished at the same time as Trapeznikova and also became a graduate student (with Viktor Bursian). Trapeznikova was already going out with Shubnikov, and when her family moved to another district she and Lev Vasilevich started to live next to each other and saw each other more often.

Trapeznikova suffered from asthma and subsequently caught an open form of tuberculosis,[22] a potentially fatal disease in any country in those days, but certainly in the Soviet Union that was still recovering from revolution and civil war. The doctors advised her to drink *koumiss* (a fermented dairy product made from mare's milk). As this was not readily available in Leningrad, Lev and Olga went in the spring of 1925 to Churakaevo in Bashkiria (also known as Bashkortostan), about 150 km from Ufa in the south of Russia, to the grass steppes. There she recovered and they married in Bashkiria. After returning to Leningrad Lev Vasilevich resumed his work, Trapeznikova got a graduate scholarship, and they lived rather well, such that she could afford to be treated by private doctors. One day Lev complained about pain on his chest. Trapeznikova was immediately worried: could it be his lungs? Could he have caught tuberculosis from her? They went to the doctor who was treating her. He examined Lev and asked: "Has not one of your ribs been broken?" It turned out that Lev Vasilevich had measured his strength with someone and had broken a rib. His age was no longer that of a gymnasium student who could simply get into a fight with someone, but he still loved to measure his strength. After their marriage they first lived with Lev's parents as Lev had done before they left for Bashkiria. It was not a comfortable arrangement, Trapeznikova writes, as the Shubnikovs were very thrifty, and they were the opposite, they only had fun and no sense at all (Ref. [1], p. 260). Timoreva and Frish were living with Frish's family in a large communal apartment. There was a spare room, which Lev and Olga took after which they started to live in the same house as the Frish family and got very close with them.

As Trapeznikova reports, Lev worked and studied all the time without taking any time off as he wanted to finish his studies and make up for the time lost. It seems that the relationship with his parents was not very good. They were constantly urging him to finish his studies and take his exams, perhaps understandable

---

[20]For more details on Shchukarev see his extensive Russian Wikipedia page.

[21]Aleksandra Vasilevna Timoreva (1902–1995) was also a graduate of Leningrad University, worked for a time at the Central Board of Weights and Measures (Mendeleev's Institute) and after the war at Leningrad State University.

[22]She may have had latent tuberculosis, which can have been activated by the asthma.

from their point of view, but such that he got completely fed up, banged the table and left. Trapeznikova does not give any further details, except that in the autumn of 1926 when he was leaving Leningrad for the Netherlands and she and his mother were accompanying him to the boat at the pier of Vasilevsky island, his mother Ljubov Sergeevna decided that she no longer wanted to see him, dropped on a bench and started to cry bitterly. There is nothing in any writings I have seen which says anything about Lev's father or about his brother and sister.

# References

1. O.N. Trapeznikova in B.I. Verkin et al. (1990), p. 256–291.
2. H. Kant, *Abram Fedorovič Ioffe, Vater der sowjetischen Physik* (B.G. Teubner, Leipzig, 1989).
3. T.K. Litinskaja, Ocherk o nauchnoj dejatel'nosti akademika I.V. Obreimova (*Essay on the scientific activity of Academician I.V. Obreimov*), *Fiz. nizk. temp.* 20 (1994) 286–295.
4. I. Obreimow, L. Schubnikow, Eine Methode zur Herstellung einkristalliger Metalle, *Z. Physik* 25 (1924) 31–36.
5. P. W. Bridgman, Certain Physical Properties of Single Crystals of Tungsten, Antimony, Bismuth, Tellurium, Cadmium, Zinc, and Tin, *Proceedings of the American Academy of Arts and Sciences*, 60 (1925), 305–383.
6. Donald C. Stockbarger, The Production of Large Single Crystals of Lithium Fluoride, *Review of Scientific Instruments* 7 (1936) 133–136.
7. B.I. Verkin et al. (1990).
8. A.V. Shubnikov, Autobiographical data and personal reminiscences, in: P.P. Ewald (ed.), *50 Years of X-ray Diffraction* (Oosthoek, Utrecht, 1962), p. 647–653.
9. Abram F. Joffé, *The Physics of Crystals* (edited by Leonard B. Loeb) (New York, 1928).
10. P. Lukirsky, S. Shchukarev and O. Trapeznikova, Élektroliz kristallov, *Zhurnal Russkogo fiziko-khimicheskogo obshchestva* LVI (1924) 453–461.
11. P. Lukirsky, S. Shchukareff and O. Trapesnikoff, Die Elektrolyse der Kristalle, *Z. Phys.* 31 (1925) 524–533.
12. I.W. Obreimow, L.W. Schubnikoff, Über eine optische Methode der Untersuchung von plastischen Deformationen in Steinsalz, *Z. Physik* 41 (1927) 907–919.

# Chapter 4
# Shubnikov in Leiden

The physics laboratory devised by Heike Kamerlingh Onnes (1853–1926) from the late 19th century in Leiden was rather different from other physics laboratories existing at that time. The low-temperature research Onnes had in mind faced him with three fundamental problems. First the low temperature must be *produced*, then it must be *measured* and thirdly it must be *utilised* in an apparatus devised in such a way that the object to be studied can be brought to the required temperature. The apparatus needed for such cryogenic work was considerably larger than usually present in physics and physics-chemistry laboratories in those days. It required a rather large establishment, with a considerable staff of trained and specialised mechanics to solve the problems of the purification of gases and liquids and to operate and maintain the various liquefiers, compressors and pumps. (Ref. [1], p. 37.) Because of its scale and approach to physical problems Emilio Segré calls Onnes's Leiden laboratory the first example of large-scale physics, physics at an industrial scale, developing interaction between research, training and industry (Ref. [2], p. 223–234).

The Leiden Physics Laboratory with Kamerlingh Onnes (Fig. 4.1) at its head was very successful and grew into the foremost laboratory in the field of low-temperature physics. The main achievements were the liquefaction of helium gas in 1908, reaching a temperature as low as 1 °K (−272 °C), and in 1911 the surprise and momentous discovery of superconductivity. In 1913 Onnes was awarded the Nobel Prize for physics for (in the words of the Nobel committee) "*his investigations on the properties of matter at low temperatures which led,* inter alia, *to the production of liquid helium*". One of the main attractions of the Leiden Physics Laboratory for low-temperature physicists was that by 1925 the production of liquid helium was a routine affair. The amount of liquid helium needed could be ordered before closing time of the laboratory and would be ready the next morning. In this respect the Leiden laboratory was unique in the world.

When Kamerlingh Onnes resigned as director in 1924, the laboratory was split into two departments.[1] One was headed by Wander de Haas, the other by Willem

---

[1]Van Delft has argued that this was an unwise decision and one of the reasons that the Leiden Physics laboratory missed several important discoveries. (Ref. [3], p. 608ff.)

© Springer International Publishing AG 2018
L. J. Reinders, *The Life, Science and Times of Lev Vasilevich Shubnikov*,
Springer Biographies, https://doi.org/10.1007/978-3-319-72098-2_4

**Fig. 4.1** Heike Kamerlingh
Onnes in 1909, painted by
Menso Kamerlingh Onnes
(*Leiden University*)

Keesom (1876–1956). In Keesom's department studies were made of the equation
of state of gases, thermometry (Shubnikov attended lectures by Keesom on thermometry), specific heat measurements and a number of other thermal properties of
matter. Keesom tried to obtain further lower temperatures by reducing the vapour
pressure of helium, using increasingly powerful pumps. Where Kamerlingh Onnes
had succeeded in obtaining 0.82 °K via this method, in the period that the
Shubnikovs were in Leiden Keesom obtained 0.71 °K. His department also had a
large glass blowing workshop, where it could be observed how large spherical
Dewars are blown, first one glass sphere and then a second one inside.

Wander Johannes de Haas (1878–1960) was one of the most important experimental low-temperature physicists of the first half of the twentieth century and has
been nominated sixteen times for the Nobel Prize. As a postdoc he had worked with
Einstein in Berlin, which resulted in the Einstein-De Haas effect, the experimental
observation of the phenomenon that a change in the magnetic moment causes a
body to rotate.[2]

De Haas was not a tall man and made the impression of being a sluggish person.
He had to be addressed as 'professor', which was unusual for Russian visitors, who

---

[2]For some more information about De Haas see the short biographical sketch by E.C. Wiersma,
*W.J. de Haas 1912–1937* (Martinus Nijhoff, The Hague, 1937).

addressed their professors with their first name and patronymic, and were not very interested in who was professor and who not. He was a very gentle, pleasant person with a soft smooth voice, without any solemnity and mentoring. In some sense he was the opposite of the laboratory's first director Kamerlingh Onnes.

It was hard to work under Kamerlingh Onnes as a director. He was always short of money and had cultivated refined ways for reducing staff salaries and curbing heating costs. In his days it was always so cold in the laboratory that peoples' hands became swollen. De Haas had a completely different style. Under him the laboratory was always well heated. He was of the opinion, although he did not put it into practice, that young people should be paid more and old professors less.

Around 1926 Wander de Haas needed a co-worker who could grow crystals of bismuth for a series of experiments he had planned. It is not very clear how De Haas got to know about Shubnikov, but it is very probable that Paul Ehrenfest (Fig. 4.2), the professor of theoretical physics at Leiden University and a close friend of Abram Ioffe since his extended stay from 1907–1912 in St. Petersburg, was involved in this. However there is nothing in their (scientific) correspondence suggesting such involvement. There is however a gap in the correspondence from Ehrenfest to Ioffe from 21 September 1925 to 17 October 1926, exactly the period that is relevant here, and a still much greater gap in the correspondence from Ioffe to Ehrenfest; there are no letters preserved from Ioffe to Ehrenfest for the whole period from 6 August 1920 to 5 April 1928.

In one of his statements to the NKVD after his arrest in 1937 Shubnikov says that "*upon finishing the institute in 1926 a foreign komandirovka was proposed to me. Ioffe made this proposal directly to me with the help of the Dutch physicist Ehrenfest, whom I had got to know at the time of his stay (if I am not mistaken in 1926) at the physico-technical institute.*" (Ref. [4], p. 244–245.) There is no reason to doubt the correctness of this statement, although it cannot be confirmed that Ehrenfest was indeed in Leningrad in 1926. A recent biography of Ehrenfest does not mention such a visit [5]. In her reminiscences Olga Trapeznikova also confirms Ehrenfest's involvement (Ref. [6], p. 260).[3] Ioffe first proposed that Shubnikov's uncle Aleksej Vasilevich go to Leiden. He, however, was a qualified crystallographer and declined to go, after which Ioffe turned to the nephew Lev Shubnikov, who gladly accepted. It should be realised that at that time Shubnikov was not a senior physicist, but at best at a level equivalent to a present-day Master of Science. Moreover he was probably more a technician than a physicist. He had just finished his studies, had no position in Leningrad or elsewhere and was actually in danger of being drafted into the military. Ioffe wrote to Mikhail Petrovich Kristi (1875–1956),

---

[3]Either in 1925 or 1926 Ioffe himself also visited Leiden, perhaps as part of his trip to Berlin, Paris and the USA from November 1925 until February 1926 (P.R. Josephson, *Physics and Politics in Revolutionary Russia* (University of California Press, 1991), p. 116) and may have discussed matters there with De Haas and/or Ehrenfest.

**Fig. 4.2** Paul Ehrenfest
painted by Menso
Kamerlingh Onnes (*Leiden
University*)

the deputy head of Glavnauka,[4] to ask his help in order to get Shubnikov out of the
army and to Leiden:

I very much ask your help in exempting the scientific worker of the Leningrad
Physico-Technical Institute Lev Vasilevich Shubnikov, born in 1901, from military service.

According to an order of the War Department/announcement of the Territorial District of
the city and province of Leningrad no. 114330 of 29 May 1926/he is subject to military
subscription.

L.V. Shubnikov is one of the most valuable workers of the institute, his work is in the field
of elastic deformations of crystals, a field I myself also work in. These issues are very
topical, both scientifically and technically; several first-class laboratories in England and
America work on them and the initiative which our laboratory has in this matter is
extremely valuable. To deprive the laboratory of one of the most experienced and talented
workers in precisely this field will put a stop to a whole series of work at least for a year,
which is equivalent to losing the initiative.

To call up L.V. Shubnikov is all the harder for the laboratory and for my work as I proposed
to send L.V. Shubnikov for a period of 6 months on *komandirovka* to Holland to the
laboratory of Kamerlingh-Onnes. There L.V. Shubnikov would continue his research at the
lowest temperatures. This *komandirovka* has already been approved by Glavnauka and has

---

[4]Main Scientific Administration of Narkompros (*Glavnoe upravlenie nauchnymi, nauchno-khu-
dozhestvennymi i muzejnymi uchrezhdenijami*) set up in December 1921; originated from the
Scientific Department of Narkompros (*Narodnyj komissariat prosveshchenija*; People's
Commissariat for Education).

a somewhat unusual character, as L.V. Shubnikov has been invited by Professor De Haas to work at the institute of Kamerlingh-Onnes, and he does not ask permission for this. On return from there he should construct a cryogenic laboratory at LFTI.

On the basis of the above I very much ask you to take all possible measures in order not to put such a valuable worker out of action.[5]

As Ioffe pointed out, Shubnikov actually did not ask permission for this *komandirovka* as he would go to Leiden at the invitation of De Haas, but it also shows that Ioffe had some sort of secret agenda as he wanted Shubnikov to build a cryogenic laboratory at LFTI after his return from Leiden. It is also clear from the letter that Shubnikov was held in high esteem by Ioffe and that he still was considered a member of Ioffe's institute (*"one of its most experienced and talented workers"*). He undoubtedly thought that Shubnikov would be able to gain enough experience in Leiden in just half a year to be able to build up a similar laboratory in Leningrad.

Ioffe's request was granted and in the autumn of 1926 Shubnikov boarded the steamship *"Preussen"*, belonging to the *"Stettiner Dampfer-Compagnie A.G."* (Stettin Steamship Company) and the same boat that in 1922 had served as one of Lenin's philosophy steamers in exiling a large part of Russia's intelligentsia [8]. It was one of the ships that made the regular trip Leningrad-Stettin.[6] From there Lev Vasilevich travelled by train to Berlin where he had to arrange a visa for the Netherlands, which at the time had no diplomatic relations with the Soviet Union.[7] He finally arrived in Leiden early in November 1926, as is clear from a letter to Vladimir Fock written from Leiden and dated 6 November (Ref. [7], p. 8). His first impression of Leiden was favourable and rather stereotypical, as he writes, among other things:

Finally the wind started to blow from where it should, and I am in Leiden. De Haas is not yet here, he is in Paris and will come back on the 8th. I saw Ehrenfest, he is very nice to me. He said that he will probably obtain a Rockefeller stipend for you.[8]

I visited the low-temperature institute; both the building itself and its content are remarkable. The young fellows that work there are also sympathetic. All in all I liked it.

The town is wonderful: ¾ of it is (taken) by the university, the rest by museums, churches, windmills, canals and pavements, cleaned with brushes and rags.

For a few days he lived in a hotel, then took furnished rooms with full board (a rather luxurious way of living) in the Witte Rozenstraat 58, opposite Ehrenfest's house in the same street.

---

[5]Part of the letter has been reprinted in Ref. [7], p. 7. Copies of the complete letter are preserved in Shubnikov's personnel file at LFTI. Its date is not very clear, probably 17 July 1926, but in any case after 29 May 1926 as this is the date of the order of the War Department mentioned in the letter.

[6]Now Szczecin in Poland, but until 1945 a major German port on the Baltic Sea.

[7]The Netherlands only recognised the Soviet Union in 1942 when it was occupied by Germany. The recognition came from the government in exile in London.

[8]And indeed he did, Fock got his stipend in 1927 and also visited Leiden.

According to Trapeznikova (Ref. [6], pp. 261–262) Lev Vasilevich worked very hard in Leiden, he studied both in the laboratory and at home. When the laboratory closed in the evenings, he went to the library, located in Ehrenfest's theory building, where he could continue working.

He liked the laboratory: "Nice guys, nice people", he wrote to his wife. The relationship between De Haas and Lev Vasilevich was exceptionally good, complete trust, esteem and friendliness. The employees of the laboratory and the service staff were very attentive to him; he liked their quiet, kind and resolute nature. However it was difficult for him to communicate and to get close to the people, and there were no Russians there. Therefore for a rather long time he felt lonely. While in Russia you can go and visit someone without an invitation, in Holland that is not customary (only one co-worker of the laboratory invited Lev Vasilevich once to dinner). On free days he sometimes rode his bicycle, going around the nearby towns, visiting museums and cathedrals; he wrote to his wife in detail about what he saw and what he liked. Unfortunately, these letters have not been preserved: they were all seized at the time of his arrest and have not reappeared. They still must be somewhere in the former KGB, now the FSB archives.

In 1927 De Haas also invited Trapeznikova, who was an assistant at LFTI, to come to Leiden, probably at Shubnikov's urgent request as he had told De Haas that he would return to Russia if Trapeznikova could not come and join him in Leiden (Ref. [6], p. 262). With help from Obreimov, who was at the time the deputy director of LFTI, she managed to get permission (after a first refusal) from Narkompros (the People's Commissariat for Education) for a *komandirovka* to Leiden to work at low temperatures, initially for three months but in the end extended to three years. In the autumn of 1927 she travelled together with Ivan Obreimov and Jakov Frenkel on the same steamship "*Preussen*", as Shubnikov had done earlier. The trip lasted several days and was not very pleasant. In Berlin it took her a week to secure a visa after which she travelled alone to the border where she was met by Lev Vasilevich. Obreimov came to Leiden later, where he worked with De Haas on the investigation of absorption spectra of molecular crystals at much lower temperatures than was possible in Leningrad. They were the first to study such spectra at the temperature of liquid hydrogen ($-253$ °C).

In spite of the harsh conditions she must have experienced in Petrograd in the years before when living in a communal apartment with just one room to themselves, she is not very positive about the living conditions they had to put up with in Leiden. They were living in the rooms Shubnikov had rented earlier, but the house at the Witte Rozenstraat was not very comfortable. They had two rooms: one on the second floor, a sort of study, and the other—on the third floor—was the bedroom. The stairs leading to the second and third floor were rather narrow; people could just pass, but carrying furniture was impossible. When they hired a piano (Trapeznikova was an eager player), it had to be dragged up via the balcony. She was struck by the absence of conveniences in the house: there was no bathroom, but on the top floor there was a water pipe, water was kept in jugs, there were no double frames in the windows. In the room below there was a small stove, lined with tiles. For some reason it was called Russian, but it stood apart. The bedroom upstairs had no heating. Lev Vasilevich worked in the downstairs room, where they also had no heating. Lev Vasilevich worked in the downstairs room, where they also had their meals. They lived on full board and the landlady brought the food, although she fed them rather poorly. The food they were given was clearly not sufficient for them. Trapeznikova writes that her head started to swim and that before her arrival Lev Vasilevich also felt hungry, especially in the evenings, such that he even couldn't get to sleep, but later got used to it. She went to the doctor, who said that she needed to eat twice as much, so they started to buy eggs and biscuits to eat in the evenings, without saying anything to the landlady (Ref. [6], p. 270).

Their experiences in that respect are rather different from those of the German physicist Walter Elsasser (1904–1991) who was hired by Ehrenfest as his assistant in the same period, after the departure in 1927 of Samuel Goudsmit and George Uhlenbeck, the discoverers of the electron spin, to the United States. In his memoirs Elsasser writes that he found an agreeable single-bedroom in a small lodging house run by a pleasant and kind middle-aged lady. Elsasser's assistantship with Ehrenfest was, however, not a success. It was rather unceremoniously and rudely cut short before the end of the semester by Ehrenfest, who sent Elsasser back to Berlin (Ref. [9], p. 83 ff).[9]

According to Trapeznikova (Ref. [6], p. 270) Shubnikov's stay in Leiden was funded by the Rockefeller Foundation. However the Rockefeller Archives do not contain any mention of this nor of Shubnikov's name. Viktor Frenkel and Paul Josephson [10] have researched the Rockefeller Archives and its connection with Russian/Soviet physicists, and they do not mention Shubnikov as the recipient of a Rockefeller grant either. This probably means that Shubnikov was not the recipient of a direct grant, but he may have been financed by Rockefeller money that the Physics Laboratory in Leiden had received for doing cryogenic research. For example, the Dutch newspapers *NRC* and *Algemeen Dagblad* reported on 8 February 1926 that the university had received $25,000 for this purpose, and still had a considerable amount left from earlier gifts.[10] From documents in the archives of the Ioffe Institute in Leningrad, as well as from Ioffe's above letter to Kristi, it is

---

[9]Elsasser was in Leiden just for the first few months of the semester, which officially ran from October 1928 to April 1929, and "stayed at the same rooming house" (Ref. [9], p. 106) as Obreimov, so he also must have met the Shubnikovs, but does not mention them in his book; nor does Trapeznikova mention him in her reminiscences. Judging from Elsasser's memoirs (in a chapter that is called *The Troubled Years*) it seems that Ehrenfest was very rude to him for no good reason and treated him very unfairly. Elsasser's opinion of Ehrenfest was consequently rather low. He wrote that the experience had taught him one thing and taught him for good, namely: "If I had ever thought, in the style of Hitler, that any one group or race of people had a monopoly on rawness, or mishandling of their fellow creatures, then the encounter with Ehrenfest disabused me of such ideas. In manner and attitudes he seemed to me still close to the Jewish ghetto." The latter comes remarkably close to Enrico Fermi's opinion on Ehrenfest. Fermi visited Leiden for some time in 1923 and later described Ehrenfest as "really very nice and wouldn't be out of place in a ghetto store for used clothing" (G. Segré and B. Hoerlin, *The Pope of Physics* (New York, 2016), p. 39). Contrary to Elsasser's experience Fermi had a very fruitful and valuable time in Leiden. Elsassser later moved to the US and became the 'father' of the geodynamic theory, as an explanation of the Earth's magnetism.

[10]Like the $100,000 received from the Rockefeller Foundation in 1924 (Ref. [3], p. 583).

also clear that while in Leiden Shubnikov still had an affiliation with the institute in Leningrad. Money was even sent to him from Leningrad. This seems a little odd as one would assume that in Leiden as an employee of the Physics Laboratory, who also took part in the teaching, as he assisted students in the physics practicum (Ref. [6], p. 264), he would be paid an ordinary salary which should be sufficient to cover ordinary living expenses. Trapeznikova writes that he was paid 200 guilders per month,[11] while she got 30 guilders per month from De Haas, so their total annual income of 2760 guilders was a bit less than half Ehrenfest's salary as a professor, which in 1927 stood at 5880 guilders.[12] This is certainly more than a junior physicist would earn nowadays at a university. They had to pay 10% in tax and paid 160 guilders to the landlady, which left them with a little less than 50 guilders for all other expenses. The Shubnikovs however enjoyed full board, which can be seen as rather extravagant and for which their salaries were indeed not sufficient. So for this reason it may not come as a surprise that they had to be supplemented by payments from LFTI. The LFTI archives contain various documents requesting that money be sent to Shubnikov in Leiden as he "is in a difficult material position".[13] They also took extensive holidays both in the Netherlands and abroad, e.g. to Paris and Brittany.

In the Leiden archives nothing has been preserved about an employee named Shubnikov. Apart from his physics papers published in the *Communications from the physical laboratory of the University of Leiden* and in the *Proceedings of the Amsterdam Academy of Sciences* there is nothing tangible left in Leiden that reminds us of Shubnikov's stay, which in the end lasted for four years.

Another interesting document from Shubnikov's personnel file in the LFTI archives is a request to the Leningrad customs board to allow that a piece of soapstone (*steatite*) be sent from LFTI to Shubnikov at the laboratory of low temperatures in Leiden as he needed it for his experiments. As soapstone is not very expensive, nor exotic, and must have been readily available also in those days, it seems odd that it had to be ordered from Russia.

---

[11] According to the International Institute for Social History (www.iisg.nl) this would be equivalent to €1511 per month in 2013, not an awful lot, certainly if you have to pay for lodgings with full board, which cost them about 80% of that amount per month. But then taking lodgings with full board was not a common thing to do. The figure of €1511 also means that Ehrenfest's salary of 5880 guilders per year in 1927 would amount to €3702 per month in 2003 (about €45,000 per year), which seems rather low, as a professor's annual salary was roughly €100,000 in 2003. In a statement to the NKVD in 1937 Shubnikov himself says that his salary was 180 guilders, which is the same amount as Trapeznikova mentions with tax of 10% already taken off (see Appendix 3).

[12] Ref. [5], p. 33. At that time the maximum annual professor's salary was actually 7500 guilders.

[13] This lack of money of the Shubnikovs sounds rather peculiar as the ledgers of the Lorentz Fund, which was funded by the International Education Board of the Rockefeller Fund, i.e. it was essentially the same pot as Shubnikov was paid from, mentions rather large payments to visiting Russian physicists, such as Igor Tamm in 1926 and 1928, amounts of up to 2000 Dutch guilders, so Tamm seems to have received half a year's salary for a short stay of a month or so. Perhaps Ehrenfest was overly generous to his Russian friends.

The only record existing of Shubnikov's stay in Leiden and of his and his wife's impressions of the physics laboratory are Trapeznikova's reminiscences in the memorial volume published in 1990 [6]. The rest of this chapter is a reworked version of her story, at some places supplemented with reminiscences from other people. The picture in Fig. 4.3 shows De Haas with Shubnikov and Trapeznikova.

## The Leiden Physics Laboratory Through the Eyes of the Russian Visitors

The department of De Haas was located on the ground floor of the building of the laboratory, occupying several large rooms. The subject matter of the work was very diverse. Work was done on superconductivity and on magnetic properties of matter, such as measurements of the magnetic susceptibility (the degree of magnetisation of a substance in response to an external magnetic field) at low temperatures and the paramagnetic properties of salts. Eliza Wiersma (1901–1944), who would later play an important role in helping Shubnikov to build up the laboratory in Kharkov, worked on the adiabatic demagnetization of paramagnetic salts and on obtaining superlow temperatures. In 1935, using this adiabatic demagnetization method with diluted chrome alum (or *Chromium(III) potassium sulphate*), De Haas and Wiersma [11] succeeded in reaching the then record temperature of 0.0044 °K Other people worked on determining the specific heat of metals. Apart from the foreigners, everybody worked on subjects proposed by De Haas and under his supervision. De Haas was also a co-author on most papers. Apart from numerous visitors who just came for a few days, there were at the time six foreigners working for longer

**Fig. 4.3**  Trapeznikova, De Haas and Shubnikov in Leiden (*from* B.I. Verkin et al. (1990))

periods at the laboratory, among whom the three Russians, Shubnikov, Obreimov and Trapeznikova.

Seminars were not arranged at the laboratory, nor were there joint discussions of work. Each collaborator of the laboratory was given a large green notebook, in which a daily record should be made of the work carried out. On Saturdays all these notebooks were collected and taken home by De Haas, and on Monday they were returned. If necessary, he called the collaborator to come and see him for a discussion. In this way he was aware of all the work. It was not necessary to write any plans. It was neither necessary for De Haas to go through the laboratory and personally check the work of his collaborators (at the Leiden laboratory they were called assistants). This was probably done by Wiersma, who until 1928 was chief assistant and later curator of the laboratory. Wiersma issued the equipment needed for making measurements, mercilessly took it away when it was not used, and gave it to someone who actually needed it. The equipment did not stand idle. Wiersma also issued the necessary materials to a collaborator who did not have a key to the storeroom. The work went on in a very organized way, without any loss of time on red tape.

The administrative staff of De Haas was very small, just one secretary. Most collaborators of the laboratory, including Lev Vasilevich, had their own key to the storeroom where the materials were kept. In the storeroom there was a notebook in which one had to write down what and how much was taken, and another notebook in which a record was made of items that had to be replenished.

On the first floor De Haas had his installation with the magnet. He always made sure that de magnet was calibrated, everything should be in order. Trapeznikova never saw De Haas angry or lose his temper. He showed his displeasure in a delicate, but rather poignant way. At one time one of the assistants carelessly went up to the magnet while it was running with a tripod in his hands. The iron tripod was attracted by the magnetic field and it bumped into the shoe of the magnet, causing a scratch. As a result the shoe had to be polished and the field to be calibrated anew. After this incident for almost a whole year De Haas asked the employee in question each time he met him how this could have happened. For the poor employee these almost daily enquiries were worse than any reprimand.

The cryogenic laboratory was a sight that had to be shown to all visitors. It was headed by Gerrit Jan Flim, who was already the head technician under Kamerlingh Onnes, and more than just a technician. Flim ensured that all the work with liquefied gases could proceed trouble free. In the cryogenic lab there was much glass, including large glass Dewars. A demonstration was given of a continuous current circulating in a lead coil in a superconducting state. The danger of working with liquid hydrogen was pointed out, as well as the fact that not a single insurance company had agreed to insure the cryogenic lab. The light-weight ceiling was shown, which in the event of an explosion of hydrogen should be the first to fly off. It was strictly forbidden to smoke in the laboratory. Even Wiersma, who usually smoked in such a way that everything and everybody was showered with sparks, and Obreimov, who never parted from his pipe, never smoked in the cryogenic lab. The only place where it was allowed to smoke was in the hallway at the entrance to the institute.

Electric measurements were carried out in a special measurement room, the *"meetkamer"*, which had a switchboard. It was equipped with a large number of highly sensitive compensators, the best at the time, with Zernike and Mollier galvanometers, and a constant temperature was maintained. In order to eliminate vibrations originating from barges passing along the nearby canal, a special foundation had been constructed on which all measuring equipment had been mounted. All rooms of the laboratory had such a foundation. In order to use it you simply had to remove the floor board at the corresponding spot, and place the equipment on the foundation curbstone; equipment was never fixed to the walls.

The cutting-edge technology used for electric measurements at the Leiden laboratory was completely new to Shubnikov and Trapeznikova. In Leningrad no experiments involving physical phenomena at low temperatures were carried out. At LFTI only a small quantity of liquid air was available, which was used for utility purposes, such as the cooling of cold traps[14] when working with a vacuum or for vapour deposition of atomized metal on a cold substrate. Not a single one of the uses made of them by the collaborators of the Leiden laboratory was known to the Russian visitors. Everything was new.

At the laboratory there were several rather large Weiss[15] magnets with which all measurements in a magnetic field were carried out. All power facilities were in the basement. In the laboratory there were only circuit breakers and resistors in oil for adjusting the necessary current. Just in this period a large new magnet was being installed and subsequently mounted. In order to calibrate the field and to set a Dewar vessel with a sample between the poles, one had to get up a ladder. The working field reached 30 kilogauss. At the time Shubnikov was the only one who worked with this magnet, which only came into full use after 1932.[16]

Throughout the building pipes had been laid collecting vapours of cooled liquids, including hydrogen and helium, in gas tanks. For measurements in the temperature range from −24 to −217 °C cascade cycles were used: methylchloride, ethylene, methane and oxygen in the temperature interval from the boiling temperature to the triple point.[17] At any intermediate temperature an accuracy to 0.01 °C was maintained. In addition, a neon cycle also operated.

There were no lab assistants at the laboratory, but there were a large number of service staff and enormous workshops with a good technician at the head. In the large mechanics workshop orders were fulfilled rapidly. There was also an

---

[14]A cold trap is a device that condenses all vapours except the permanent gases into a liquid or solid (*Wikipedia*).

[15]Named after the French physicist Pierre-Ernest Weiss (1865–1940). He developed the domain theory of ferromagnetism in 1907. Weiss domains and the Weiss magneton are named after him.

[16]And in May 1933, using this new electromagnet, De Haas and Wiersma managed to reach a temperature of 0.27 °K, using adiabatic demagnetization, and went still further down to 0.08 °K somewhat later (Ref. [3], p. 607) and the already mentioned record temperature mentioned above.

[17]In thermodynamics, the triple point of a substance is the temperature and pressure at which the three phases (gas, liquid, and solid) of that substance coexist in thermodynamic equilibrium (*Wikipedia*).

electromagnetic workshop and a small glass-blowing workshop with one techni-cian, who carried out the work immediately without any formal orders. In the workshop a large number of young apprentices worked. The training program included plumbing, electromagnetic and glassblowing specializations. The apprentices rendered services to the collaborators of the laboratory: they mounted samples for measurements, checked electrical circuits, washed Dewar vessels and checked connections for leaks, in order that there were no leaks of hydrogen and helium. During the measurements carried out by the assistants the apprentices kept the temperature constant, by regulating the vapour tension above the cooling liquid by means of a differential oil manometer. In this way they spent the entire day.

When joining the laboratory each collaborator was issued with a box containing a set of necessary instruments (screwdrivers, keys, jars with rubber cement etc.). For all this a small amount of money was deposited as security. When the box was returned upon leaving the laboratory, the security was repaid, minus the value of any instruments lost.

There was a special room (the "*rekenkamer*") for doing calculations, writing papers, drawing figures and suchlike. Handbooks were also kept there, there were drawing boards as well as sets of flexible patterns and other necessary appliances.

In the summer there was vacation and the collaborators left. In one of her letters to Aleksandra Timoreva in Leningrad Trapeznikova described this time as follows: "Today there is a very pleasant event here, we break up for a one month holiday, for August. Actually, except for Lev and Ivan Vasilevich, nobody has worked the last few days, but officially today is the end. We have to go to the two professors, De Haas and Keesom, and wish them a pleasant holiday, and subsequently go around all the workshops, shake hands with the most senior technicians and make a nice bow to the apprentices, then make a visit to the liquid hydrogen and helium in the person of Flim. It is also a must to shake Crommelin's hand.[18] Finally you have to stay in your room and wait for the visits of the assistants who will surely come to shake your hand and wish you a pleasant holiday. This is very much the custom here; even before Christmas and Easter when we break up for 1.5 weeks, all go round to greet each other."

## The Shubnikovs in Leiden

In spite of the intensive work Shubnikov always was in a good and cheerful mood. Even when things went wrong, he did not get upset, and jokingly assured in Dutch that we should "*thee drinken en naar buiten kijken*" (go and drink tea and look out

---

[18]Claude August Crommelin (1878–1965) had been trained by Kamerlingh Onnes and was supervisor (from 1907) and assistant director (from 1924) of the laboratory (Ref. [3], p. 210). See also http://www.crommelin.org/familytree/1878ClaudeAugust.htm. Crommelin visited Moscow and Leningrad in 1924, where he gave a few lectures and Shubnikov and Trapeznikova may have met him.

of the window). He very much loved to sing. The morning always started with songs. Then they had breakfast and left for the laboratory. At 10 o'clock the doors of the laboratory were closed. In order to get in after ten, you had to ring the bell, the foreman came from the mechanics workshop to open the door. For Trapeznikova this was very unpleasant and she tried not to be late. She went to work together with Wiersma, who came round as he lived nearby. Lev Vasilevich was always later than ten at work. He was perhaps the only collaborator who took such liberty. But this did not cause any displeasure; everybody knew how much he worked. Lev Vasilevich had his own key to the entrance door, so he did not have to disturb anyone. One wonders why the Shubnikovs could not go together, so that he could let her in.

In the evenings after the closure of the laboratory, as a rule Lev Vasilevich went to the library to work. He was always very busy, neglecting his wife, although he had specifically demanded that she join him in Leiden, about which she complained in her frequent letters to her friend Aleksandra Timoreva. She wrote very often, while Lev, on the contrary, did not like writing letters, at least not to his parents.

Lev was also good at riding a bicycle. When they finally had also bought one for her, they went on cycling tours through the country.

For everybody it was a large event when it started to freeze and the canals became frozen. For two to three days the laboratory was closed and all collaborators and apprentices of the workshops went skating. Special shoes were not needed for the skates. Old fashioned Dutch skates were very different from modern ones. They consisted of little wooden foot-plates with metal runner blades underneath and could be tied with ropes or leather straps to any shoe, even to shoes with heels. According to Trapeznikova, all Dutchmen were able to skate well, only they, the two Russians, could not, but they were persuaded to have a try. Like always Lev Vasilevich immediately started to develop a great speed, assuming (not without reason) that speed gives stability, sometimes bumping into people who happened to be in his way. She very soon refrained from learning to skate.

In due course they got to know the collaborators of the laboratory better. Several times they were invited by De Haas for a cup of tea. His wife served them while De Haas asked questions about Russia and said that he would like to visit it. They became close friends with Eliza Wiersma and his wife Ali[19] (Fig. 4.4) and spent lots of time with them outside the laboratory, as well as with Ehrenfest's family. It was a strong friendship, according to Trapeznikova, also reflected in the fact that Wiersma came to visit the Kharkov laboratory almost every year in the early thirties, until entrance visas to foreigners were refused. Lev Vasilevich established a very good relationship with Wiersma. They happened to have been born in the same year and on the same day, 29 September 1901, and were both destined to die young, both by the actions of an authoritarian regime, Wiersma under and at least

---

[19]Alida Petronella Jöbsis, who was a chemist by education, but after her marriage became a housewife.

**Fig. 4.4** The Shubnikovs with the Wiersmas (*from* B.I. Verkin et al. (1990))

partly due to the German occupation of the Netherlands and Shubnikov by Stalin's butchers.

They also visited the Wiersmas on weekdays. Ehrenfest went there too. There was a lot of talk about physics. Here Trapeznikova experienced for the first time how Lev Vasilevich could question Wiersma for hours, for instance, about details concerning the construction of the electromagnet and much else. After such conversations Wiersma complained that he felt like a squeezed lemon. Lev Vasilevich himself was a restless person who did not know fatigue. The more difficult the issue, the better it seemed was his mood.

The Ehrenfests lived opposite them in the Witte Rozenstraat. Their house was built according to a design of Tatjana Alekseevna Afanaseva, Ehrenfest's Russian wife, and was rather unusual among the standard three-storeyed houses in Leiden. Towards the street it had a blind wall with a small viewing window below. The entrance was through the garden. Tatjana Alekseevna could not find any work in Leiden; she therefore spent much time in Simferopol in the Crimea, where she gave lectures on thermodynamics and statistical physics at the university.[20] She spent only two to three months per year in Leiden. Her mother Ekaterina Uljanovna stayed with Ehrenfest and the children. Trapeznikova thought Ehrenfest an amazing person. You opened the door and saw his shining eyes; everything around him lit up. It was a joy to see his shining eyes. But still at times Ehrenfest suffered from the most severe depression. Trapeznikova remembers that they, he, Lev Vasilevich, Wiersma and herself, made a walk and he talked about what worried him. In 1933, in order not to burden his other children with his youngest son who had Down syndrome, he killed the son and then himself. They could then not imagine that he would do such a thing.

Ehrenfest played a tremendous role in Leiden scientific life. There were always a dozen theorists working with him. He was a friend of De Haas and, although he

---

[20]The Taurida National V.I. Vernadsky University. Since the Russian annexation of Crimea in 2014 it is called the Crimean V.I. Vernadsky Federal University.

seldom visited the laboratory, he was familiar with all the results. And most importantly, Ehrenfest held famous theoretical seminars, attended by scientists from all over the world. The seminars took place in the auditorium of the theory building. First there was a lecture from one of the visiting celebrities, followed by a tea party, for which Trapeznikova made the tea and bought the biscuits, and a discussion without any time restriction. Shubnikov, Wiersma and Trapeznikova attended all Ehrenfest's seminars. At various times they listened to Albert Einstein, Max Planck, Max Born, Arnold Sommerfeld, Wolfgang Pauli, Erwin Schrödinger, Percy Bridgman (1882–1961),[21] Peter Debye, Dirk Coster (1889–1950),[22] Adriaan Fokker (1887–1972) and others, but never to Niels Bohr and Ernest Rutherford. In this way they met many famous physicists (Fig. 4.5).

In 1928 Pëtr Kapitsa came to Leiden from England. Abram Ioffe and his wife, Anna Vasilevna, visited once, as did other Russians. The arrival of Sergej Frish was a momentous occasion for the Shubnikovs. He was on *komandirovka* in Groningen in the Netherlands, where he worked with Dirk Coster, and came to Leiden for four days.[23]

Many theorists visited Ehrenfest. Einstein and Pauli always stayed at Ehrenfest's home. But Einstein did not want to eat the canteen food that was brought to Ehrenfest. Wiersma lived opposite and Einstein went to have dinner at his place. Ali was a very good housewife and a very good cook. Einstein subsequently wrote her a long thank-you poem.

On Ehrenfest's invitation Igor Tamm presented Dirac's relativistic electron theory at Ehrenfest's seminar. Tamm and Paul Dirac became friends when the latter also visited Leiden in the same summer of 1928; a friendship that aroused considerable surprise among physicists as the two men had completely opposite characters—Dirac being extremely introvert and Tamm extrovert.[24]

When it became known that Lev Landau would come to Leiden from Copenhagen, Ehrenfest became very worried. He had heard from Ioffe[25] about Landau's alleged bad temper, and feared unpleasantness terribly. Indeed Landau sometimes behaved provocatively at the seminars at LFTI. In Trapeznikova's

---

[21]Percy Williams Bridgman: American physicist who won the Nobel Prize in 1946 for his work on the physics of high pressures. Bridgman committed suicide by gunshot after living with metastatic cancer for some time. His suicide note read in part, "It isn't decent for society to make a man do this thing himself. Probably this is the last day I will be able to do it myself." As recalled in Chap. 3, he developed a method for growing crystals which was very similar to the earlier Obreimov-Shubnikov method.

[22]Student of Ehrenfest who got a PhD in 1922 and became professor in Groningen. Also worked in Copenhagen with Niels Bohr.

[23]He wrote a little piece in the popular scientific journal *Vestnik znanija* (no. 17–18 (1929) 621–622) about the low-temperature laboratory in Leiden, as a letter from "Germany" under the heading *Avanposty mirovoj nauki* (Outpost of world science).

[24]See e.g. the excellent recent biography of Dirac by Graham Farmelo (G. Farmelo, *The Strangest Man: The Hidden Life of Paul Dirac, Quantum Genius* (Faber and Faber, London, 2009)).

[25]This was in 1928 or a little earlier. So already at that time the relationship between Ioffe and Landau was not very good, it seems.

**Fig. 4.5** Trapeznikova walking along a canal near Leiden with Obreimov and Kapitsa (*from* B.I. Verkin et al. (1990))

opinion he was in general a shy person and out of shyness behaved rather awkwardly. Ehrenfest charged Trapeznikova, who unlike Shubnikov knew Landau fairly well from Leningrad University, with keeping Landau in check, so that he could not cause any scandal. Shubnikov and she were very happy to see Landau. They tried to entertain him, went to the cinema, and immediately started to address each other with the informal "ty", which was rather uncommon in Russia (at the university they all addressed each other with the formal "vy"). Subsequently they went to the cinema for a second time, taking Wiersma along. Landau's character was somewhat prickly, but here he met people who were very cordial to him and took a fancy to him. He was very sweet, at the seminar everything turned out very well, Landau behaved himself and everybody liked him, in particular Ehrenfest. Thus the friendship with Landau started, which continued and became stronger at Kharkov.

The pride of the Dutch was Hendrik Lorentz (Fig. 4.6), Ehrenfest's predecessor as professor of theoretical physics in Leiden and in 1902 the second winner of the Nobel Prize in physics; he enjoyed the love and esteem of the entire Dutch nation. In the last years of his life Lorentz lived in Haarlem, but he could often be seen in Leiden, where he still lectured at the University.

De Haas was married to Lorentz's daughter (she was also a theoretical physicist), therefore when Lorentz got ill, the people at the laboratory were constantly informed about his illness and expressed their sympathy with De Haas. When he died on 4 February 1928, Haarlem was in mourning. The death of Hendrik Lorentz

**Fig. 4.6** Hendrik Antoon Lorentz, in 1916, painted by Menso Kamerlingh Onnes (*Leiden University*)

brought grief to the entire population. The funeral was on 9 February, and on the route of the procession all lampposts in the streets were covered in black crêpe. The crowd accompanying the catafalque was very large, but only those invited by the Lorentz family were admitted to the cemetery. The Shubnikovs also obtained an invitation and were present at the internment. They urgently had to obtain a hat for Lev; the cap he always wore was not a suitable headdress. Einstein, Ehrenfest and many others attended.

In April 1928 during the Easter holidays they made a bicycle trip through Holland, possibly on the initiative of Igor Tamm.[26] In the summer holidays of the same year, in August, the Shubnikovs went to France. They took their bicycles with them. They stayed in Paris and Versailles, and from there travelled to Brittany. Shubnikov was not only interested in the old cathedrals of Brittany, but also in prehistoric structures: dolmens, menhirs, etc. constructed of stones of enormous dimensions.

They very much wanted to go to England. Pëtr Kapitsa had invited them to Cambridge. But life there was very expensive, and they could not afford such a trip. In the next summer holidays, in 1929, they went to Germany to Cologne, and then by boat along the Rhine from Cologne to Mainz. For the return journey to Leiden, they decided to take a plane; their first flight in an airplane, first in a four-seater from Cologne to Essen and from there in a twelve-seater to Amsterdam.

---

[26]This trip is also mentioned by Ehrenfest in his letter of 13 April 1928 to Abram Ioffe. He also writes that the trip was organized thanks to Tamm's energy. From postcards sent by Ehrenfest's daughter Tanechka to her father he knew that the trip was a great success.

The time in Leiden was very fruitful for Shubnikov and Leiden played an enormous role in his formation as a physicist. He started off as a junior physicist, but left as a very experienced experimentalist, who had become aware of his vocation and was working at the cutting edge of physics research. He benefited from the high level of research carried out in Leiden, from Ehrenfest's famous seminar and from the many visitors, theorists and experimentalists, from all over the world including Russia who came to Leiden. He had the opportunity to associate with the world's most outstanding theorists and experimentalists, and brilliantly used the opportunities presented to him. Shubnikov found himself in a first-class school of physicists and in four years he grew into one of its brightest representatives.

De Haas had wanted Shubnikov to stay on for some more years in order to finish experiments at still lower temperatures with the crystals he had prepared, and he and Ehrenfest made an effort through Ioffe to achieve this, but his *komandirovka* could not be extended (Ref. [7], p. 11). According to Landau Shubnikov returned to Russia voluntarily, but according to Trapeznikova when he went to Berlin in the spring of 1930 to renew his visa he was told that it was time to return (Ref. [6], p. 277). This can hardly be true, however, for once in the Netherlands there was no need to go abroad to have a visa extended; that could be done with the local police. Trapeznikova neither says that she herself had to go through the same procedure, which would have been logical. If Shubnikov had to go there, then she too, one would think. I believe that he actually had to go to Berlin to visit the Soviet embassy in order to have his *komandirovka* extended, which initially had been for only six months (as stated in Ioffe's letter to Kristi quoted at the beginning of this chapter). The archives of the Ioffe Institute contain documents in which Shubnikov requests an extension of his *komandirovka*, e.g. a letter by Shubnikov to the administration of LFTI dated 10 December 1927 requesting an extension until 1 June 1928, and a further letter from the deputy director of LFTI to Glavnauka asking permission for such an extension, since "the work has not yet been finished". It is quite possible that this procedure had to be followed every six months and that Shubnikov had to go to Berlin to explain at the embassy why an extension was needed, but it is still rather odd as it is unclear why this could not be arranged via a simple exchange of letters between Leiden and Russia. There is also a rather cryptic remark in a postscript to Ehrenfest's letter of 1 May 1928 to Ioffe: "De Haas is more than surprised that there is no *news whatsoever* from Shubnikov who should have been here on *1 May*. Maybe he intends to send a telegram after his arrival?" (Ref. [12], pp. 199–200), which only makes sense if Shubnikov had gone to Russia and not to Berlin. How else could it have had anything to do with Ioffe? At any rate no answer from Ioffe is known.

Ioffe's 1926 letter to Kristi referred to above has also revealed that after his return from Leiden Ioffe intended to charge Shubnikov with setting up a cryogenic lab in Leningrad. So, Ioffe saw Shubnikov's trip to Leiden as a useful means for getting a Soviet scientist to acquire expertise on low-temperature physics, with the aim to exploit this later in the Soviet Union. That Shubnikov's fate had indeed already been decided earlier is also revealed by a letter dated 6 January 1929 (i.e. a full year before Shubnikov allegedly went to Berlin to try to renew his visa or passport) from Ivan Obreimov in Leningrad to Pëtr Kapitsa in Cambridge. In this letter Obreimov stated that Shubnikov would be one of the scientists to work at the new institute in Kharkov, which for that matter still had to be built and of which assignment Shubnikov was at that time perhaps not yet aware (Ref. [13], p. 553).

In the spring of 1930 he was told that it was time to return to the Soviet Union, which must have been around the time when he wrote the following report (in the concise style typical of Shubnikov) about his work of the six months from October 1929 to March 1930 to Ioffe:[27]

For approval! 12.IV.30

A. Ioffe

Report on scientific *komandirovka* October-March 29/30

L. Shubnikov, Natuurkundig Laboratorium Leiden, Holland

In the said time the following has been carried out:

§1. Published a method of obtaining Bi crystals of any form and orientation, L. Shubnikov, Proc. Amst. Acad. 1930.

§2. Discovered a new phenomenon of the electro-resistance of crystals of Bismuth in a transverse magnetic field and a temperature of 14 °K. Preliminary publication:

L. Shubnikov and de Haas, Proc. Amst. Acad. At the end of the month a detailed paper will be published in Proc. Amst. Acad.

§3. Measured the electro-resistance of Bi crystals (without a magnetic field) at temperatures down to 1 °K inclusive. Found the cause of the repeatability of the electro-resistance. Paper is ready to be printed and will be published in April.

Currently the following is being worked on:

(1) I am preparing the equipment for continuing the work of §2 to higher fields and at temperatures of liquid helium (1.5–4.2 °K).
(2) Have started an investigation similar to §2, but in a longitudinal magnetic field.

I hope to finish the work indicated under (1) and (2) by 1 July 1930.

3 April 1930 L. Shubnikov

The new phenomenon, casually mentioned in §2 of the above letter, is the Shubnikov-De Haas effect. At that time, April 1930, Shubnikov was certainly very familiar with the plans of Ioffe and others to establish a new institute in Kharkov in the Ukraine. Kapitsa had been asked to become the head of a department of this institute, and Obreimov had discussed questions of the organisation of the institute with Shubnikov as is evidenced by the following letter from Shubnikov to Kapitsa (Ref. [7], p. 11).

Leiden, 8 December 1929

Much-esteemed Pëtr Leonidovich!

---

[27]Archive AN SSSR, Leningrad, folio 1034, inv. 3, quoted in Ref. [7], p. 11. The fact that Shubnikov wrote such reports also shows that he still had a strong affiliation with LFTI.

I have heard from Ivan Vasilevich that you will shortly organise a department of our laboratory in Kharkov.

Upon arrival in Kharkov (VII–VIII 30) I thought to engage there in roughly the same topics as I worked on here in Leiden. This field is closely related to your recent work, therefore I would like to continue my work in your laboratory.

Please be so kind as to inform me how soon you will set up your laboratory and whether I can expect to work at your laboratory.

Sincerely yours,

Lev Shubnikov.

Could you please send me a reprint of your recent work on the electro-resistance in a magnetic field?

So already in 1929 Shubnikov expected to go to Kharkov in July or August 1930, immediately after finishing his work in Leiden. For his part Obreimov also very much expected to attract Shubnikov to work at Kharkov, as he wrote to Kapitsa (Ref. [7], p. 11):

24.04.1930

Dear Pëtr Leonidovich!

… You proposed to establish a department of your laboratory with us. Do you not think that the best start for this would be if you allowed Sinelnikov to continue the work started with you here, together with Shubnikov, who also will soon return? They would form a great core group which could form the basis of your institute…

Yours I. Obreimov

How this would develop further will be related in Chap. 6.

The result was that although De Haas was very much interested in carrying out research on galvanomagnetic phenomena at low temperatures, and wanted to assign a leading role in this work to Shubnikov, this could not go ahead due to the other plans the Russian had made.

And so in the summer of 1930 the Shubnikovs left hospitable Leiden and would never return.

De Haas and Keesom were convinced that after having returned to Russia Shubnikov would establish and develop low-temperature physics, and would manage a cryogenic laboratory. They promised to help where possible, and subsequently fulfilled their promises. Behind all this was the personal relation towards Shubnikov. And there were grounds for this. Shubnikov was a wonderful experimentalist. He understood the structure of everything, he knew and felt the machines. Trapeznikova, who apparently only got to know the scientific skills and talents of her husband in Leiden, was amazed about his purposefulness, his energy, his very clear understanding of what and how to do something, and she wonders how he learned all this.

Shubnikov's scientific work, as well as Trapeznikova's scientific activities, in Leiden will be discussed in the next chapter.

# References

1. M. and B. Ruhemann, *Low Temperature Physics* (Cambridge University Press, 1937).
2. Emilio Segré, *From X-Rays to Quarks* (Dover Publications, 1980).
3. Dirk van Delft, *Freezing Physics: Heike Kamerlingh Onnes and the quest for cold* (Koninklijke Nederlandse Akademie van Wetenschappen, Amsterdam, 2007).
4. Ju.V. Pavlenko, Ju.N. Ranjuk and Ju.A. Khramov, *"Delo" UFTI 1935-1938* (The "UFTI" Case 1935–1938) (Feniks, Kiev, 1998).
5. M.J. Hollestelle, *Paul Ehrenfest—Worstelingen met de moderne wetenschap, 1912–1933* (Leiden University Press, 2011).
6. O.N. Trapeznikova in B.I. Verkin et al. (1990), p. 256–291.
7. B.I. Verkin et al. (1990).
8. L. Chamberlain, *Lenin's Private War, The voyage of the philosophy steamer and the exile of the intelligentsia* (St Martin's Press, New York, 2006).
9. Walter M. Elsasser, *Memoirs of a Physicist in the Atomic Age* (Science History Publications, Adam Hilger, Bristol, 1978).
10. V.Ja. Frenkel and P. Josephson, Sovetski Fiziki—Stipendiaty Rokfellerovskogo Fonda, *Usp. Fiz. Nauk* 160 (1990) 103–134.
11. W.J. de Haas and E.C. Wiersma, Adiabatic demagnetization of some paramagnetic salts, *Physica* 2 (1935) 335.
12. N.Ja. Moskovchenko and V.Ja. Frenkel, *Ehrenfest-Ioffe Nauchnaja Perepiska* (Ehrenfest-Ioffe Scientific Correspondence) (Akademija Nauk SSSR, 1990).
13. P.E. Rubinin, P.L. Kapitza and Kharkov. Chronicle in letters and documents, *Low Temp. Phys.* 20 (1994) 550–578.

# Chapter 5
# Shubnikov's Scientific Work in Leiden: Shubnikov—De Haas Effect

Bismuth, the substance Shubnikov was going to work with, had for some time been a favourite material of experimentalists. At the time of Kamerlingh Onnes it had been used in many experiments carried out in Leiden and elsewhere, where the problem had always been to get sufficiently pure samples. At ambient temperatures bismuth has a layered structure and shows an extremely strong fluctuation of the resistance in a magnetic field, as well as substantial diamagnetism, meaning it is repelled by a magnetic field, as an applied magnetic field creates an induced magnetic field in the opposite direction. Also characteristic is a reduction of volume (the crystal structure changes) and of the resistance under pressure. These phenomena occur in other metals too, but are most pronounced in bismuth. Already in 1895 Pierre Curie had observed that the diamagnetic properties of bismuth did not vanish at melting temperature as is usually the case, but at a somewhat lower temperature. In the same temperature region Bridgman had observed a change of the sign of the temperature coefficient of the resistance, while Philip Lennard had observed anomalies in the resistance at an alternating current. A number of authors even proposed the existence of two modifications of bismuth.

At the time when De Haas decided to embark on the investigation of the behaviour of bismuth in a magnetic field, it was already known that its magneto-resistance increases much faster with an increase of the field than for other metals, and that for mono-crystals this phenomenon depends on the orientation of the field to the crystal axes. It was also known that the resistance of bismuth in a magnetic field increases rapidly with decreasing temperature,[1] but different samples of bismuth gave different results. It was therefore not surprising that De Haas decided to extend the investigation of the magneto-resistance of bismuth to hydrogen and helium temperatures (i.e. down to a few degrees Kelvin). It is possible that the interest of Paul Ehrenfest also played a stimulating role in this respect, as he had devoted a few papers to the problem of the diamagnetism of bismuth [1].

At Cambridge Kapitsa had also performed experiments on bismuth, but at higher temperatures and with less pure crystals, hypothesizing that some of the anomalies

---

[1]This property of bismuth was actually often used to measure magnetic fields, but its behaviour was still far from understood.

© Springer International Publishing AG 2018
L. J. Reinders, *The Life, Science and Times of Lev Vasilevich Shubnikov*,
Springer Biographies, https://doi.org/10.1007/978-3-319-72098-2_5

he had observed had to do with imperfections in the crystal. These anomalies included cracks (defects) in the mono-crystals of bismuth he had used, a problem which would later be studied by Trapeznikova (see later in this chapter). Kapitsa had observed that in weak fields the electric resistance is proportional to the square of the field, and in strong fields linear with the field [2]. In this "bismuth-fever" each side had its own advantage: Leiden had low temperatures, Cambridge strong magnetic fields. Leiden worked at magnetic fields that were a few orders of magnitude lower (with a permanent magnet of up to 30 kilogauss), while Cambridge had only temperatures in the nitrogen range (about −200 °C) at its disposal. After completion of the experiments it became clear that for this problem low temperatures were the more important (Ref. [3], p. 22).

De Haas also suspected that imperfections in the crystals were the reason for the inconsistency in the results. Therefore, Shubnikov's first task in Leiden was to prepare very pure single crystals of bismuth. He started from the purest bismuth commercially available, purified it further by chemical means, and used this very pure bismuth for growing the single crystals he needed. For this growing process he used a method that was very similar to the one developed by Kapitsa and independently by Goetz [4]. The method permits the production of single crystals of metals, with practically no limit as to size or desired orientation, thus indicating that all external mechanical influences are avoided. Furthermore, the method permits the zone of formation in a growing crystal to be subjected to a strong magnetic field.

All other methods, including Shubnikov's own with Obreimov, had to be rejected or substantially modified since they were not suitable for growing the crystals in the geometric forms he needed for the experiments. His conditions for the crystals were that they should have a pre-determined geometric form and that it should be possible beforehand to fix the position of the crystallographic axis.[2] To this end an intricate piece of equipment containing two air-tight cavities, similar to the one used by Kapitsa, was constructed. In the larger cavity the bismuth was melted and subsequently cautiously pressed through a narrow gap connecting the two cavities into a narrower, needle-like cavity. There it was brought into contact with a seed crystal in the form of a rectangular rod of dimension 2.5 × 2.5 × 22 mm. The seed crystal could be introduced in such a way that the direction of the main crystallographic axis of the crystal to be grown would coincide with the axis of the seed crystal and the latter axis could be manipulated in the device to have a fixed orientation with respect to the outside form of the crystal to be grown.

Shubnikov continued to measure the residual resistance of bismuth as a function of the number of recrystallizations and its dependence on the conditions of growth and deformation. After a year, when Trapeznikova had arrived in Leiden, she assisted him in this work. To make sure that the crystals grown were as pure as possible the crystallization process was repeated a number of times in order to

---

[2]Kapitsa had also tried to use the Obreimov-Shubnikov and the Bridgman methods to obtain crystal rods of about 3–5 mm in length with the crystallographic axis parallel and perpendicular to the axis of the rod, i.e. with perfect cleavage planes perpendicular or parallel to the length of the rods, but had not succeeded and developed an alternative method.

remove at each subsequent crystallization contaminated parts of the crystal. The presence of contamination could be deduced from the fact that the value of the residual resistance continued to decrease. After seven or more crystallizations the residual resistance did no longer change and did not differ in value from the residual resistance of other very pure metals, indicating that no further increase in purity could be achieved.

It was important that no tension arose in the crystallization process. To this end a little mould of *steatite*[3] was used which expanded during the growth of the crystals. The crystallization took place in vacuum. This work had great practical significance, especially for the Leiden laboratory.

The purity was verified by spectroscopic analysis which showed that the decrease of the residual resistance at recrystallization is uniquely connected with the decrease of the intensity of the lines of silver and lead, which are therefore the most significant impurities that remain after recrystallization. They were removed by chemical means. It took Shubnikov a whole year to obtain crystals of the required, for that time unique, purity, which were subsequently exposed to high magnetic fields at low temperatures.

Shubnikov wrote a paper about the procedure he had applied. Contrary to the other papers he published from Leiden, it had only him as an author [5], as it was very much his own work. An extended version of this paper, including also a survey of the research carried out on the crystals was published with De Haas in the Journal of the Russian Physico-Chemical Society [6], clearly meant for a Russian audience.

Shubnikov was always very precise in references to work of other authors. In the above-mentioned paper on crystallization he cited many crystallization methods of other authors, but there is no reference to a purification method. It is obvious that he would not have started the investigation of the dependence of the residual resistance on the number of crystallizations if it had already been done before by somebody else. Shubnikov was the first to use this method. Nowadays it is well known that measuring the residual resistance at low temperatures is one of the most sensitive methods of determining the purity and perfection of a metal. However few people know that Shubnikov and De Haas were the first who proposed to use the ratio of the resistance at room temperature and at helium temperatures as a criterion for the perfection of a sample—a measure for the degree of purity and the absence of stresses.

The measurements carried out by Shubnikov on the residual resistance of samples of "pure" bismuth, supplied by various firms, showed that at nitrogen temperature all samples behave roughly the same, but at the temperature of boiling

---

[3]This must have been the *steatite* brought over from Leningrad as mentioned before.

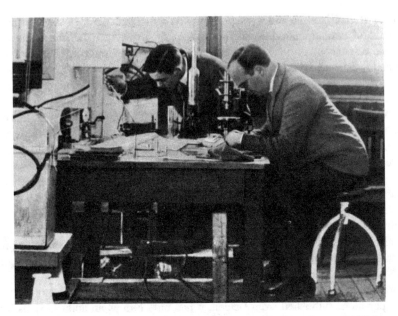

**Fig. 5.1** Shubnikov and Blom at work in the laboratory (*from* B.I. Verkin et al. (1990))

hydrogen the bismuth of the firm Hartmann and Braun,[4] used by Kapitsa in Cambridge, showed a residual resistance that was orders of magnitude higher than the bismuth of the firm Hilger,[5] indicating that the latter was much purer. The Hilger bismuth served as the original raw material for the Leiden measurements. However, even samples grown from the best Hilger bismuth displayed a significant variation in the residual resistance. This was systematically investigated further. First the mono-crystalline nature of the crystals was verified: the samples were polished and examined under the microscope to make sure there were no visible traces of impurities, like grains or twins (Fig. 5.1). Subsequently they were

---

[4]The firm was founded in 1879 in Germany by Eugen Hartmann, who was joined in 1882 by Wunibald Braun forming the firm E. Hartmann & Co and continuing as Hartmann and Braun from 1901. By the turn of the century it had become a leading supplier of instrumentation devices. After several takeovers in the post-war years, it was finally taken over in 1999 by Asea Brown Boveri (ABB) AG and fully integrated into the ABB Group.

[5]Hilger was founded in 1874 by two German precision optical instrument technicians Adam and Otto Hilger. They had fled Germany a few years previously to escape religious persecution and set up manufacturing of precision optical and mechanical instruments in London. The company prospered and also started to make high quality synthetic crystals since crystals that had previously been used did not keep pace with the increasing demands on optical components for infrared spectroscopy. Hilger started to grow and process synthetic crystals in house, which implied the birth of crystal growth at Hilger. In 1948 Hilger amalgamated with the metrology instrument manufacturer E.R. Watts and the company became known as Hilger and Watts. In July 2010 Hilger was acquired by the US corporation Dynasil. (Information from the website www.hilger-crystals.co.uk.)

inspected by X-rays. It was established that along their whole length the samples had the identical orientation and were mono-crystals to a very high degree.

The first measurements of the magneto-resistance of bismuth carried out on samples from the firm Hilger showed that the simple dependence of the resistance on the magnetic field, observed by Kapitsa at room and hydrogen temperatures, is not satisfied at lower temperatures: the curve of the increase of the dependence of the resistance on the field shows a much more complicated behaviour, showing clear signs of non-monotonic behaviour. The same experiments observed an extremely strong sensitivity of the magneto-resistance on mechanical stresses and on the presence of very small quantities of impurities (at the level of $10^{-4}$ %), which was a completely new factor and was in itself already an important physical result (Ref. [3], pp. 22–23).

Shubnikov and De Haas wrote two papers on the development of methods for purifying and obtaining crystals of bismuth, as well as on the dependence of the residual resistance of monocrystals of bismuth on the purity of the metal in the temperature range from 0 °C to 1.3 °K. Their research did not only involve bismuth of the firm Hilger, but also of the firms Hartmann and Braun and Kahlbaum.[6] These papers deserve special attention, as the significance of the work invested by Shubnikov in the process of obtaining unique crystals of bismuth can only be appreciated by reading them. Until Shubnikov's work nobody had expected that a tiny pollution such as the presence of a single foreign atom on a million bismuth atoms would have such a pronounced effect. In their paper Shubnikov and De Haas wrote that the influence of the purification is so great that at 20 °K the extra purified bismuth shows a residual resistance which is about three times smaller than of the already very pure Hilger bismuth and about twelve times smaller than that of the purest bismuth of Kahlbaum. The residual resistance measured in these mono-crystals was fully comparable with that of the purest metals [7].

After this preliminary work on purification and on growing pure crystals, they could now proceed to investigate the magneto-resistance.

The Dutch physicist Hendrik Casimir (1909–2000), who at the time was a graduate student with Ehrenfest, writes in his memoires about this work (Ref. [8], pp. 335–336): "*Around 1930 the Russian physicist Shubnikov, together with his charming wife A.[7] Trapeznikova, spent quite some time at Leiden and did remarkable work on the magneto-resistance of bismuth. It was known since the end of the nineteenth century that the electrical resistance of bismuth increases considerably in a magnetic field and it was also known that this effect becomes more pronounced at low temperatures (...) but I do not think anybody had expected an increase by a factor of several hundred thousand in a field of 30,000 Gauss at a temperature of 14 °K. Such spectacular results were obtained with extremely pure crystals: Shubnikov started from the purest metal he could get and recrystallized it numerous times.*

---

[6]A German firm in Berlin. The Firma Kahlbaum Laborpräparate was founded in 1890, merged in 1927 with Firma Schering to form Schering-Kahlbaum AG, which was acquired in 2006 by Bayer AG.

[7]Casimir cannot have known the Shubnikovs very well, else he would certainly have remembered that her first name was Olga; so it is also doubtful that he knew her as being charming.

*Post-World War II semiconductor technology has familiarized us with the fact that minute quantities of "dope" may drastically modify physical properties, and the art of growing pure crystals has reached a high level of perfection. In a way, Shubnikov's results anticipated these developments".*

As is clear from the quote from Casimir's book, it was not unexpected that at hydrogen temperature (about 20 °K (−253 °C), the temperature at which hydrogen becomes liquid) the increase of the resistance when exposed to a strong magnetic field would be larger than already measured at higher temperatures, but nobody had expected that at 14 °K (−259 °C), the temperature at which hydrogen becomes solid, the resistance would increase by a factor of several hundred thousand in a magnetic field of 30,000 Gauss. It also oscillates as a function of the magnetic field, the first observation of such oscillation of a physical quantity. Such results were obtained with extremely pure single crystals. The results of the measurements, which took most of Shubnikov's remaining three years in Leiden, caused considerable excitement and were published in a number of papers, including a publication in *Nature* [9] (Fig. 5.2).[8] This quantum mechanical effect (i.e. the oscillation of the resistance of a substance at very low[9] temperatures in a high magnetic field) is now known as the Shubnikov-De Haas effect[10] and is the main result of Shubnikov's four-year stay at the Leiden laboratory. A long time passed before the effect got the recognition it deserved. For some time the results of De Haas and Shubnikov were considered an exotic anomaly only applying to bismuth—a semimetal with a small number of charge carriers.

The experiments also established that the magneto-resistance is very sensitive to the orientation of the crystal. This confirmed the necessity of working with exceptionally pure mono-crystals with a fixed orientation of the axes. Shubnikov solved the problem of obtaining mono-crystals with fixed geometric characteristics by improving Kapitsa's method and combining it with the Obreimov-Shubnikov method.

The good reproducibility of the results on samples obtained in this way when increasing and decreasing the magnetic field, as well as the complicated field dependence that emerged when using the most perfect crystals with the lowest residual resistance, was reason to assume that a real physical effect lies at the basis of this dependence.

Shubnikov and De Haas also studied the effect of annealing (to remove residual internal stresses) over several days at pre-melting temperature, the effect of small plastic deformations and the thermal conditions of growth, which showed that only the latter significantly affect the quality of the crystals. They suggested that

---

[8]And six publications in *Proc. Roy. Acad. Amsterdam* 33 (1930) 130–133; 327–331; 351–362; 363–378; 418–432; 433–439, which were also published as *Leiden Communications* (207a (1930) 3; 207b (1930) 9; 207c (1930) 17; 207d (1930) 35; 210a (1930) 3; 210b (1930) 21).

[9]At nitrogen temperatures (about −200 °C) the quantum effects already disappear.

[10]It was apparently first called as such in 1950 by B.I. Verkin, B.G. Lazarev, N.S. Rudenko, Periodicheskaja zavisimost' vospriikchivosti metallov ot polja pri nizkikh temperaturakh (Periodic dependence of the susceptibility of metals on the field at low temperatures), *Zhurn. eksperim. i teoret. fiziki* 20 (1950) 93–97. This shows that Shubnikov's name was not completely shunned by physicists even before Stalin's death.

500 *NATURE* [OCTOBER 4, 1930

### Letters to the Editor.

*[The Editor does not hold himself responsible for opinions expressed by his correspondents. Neither can he undertake to return, nor to correspond with the writers of, rejected manuscripts intended for this or any other part of NATURE. No notice is taken of anonymous communications.]*

**A New Phenomenon in the Change of Resistance in a Magnetic Field of Single Crystals of Bismuth.**

FROM many investigations it is well known that bismuth shows very variable behaviour. We have investigated a very pure specimen; Hilger's bismuth was purified still further. The crystals made from this material proved to be excellent. With X-rays they show very sharp interference spots or lines, and when compressed nothing could be observed of the phenomenon of 'cracks' as described by Borelius, Lindh, and Kapitza (*Proc. Roy. Soc.*, A, vol. 119, p. 366; 1928). From these crystals we measured the change of resistance in the magnetic field at different temperatures.

First we determined the change in resistance of several crystals having the principal axis parallel to their length. The current flows in the length direction of the crystal. The rod is put in the magnetic field with its length (that is, principal axis) at right angles to the lines of force of the field, and it is possible to turn it round an axis coinciding with the principal axis.

We determined the curves giving the change of resistance as a function of the intensity of the magnetic field, when one of the binary axes was either parallel to or at right angles with the field. These curves show a very complicated form, extremely so if the binary axis is at right angles with the field.

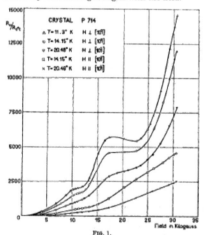

FIG. 1.

In Fig. 1 these curves are given for the temperatures 20·48° K., 14·15° K. and 11·3° K. The abscissæ are the intensities of the magnetic field; the ordinates are the values of $R_H/R_{0\,C}$. $R_H$ is the resistance in the magnetic field at low temperatures; $R_{0\,C}$ the resistance without a field at 0° C. It will be seen that the curves do not show a parabolic part in the beginning which

No. 3179, VOL. 126]

gradually changes into a linear part at higher field strengths. It has been found that the whole phenomenon strongly depends on temperature : at higher temperatures the curves become more and more simple. This can already be seen at 20·48° K. Here the first flat part found at about 9·5 kilogauss, and prominent at 11·3° K., has nearly disappeared. Measurements at higher temperatures, for example, 64·25° K. and 77·40° K., show a very simple curve, just as has been found hitherto at all temperatures.

In order to investigate the phenomenon more thoroughly we measured the change of the resistance, keeping the field constant, but changing gradually the angle between a binary axis and the lines of force, and

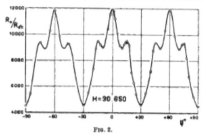

FIG. 2.

reading the resistance and the angle after each displacement. In Fig. 2 the abscissæ are the angle $\phi$, between the crystallographic direction [112] in the crystal and the lines of force, ordinates are the values of $R_H/R_{0\,C}$ at those different angles, in a field of 30·650 gauss at a temperature of 14·15° K. This curve does not show cosine form, but gives a much more complicated relation of the resistance to small changes of the angle. Simple cosine curves have been found only at very low field strengths. At higher temperatures we do not find the complicated form.

We are now investigating some crystals having two different orientations. Both these orientations have the principal axes at right angles with the length of the crystal. For the first orientation, the length coincides with the direction of a binary axis (and with the axis round which the crystal can be turned, it being also at right angles with the lines of forces and coinciding with the direction of the current). For the second one all this is the same, but the length coincides now with the direction of a bisectrix of two binary axes. Here, too, we investigated the change of resistance with temperature, field strength, and angle of the principal axis with the field. The most important result of these investigations is that the curves have much in common with those given above for the other orientation (Figs. 1 and 2).

Here, too, we find at low temperatures that the resistance in the field changes rapidly with small changes of $\phi$. This phenomenon disappears only when we pass to high temperatures and to weak magnetic fields. Of course the form of the curve giving $R_H/R_{0\,C}$ as a function of $\phi$ is in this case quite different from the one given in Fig. 2.

The results are very much influenced by the purity of the material used for the crystals. As an indication of this purity, it may be stated that our crystals show at 1·3° K. a resistance having a value of some thousandths of that at 0° C. At 11·3° K. the resistance in a magnetic field of 31 kilogauss is 922·000 times higher than that without the field.

L. SCHUBNIKOW.
W. J. DE HAAS.

University of Leyden.

**Fig. 5.2** Paper in *Nature* by Shubnikov and De Haas

impurities, and their different distribution at different growth regimes, play a decisive role in this respect. Before Shubnikov did this work, there were few concerns about the purity of the samples used. At any rate it was not given due consideration, as can be judged from Kapitsa's work of 1928 ([2], quoted above), for instance, who just used commercially available bismuth and not even the purest one, although he seems to have been aware of the fact that imperfections and impurities played an important role in his measurements. How high the degree of perfection of mono-crystals obtained by Shubnikov's method was can be judged from work carried out in the mid-fifties, which essentially still used the same method [10].

In their fourth paper, in 1930 Shubnikov and De Haas reported the results of an investigation of the oscillating dependence of the electro-resistance on the orientation of the magnetic field in a plane perpendicular to the direction of the current. The observation of this dependence, reflecting an anisotropy in the mobility of the charge carriers, made an even stronger impression on the authors than the oscillation at a variation of the field: "*These remarkable curves of the change of the resistance in a magnetic field which we have observed here have never before been seen. For this there are probably two reasons: firstly, crystals were never investigated in this way at temperatures of liquid hydrogen, and secondly in order to obtain these curves it is absolutely necessary that very pure material be used for preparing the crystals*" (Ref. [11], p. 369–370). Not only Shubnikov and De Haas, but also Ehrenfest understood that precisely these oscillations were most important. At first they were asymmetrical. It was necessary to learn how to grow crystals of bismuth with a fully determined orientation in order that the direction of the growth coincides with the crystallographic axis. Then finally Shubnikov saw distinctly expressed symmetric oscillations. All these measurements were made at nitrogen and hydrogen temperatures, and the new large magnet was used. The total cycle of investigations took four days (Ref. [12], p. 267).

The investigation at hydrogen temperatures was subsequently continued for a purer and more precisely oriented mono-crystal [13]. The results of this work look even more impressive. The authors understood that they were dealing with a principally new phenomenon, but in none of the papers of the series did they express themselves on its nature. The only indication that such attempts were undertaken is a reference to a suggestion by Ehrenfest to analyse the field dependence of the coefficients of the Fourier expansion of the angular dependences.

The measurements carried out at nitrogen temperatures showed that at such temperatures the magneto-resistance does not oscillate [14]. Only when cooling down to 64 °K a small kink appears in the dependence of the resistance for a field of 15 kilogauss. Thus this work demonstrated that quantum effects already vanish at nitrogen temperatures.

The observation of the oscillations of the magneto-resistance also stimulated the search for oscillations of the magnetic susceptibility of bismuth. Already in 1929 when Shubnikov was still in Leiden, his crystals were used by Pieter van Alphen[11]

---

[11]Pieter Martinus van Alphen (1906–1967). Obtained his PhD with De Haas in 1933 and later went to work for the Philips company in Eindhoven (Netherlands).

to measure the magnetic susceptibility. While in the Shubnikov-De Haas effect oscillations of the electrical resistance of a crystal as a function of an external magnetic field are measured, in the work of Van Alphen it concerns oscillations in the magnetic moment (i.e. in the torque the crystal will experience in a magnetic field) of a pure metal crystal when the intensity of the applied magnetic field is increased. Such oscillating behaviour was indeed discovered in the same year 1930 by De Haas and Pieter van Alphen and is known as the De Haas-Van Alphen effect [15]. A theoretical explanation of the effect has been given by Lev Landau [16], who showed it to be a direct consequence of the quantization of closed electronic orbits in a magnetic field. It is therefore a direct observational manifestation of pure quantum mechanics.

The discovery of the effect became possible thanks to Shubnikov's work and was very directly connected with it: *"The most detailed investigation was that of the bismuth single crystals. First, because we had at our disposal a single crystal of extremely pure bismuth and second because of the desirability of the examination of bismuth single crystals in connexion with the anomalous results of the resistance measurements by L. Schubnikow and W. J. de Haas with these crystals. Because of the evident correlation of the diamagnetic susceptibility with the change of resistance we were inclined to expect a dependence of the susceptibility on the field analogous to that found for the resistance. Further on we shall see that our expectations were justified... For the above reason L. Schubnikow prepared for us a small single-crystal (5 × 5 × 5 mm)... The material for this crystal was chemically pure bismuth which was further re-crystallized 12 times. The bismuth was therefore extremely pure."* (Ref. [15], pp. 1107–1112.)

Shubnikov had planned to carry out further investigations of the magneto-resistance at low temperatures for other metals in Kharkov, but this work was cut short by his arrest. The work was continued by others and in 1939 B.G. Lazarev, N.M. Nakhimovich and E.A. Parfenova observed the Shubnikov-De Haas effect in mono-crystals of zinc [17]. Neither this paper, nor an accompanying paper in *Doklady Akademii Nauk SSSR* [18], contains any reference to Shubnikov (his work is just referred to as W.J. de Haas and others without specifying the 'others'). They probably did this in order not to expose themselves to criticism in line with the policy that 'enemies of the people' should be treated as if they had never lived. Even for the preparation of crystals they referred to Bridgman's paper and method, completely ignoring Shubnikov's Leiden paper, where he described the preparation of crystals for his experiments on bismuth or the earlier method developed by their fellow countrymen Obreimov and Shubnikov. At the time they were both 'enemies of the people' as Obreimov was kept in prison until 1941.

From about 1947 oscillation effects for a wide range of metals were studied at the cryogenic laboratory in Kharkov. It was established that such effects are characteristic for all metals and that the problem of observing them in concrete matter depends on the possibility of creating sufficiently strong magnetic fields, and also of obtaining perfect mono-crystals of metals free of impurities.

In due course it has turned out that not only the conductivity and the magnetization, but also many other physical properties of metals oscillate under a changing magnetic field, for instance the speed and absorption coefficient of sound, the reflection coefficient of light, the thermal conductivity, the specific heat, the chemical potential and a series of other thermodynamic and kinetic properties of metals. For a detailed discussion see Ref. [19], p. 133 ff.[12]

The realization that the effect he had discovered is a general property of metals came only after Shubnikov's death. Much research into these and other related phenomena has since been carried out. The importance of these effects for condensed matter physics only became clear in the fifties and sixties, when the large role these oscillations play in determining the electron energy spectrum of metals was clarified (Ref. [3], p. 25). Nowadays it is one of the most important tools for investigating quantum electron properties of matter. It has essentially laid the foundation for the contemporary physics of electronic properties of semiconductors—an area of investigation which later was called 'fermiology' (the study of the shape and size of Fermi surfaces[13]). The effect was and is used for determining the Fermi surfaces of virtually all metals and semimetals, as well as of many semiconductors. To this day the Shubnikov-De Haas and the De Haas-Van Alphen effects and phenomena related to these effects are still used to obtain vast amounts of information on the cross sections of Fermi surfaces, cyclotron masses, the g-factor and other properties. The Shubnikov-De Haas effect, which in present-day experiments can easily be observed, has found broad application in investigating magnetic breakdown, non-equilibrium processes in semiconductors, and also in the restructuring of electronic spectra under extreme conditions (at high pressure, strong electric fields), when the observation of other oscillation effects is significantly impeded. At present it is one of the main instruments for investigating quantum electronic properties of solids and the area of its application is continuously expanding. In the future the Shubnikov-De Haas effect will undoubtedly find more and new applications (Ref. [3], p. 25).

Lev Vasilevich wrote all the papers on the Shubnikov-De Haas effect in German, and subsequently took them to De Haas. Shubnikov was the first author for all of the papers. This was unusual, but it was done by De Haas to stress the decisive role of Shubnikov in the discovery of the effect.

---

[12]David Shoenberg (1911–2004) carried out many experiments on magnetic oscillations, at Cambridge and at Kapitsa's institute in Moscow. "In his pioneering experiments of the 1960's, Shoenberg revealed the richness and deep essence of the quantum oscillation effect and showed how the beauty of the effect is disclosed under nonlinear conditions imposed by interactions in the system under study." (V.M. Pudalov, David Shoenberg and the beauty of quantum oscillations, *Low Temp. Phys.* 37 (2011) 8–18, p. 8.)

[13]The Fermi surface is an abstract boundary (i.e. not in real space) useful for predicting the thermal, electrical, magnetic, and optical properties of metals, semimetals and doped semiconductors. The shape of the Fermi surface is derived from the periodicity and symmetry of the crystalline lattice and from the occupation of electronic energy bands. Electrons in crystals are arranged in energy bands separated by regions in energy which cannot be occupied by electrons (energy gaps). The existence of a Fermi surface is a direct consequence of the Pauli exclusion principle (*Wikipedia*).

A further proposal was to carry out the same measurements at helium temperatures, but Shubnikov did not manage to do this during his stay in Leiden. De Haas and Ehrenfest were well aware of the importance of this work and on 25 April 1931, well after Shubnikov's departure from Leiden, Ehrenfest wrote to Ioffe with a request to help Shubnikov return to Leiden:

> ... You know that when staying here Shubnikov achieved such important and outright beautiful results in measuring the resistance of bismuth crystals at the temperature of liquid hydrogen and that agreement has been reached that he should continue these measurements at helium temperatures. After all, you know how interesting this transition to helium temperatures promises to be. Now it turns out, if I correctly understood the matter, that it is unlikely that Shubnikov will come to Leiden in the near future, for reasons I do not know in detail. Could I ask you to ask Obreimov how it stands with this matter? I know that Professor De Haas will be very, very sorry if Shubnikov cannot come to Leiden in the near future for such a period as is necessary for carrying out the measurements at liquid helium temperatures. After all, you very well know how very much De Haas is always prepared to help Obreimov with the development of cryogenic work at Kharkov. And I know that after the splendid successes of Shubnikov in measuring at hydrogen temperatures, De Haas very much would want Shubnikov himself to do the measurements with helium. And now it suddenly turns out that this is impossible, and I absolutely do not know why. I am afraid that from this not only this important research will suffer, but that also the promising collaboration between the Leiden laboratory and Russian physicists, which is very dear to me, can come under serious threat. After all, you yourself also see this very well so that I do not have to write in detail about this. I in any case asked De Haas to have a little more patience and proposed to him to resort to your mediation in talks with Kharkov. Please ask Obreimov about this and promote that De Haas will as soon as possible receive a favourable answer.

> Forgive me that to the thousands of your worries and heavy duties I add this one. But it is very important that Shubnikov will as soon as possible finish his work which he has so splendidly started here!... (Ref. [20], pp. 222–223)

In the Ehrenfest-Ioffe correspondence no reply from Ioffe to this letter can be found, but whatever it was it had no positive effect, which is all the more regrettable as at that time Shubnikov was sitting idle in Kharkov where Obreimov was in no hurry to appoint him head of the cryogenic laboratory (see Chap. 7).

In Leiden Shubnikov had already collaborated with Jan Wilhelmus Blom,[14] who when a student had been Shubnikov's helper in his work on the magneto-resistance of bismuth, and naturally he taught him how to purify and grow crystals. So several years later, in 1935, it was Blom, who, using Shubnikov's crystals, finally took up the work De Haas had still wanted Shubnikov to do and measured the dependence of the resistance of bismuth on the magnetic field at helium temperatures and oscillations of the resistance as a function of the angle of rotation of the binary axes of the crystal with respect to the magnetic field. It transpired that the effect of the field on the resistance at a temperature of 4.22 °K is larger than at 14.15 °K. Further

---

[14]J.W. Blom (1908–1975): studied in Leiden, gained his PhD with De Haas in 1950 on the *Magneto-resistance for crystals of gallium*, but always wanted to become a teacher. From 1944 he taught at the Rembrandt Lyceum in Leiden, later became director of a secondary school in Den Helder and in 1961 of his former school, the Rembrandt Lyceum in Leiden.

cooling to 1.35 °K did not result in any change. Moreover, the structure of the oscillations at helium temperature turned out to be more complicated than at hydrogen temperature, but did not give any qualitatively new results. The results were published in 1935 with three authors on the paper: De Haas, Blom and Shubnikov [21].

—

So what is this Shubnikov-De Haas effect? The theory of the effect is actually quite complicated as it involves the detailed problem of electron scattering in a magnetic field.

In 1928 when Lev Landau visited Kapitsa, who had just discovered an anomalous property of the electric conductivity of bismuth in a strong magnetic field, at the Cavendish laboratory in Cambridge he suggested that Landau theoretically investigate the diamagnetism[15] of bismuth. This led to one of the highlights of Soviet physics in the 1920s: Landau's paper on diamagnetism [16]. Landau described a quantum-mechanical electron in a uniform magnetic field by an harmonic oscillator and then used the statistical mechanics of a degenerate free electron gas in a magnetic field to predict a weak steady diamagnetic susceptibility,[16] which should be exactly one third of the electron spin paramagnetic susceptibility (but with the opposite sign) already calculated by Wolfgang Pauli [22]. One of Pauli's assumptions was that the spatial motion of the electron is not affected by the magnetic field. By relaxing this assumption and calculating the effect on the electron wave function Landau found for free electrons that this causes a diamagnetic moment.

Another important idea in Landau's paper was the prediction that at low temperatures the magnetization should oscillate, which is essentially a prediction of the De Haas-Van Alphen effect. He dismissed it however as essentially impossible to observe, although within two months of Landau's paper De Haas and Van Alphen actually observed the phenomenon. They refer to Landau's paper as a source of information on diamagnetism, but either overlooked Landau's remark about the oscillatory behaviour or did not appreciate its relevance (Ref. [19], pp. 2–3).

Shubnikov-De Haas oscillations are not oscillations of the magnetic susceptibility, but of the specific resistance in an applied magnetic field parallel to the flow of the current in the edge states (states at the boundary of a sample) of a 2-dimensional electron gas. Their origin is, however, the same.

---

[15]Diamagnetism is the phenomenon that, when an external magnetic field is applied to a material, an induced magnetic field is created in a direction opposite to the external field. It is a (rather weak) quantum mechanical effect occurring in all materials. When it is the only contribution to the material's magnetism, the material is called a diamagnet. It was first discovered in bismuth and antimony by the Dutch botanist and physician Sebald Justinus Brugmans (1763–1819). In 1845 Michael Faraday (1791–1861) demonstrated that every material reacted (in either a diamagnetic or (para)magnetic way) to an external magnetic field and coined the name diamagnetism, contrary to substances that are called magnetic as they are attracted by a magnet. The magnetic group was later subdivided into ferromagnetic and paramagnetic substances.

[16]In the classical theory it would be zero.

Why do such oscillations occur? In 1933, on the basis of Landau's formulation of the energy levels in a magnetic field, Rudolf Peierls formulated a quantitative theory of such oscillations, treating a metal as an isotropic gas of electrons [23].[17] At sufficiently low temperatures and high magnetic fields, the free electrons in the conduction band of a metal or other substance will behave like simple harmonic oscillators, and consequently the energy-eigenvalues of the circular motion are the same as for an harmonic oscillator. When the magnetic field strength is changed, the oscillation period of the simple harmonic oscillators changes proportionally. When solving the Schrödinger equation for free electrons in a magnetic field, one obtains the energy-eigenvalues of the Landau-levels, the quantised levels of the cyclotron orbits of charged particles in magnetic fields. Increasing the magnetic field B entails two effects. On the one hand, the energetic separation of two Landau-levels increases linearly with B. The gap between the Landau-levels gets bigger. On the other hand, the degeneracy of a Landau-level also increases linearly with B; more and more electrons can fit into a given Landau-level. This leads to a redistribution of electrons among the Landau-levels to always occupy the energetically most favourable states. The occupation of the highest Landau-level ranges from completely full to entirely empty, leading to oscillations in various electronic properties. The oscillations occur when these Landau-levels pass through the Fermi-level[18] and become depopulated.

In the more intuitive picture given below we can also view the process as follows: when applying a magnetic field to a 2-dimensional electron gas, the electrons in the bulk regions perform circular motions. In the border (edge) regions of the sample, the electrons cannot perform full circular motions, because they get scattered back from the interface (Fig. 5.3). These scattering events give these electrons a higher energy. The magnetic field forces scattered electrons to move in the forward direction so that the current flows without specific resistance in the edge channels.

If the highest Landau-level is far away from the Fermi-energy, there are no states available for scattering and the Shubnikov-De Haas specific resistance goes to zero. In that case the electron transport in the edge states is ballistic. When increasing the magnetic field, the highest Landau-level gets close to the Fermi-energy. Now there are states available for scattering in the bulk regime. This scattering causes the peaks in the Shubnikov-De Haas oscillations.[19]

---

[17]The actual predictions of his theory indeed qualitatively resemble the experimental data, but the magnetisation came out much too small. This was later rectified. It goes beyond the scope of this book to treat this here in any detail.

[18]The Fermi-level or Fermi-energy is the topmost energy filled by an electron at absolute zero, where it is equal to the chemical potential. In a band structure picture, the Fermi level can be considered to be a hypothetical energy level of an electron, such that at thermodynamic equilibrium this energy level would have a *50% probability of being occupied at any given time.*

[19]https://lampx.tugraz.at/ ~ hadley/ss2/problems/shubnikov/s.pdf.

electrons can move along edge (conducting)

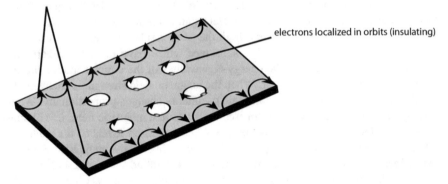

electrons localized in orbits (insulating)

**Fig. 5.3** Motions of electrons in a 2-dimensional electron gas

Shubnikov's wife Olga Trapeznikova started work at the Leiden laboratory around a month and a half after her arrival in 1927. She was very impressed by the Leiden Physics laboratory, by the special room for measuring (the "*meetkamer*"), by the numerous assistants and employees of the workshop who did everything for the scientific staff, by the value attached to the scientific staff and the respect for scientific work. The style of the laboratory implied that measurements should be very precise, to be carried out on pure samples. The spirit was completely different from Leningrad University, and also from the Physico-Technical Institute. Moreover, there was much better equipment. What impressed her most was the accuracy with which measurements were made. She saw for the first time how important it was or can be to measure accurately! In this respect she actually states that at Leningrad University, and also at the Physico-Technical Institute no value was attached to quantitative measurements of such precision (Ref. [12], p. 263), a remark that throws a refreshing light on the physics pursued in Leningrad at that time.

It is not clear which stage she was in with her postgraduate work. In principle she was a more experienced physicist than her husband, who only just had finished his diploma work, while she had already done a few years of postgraduate work with Lukirsky, but she was much less talented, although quite competent, as is also clear from the following.[20] Initially she did not have a definite subject and she acquainted herself with new methods of working with a platinum thermometer. Her teacher-"torturer", as she described him, was her husband. At first he frightened her so much that she was afraid to touch the switch of the compensator, and then he became angry because of the confusion she showed, but she suffered everything and learned a lot. And learning from Lev Vasilevich was quite something. The way he worked was absolutely amazing; he had golden hands, and somehow grasped things instantly. This was especially striking when he was measuring. He did everything

---

[20]There can also be no doubt that as a woman she was in a disadvantageous position, both in Leningrad and in Leiden, but especially in Leiden.

quicker and better than others. As regards setting up and constructing equipment he also did very well, so much so that the chief technician of the workshop always asked him for advice (Fig. 5.4).

When Trapeznikova had learned to use the platinum thermometer, she started to help Lev Vasilevich with his measurements. Mostly she had to regulate the magnetic field. All rooms were connected with each other and with the measurement room. Depending on the signals she was given, she increased or decreased the magnetic field. The temperature and pressure of the vapour above the cooling liquid were also regulated in the same way; this was normally done by apprentices. Another person sat in the measurement room and measured the temperature with a platinum thermometer. Shubnikov himself measured the resistance of the sample.

After a couple of months Obreimov also arrived in Leiden. He was going to measure the vibration spectra of some organic crystals, such as azobenzene, and also of iodine crystals. He was interested in the change of their absorption spectra at low temperatures. Trapeznikova was also charged with assisting him. She grew the organic crystals for him, and subsequently helped to mount them in the apparatus, and also developed the plates.

After this she was placed at the disposal of De Haas. He proposed that she measure the magnetization curve of an iron bar sent by the Philips company from Eindhoven as a specimen with a large coercive force for the construction of a permanent magnet. However while she was busy with this, it became gradually clear to her that she was dealing with extremely soft iron. De Haas was enquiring all the time how the work with the Philips specimen was going. He teasingly called her the "kleine Shubnikov" (little Shubnikov).[21] She carried out measurement after measurement, but got no result. In order to reduce the effect of demagnetization, she employed all kinds of tricks, but the outcome was the same. Then one day a letter arrived from Philips containing apologies. It transpired that by mistake the incorrect specimen had been sent, indeed very soft iron. De Haas then said that from now on he would always believe her. But later when he asked her about Russia and she showed herself a fierce patriot, he said: "I believe you in everything, but it is still better that I go myself and have a look."

In one of his papers [24] Kapitsa had reported on an inherent imperfection in crystals of bismuth, and the presence of cracks in such crystals. The paper included curves of the dependence of the resistance of bismuth crystals on the deformation. Kapitsa had used bismuth from Hartmann and Braun and from Kahlbaum (while Hilger bismuth was actually purer). At that time Shubnikov had already obtained uniquely pure crystals of bismuth, in which spectroscopically no impurities could be discovered, so much better than any of the crystals used by Kapitsa or commercially available. De Haas proposed that Trapeznikova repeat the Kapitsa experiment with a similar set-up, but with Shubnikov's crystals. The measurements showed that there were no cracks in these crystals, confirming his suspicion that they were not inherent

---

[21]This shows that as a woman she was not taken as seriously as her husband and even belittled; it would have been unthinkable for Shubnikov to have been given such a name.

**Fig. 5.4** Trapeznikova and Joost van den Handel, a PhD student of De Haas, at work in the laboratory (*from* B.I. Verkin et al. (1990))

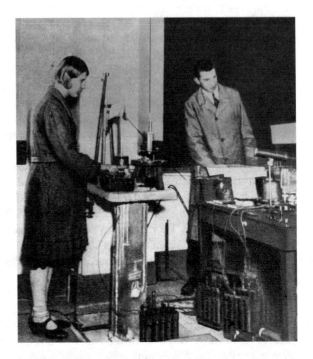

to such crystals, but due to impurities. This work was printed in the Journal of the Royal Academy of Sciences and in the Leiden Communications [25].

Trapeznikova also was involved in a study of the "Temperature dependence of the torsion modulus of lead, bismuth, cadmium and zinc at low temperatures", which was directed by De Haas. This work was finished and prepared for the printer with De Haas and herself as authors. But when she was later in Kharkov the almost total absence of the dependence of the torsion modulus of cadmium and zinc on the temperature started to worry her and she wrote to De Haas asking for the work not to be printed. She subsequently regretted this very much as similar results were obtained and later published by others.

# References

1. P. Ehrenfest, Bemerkung über den Diagmagnetismus von festem Wismut, *Physica* 5 (1925) 388–391; *Z. Phys.* 58 (1929) 719–721.
2. P.L. Kapitza, The study of specific resistance of bismuth crystals and its change in strong magnetic fields and some allied problems I-III, *Proc. Roy. Soc.* A 119 (1928) 358–443.
3. B.I. Verkin et al. (1990).
4. A. Goetz, On Mechanical and Magnetic Factors Influencing the Orientation and Perfection of Bismuth Single-Crystals, *Phys. Rev.* 35 (1930) 193–207.
5. L. Schubnikow, Über die Herstellung von Wismuteinkristallen, *Proc. Roy. Acad. Amsterdam* 33 (1930) 327–331.

6. L.V. Shubnikov and W. de Haas, Izgotovlenie i issledovanie kristallov Vismuta (*Preparation and investigation of Bismuth crystals*), *Zhurnal Russkogo fiziko-khimicheskogo obshchestva* LXII (1930) 530–537.
7. L. Schubnikov, W.J. de Haas, Die Abhängigkeit des elektrischen Widerstandes von Wismutkristallen von der Reinheit des Metalles, *Proc. Roy. Acad. Amsterdam* 33 (1930) 350–362, p. 350.
8. H.B.G. Casimir, *Haphazard Reality: Half a Century of Science* (Harper & Row, New York, 1983).
9. L.V. Shubnikov and W.J. de Haas, A new phenomenon in the change of resistance in a magnetic field of single crystals of bismuth, *Nature* 126 (1930) 500.
10. P.B. Alers and R.T. Webber, The Anomalous Magnetoresistance of Bismuth at Low Temperatures, *Phys. Rev.* 84 (1951) 863–864; The Magnetoresistance of Bismuth Crystals at Low Temperatures, *Phys. Rev.* 91 (1953) 1060–1065.
11. L.W. Schubnikow, W.J. de Haas, Neue Erscheinungen bei der Widerstandsänderung von Wismuteinkristallen im Magnetfeld bei der Temperatur von Flussigem Wasserstoff I, *Proc. Roy. Acad. Amsterdam* 33 (1930) 363–378.
12. O.N. Trapeznikova in B.I. Verkin et al. (1990).
13. L.W. Schubnikow, W.J. de Haas, Neue Erscheinungen bei der Widerstandsänderung von Wismuteinkristallen im Magnetfeld bei der Temperatur von Flüssigem Wasserstoff II, *Proc. Roy. Acad. Amsterdam* 33 (1930) 418–432.
14. L. Schubnikov, W.J. de Haas, Die Widerstandsänderung von Wismuteinkristallen im Magnetfeld bei der Temperatur von Flüssigem Stickstoff, *Proc. Roy. Acad. Amsterdam* 33 (1930) 433–439.
15. W.J. de Haas and P.M. van Alphen, The dependence of the susceptibility of diamagnetic metals upon the field, *Proc. Roy. Acad. Amsterdam* 33 (1930) 1106–1118.
16. L. Landau, Diamagnetismus der Metalle, Diamagnetismus der Metalle, *Z. Phys.* 64 (1930) 629–637.
17. B.G. Lazarev, N.M. Nakhimovich and E.A. Parfenova, Vlijanie magnitnogo polja na elektricheskoe soprotivlenie monokristallov tsinka i kadmija pri nizkikh temperaturakh (Effect of a magnetic field on the electrical resistance of mono-crystals of zinc and cadmium in a magnetic field at low temperatures), *Zhurn. Eksperim. i Teoret. Fiziki* 9 (1939) 1169–1181.
18. B.G. Lazarev, N.M. Nakhimovich and E.A. Parfenova, Elektricheskoe soprotivlenie monokristallov tsinka i kadmija v magnitnom pole pri nizkikh temperaturakh (Electrical resistance of mono-crystals of zinc and cadmium in a magnetic field at low temperatures), *Dokl. Akad. Nauk SSSR* XXIV (1939) 855–859.
19. D. Shoenberg, *Magnetic Oscillations in metals* (Cambridge University Press, 1984).
20. N.Ja. Moskovchenko and V.Ja. Frenkel, *Ehrenfest-Ioffe Nauchnaja Perepiska* (Ehrenfest-Ioffe Scientific Correspondence) (Akademija Nauk SSSR, 1990).
21. W.J. de Haas, J.W. Blom and L. Shubnikov, Über die Widerstandsänderung von Wismutkristallen im Magnetfeld bei tiefen Temperaturen, *Physica* 2 (1935) 907–915.
22. W. Pauli, Über Gasentartung und Paramagnetismus, *Z. Phys.* 41 (1927) 81–102.
23. R. Peierls, On the Theory of the Diamagnetism of Conductions Electrons II. Strong magnetic fields, *Z. Phys.* 81 (1933) 186–194.
24. P. Kapitza, Specific Resistance of Bismuth Crystal and its Change in Strong Magnetic Fields, *Proc. Royal. Soc. A* 119 (1928) 358–443.
25. O. Trapeznikowa, Untersuchung über den Einfluss van Druck und Zug auf den spezifischen Widerstand van Wismut Einkristallen, *Proc. Roy. Acad. Amsterdam* 34 (1931) 84–843; *Leiden Comm.* no. 215c.

# Chapter 6
# Founding of the Ukrainian Physico-Technical Institute in Kharkov

When Shubnikov returned to the Soviet Union in 1930, he found a country that was very different from the one he had left in 1926, during the heyday of Lenin's New Economic Policy (NEP). At the end of the twenties, in a spectacular and dramatic change of policy decreed during November-December 1929, the Bolsheviks made an end to the concessions of the NEP period and announced the Great Break (*Veliky Perelom*), which was to culminate in the collectivization and industrialization campaigns of the 1930s. The first five-year plan had been drawn up, also for science, and the entire country was being mobilized for the industrialization effort.

The character of physics had consequently also changed; it now had to make its contribution to the industrialization effort and Shubnikov's practical expertise was eminently suited for this. While still in Leiden Shubnikov had been invited by Obreimov, with whom he had collaborated in Petrograd, to come to Kharkov to help build up a new institute and he very much wanted to go there.

According to Trapeznikova's recollection (Ref. [1], p. 277), the Shubnikovs left Leiden in the summer of 1930. When they arrived in Leningrad, their former room in the apartment where they had lived together with Sergej Frish and his wife Aleksandra Timoreva was occupied and they were forced to live somewhere else. Although Trapeznikova was very reluctant to leave Leningrad, she realized that there was nothing left for her in that city. She had finished graduate school and had cut her ties with the university. They were still both registered as collaborators of the Leningrad Physico-Technical Institute (LFTI), so to obtain a transfer to Kharkov was easy and she proposed that they go away. Shubnikov's personnel file at the Ioffe Institute contains a hand-written request by Shubnikov dated 8 August 1930 to the LFTI office to apply for a reprieve of military service. Later in August the Shubnikovs attended the physics conference in Odessa, held from 19 to 24 August 1930. The file also contains another document dated 27 October 1930 in which Shubnikov states to go to Kharkov and asks for his name to be deleted from the list of employees of LFTI, as well as a certificate dated 30 October stating that he has now been included into the list of employees of the Ukrainian Physico-Technical Institute and consequently must move to Kharkov.

© Springer International Publishing AG 2018
L. J. Reinders, *The Life, Science and Times of Lev Vasilevich Shubnikov*,
Springer Biographies, https://doi.org/10.1007/978-3-319-72098-2_6

The Ukrainian Physico-Technical Institute (UFTI) in Kharkov was the first of a string of institutes that originated from Abram Ioffe's efforts to create a network of physics institutes beyond Moscow and Leningrad. Other important institutes were founded in Dnepropetrovsk (also in the east of Ukraine, about 200 km to the south from Kharkov), in Sverdlovsk (the current Yekaterinburg) in the Urals and in Tomsk in Siberia. It started off with a letter Ioffe wrote on 23 April 1928 to Veniamin Mikhajlovich Sverdlov,[1] the head of the Scientific-Technical Administration of Vesenkha,[2] in which he set out his ideas about starting a branch of the Leningrad Physico-Technical Institute in Kharkov. In his view such an institute was necessary in order "*to unify the local, currently scattered forces in Kharkov (...) and to strengthen the Kharkov group with a number of persons specialised in other branches of physics,*" who could for instance be sent down from his laboratory. In his letter Ioffe proposed to keep the construction and equipment of the institute in the hands of local Kharkov people supplemented with one or two representatives from LFTI. After a year and a half or so, when all the preparatory work had been done and the institute was fully equipped and running, people could then be sent from Leningrad.

A few months later in 1928, at a meeting in Kharkov Ioffe pointed out that Germany's strength in science and technology was for a large part due to its decentralisation. Centralization of scientific efforts in Moscow and Leningrad would be dangerous.[3] It was important to spread scientific centres throughout the country as it would raise the general level of culture and fertilize science. An institute in support of industry should be where the factories are. The foundation of

---

[1]Veniamin Mikhajlovich Sverdlov (1886–1939), the younger brother of Jakov Mikhajlovich Sverdlov (1885–1919), who was chairman of the All-Russian Central Executive Committee from 1918 until his death in 1919 and allegedly ordered the murder of the Romanov family. Veniamin was arrested in 1938 as a "Trotskyist terrorist" and executed in 1939.

[2]Supreme Soviet of the National Economy or All-Union Economic Council (*VSNKh, Vysshy Sovet Narodnogo Khozjajstva*). Already set up in December 1917 and in operation until 1932. Its stated purpose was to "plan for the organization of the economic life of the country and the financial resources of the government". In August 1918 the Scientific-Technical Department of Vesenkha (*Naucho-tekhnichesky otdel; NTO*) was organised. It was later named the Scientific-Technical Administration (*Naucho-tekhnichesky Upravlenie; NTU*). The idea behind this was to mobilise science for industrial production. After the creation of the Soviet Union in 1923 it was transformed into the joint all-Union and republican People's Commissariat. In 1932, it was reorganized into three People's Commissariats of heavy industry, light industry and forestry.

[3]Something Ioffe was at first not convinced of, but in his letter of 13 April 1928 (Ref. [3], p. 195–196) Ehrenfest pointed out to him that centralization of all physics in Leningrad and Moscow, or actually as far as Ioffe was concerned in Leningrad since Ioffe also wanted the Moscow physicists Igor Tamm and Leonid Mandelshtam to come to Leningrad, would be a bad idea as was shown by the extreme centralization of science in Paris caused by the French revolution. "One of the deadliest poisons of such centralization is that a bunch of older people owns everything. In the best case these are people who have done something quite remarkable in their youth, but often that is not so. The life of the young people then becomes hell. A *healthy* development can only be obtained in *decentralized* places, filled only for one half with staff. This *evil* is dangerous as it appears only very slowly. But it is a lethal poison".

an institute in Kharkov, where a huge industry was being built up, was therefore expedient. The huge industrial complexes being constructed in Kharkov at the time included the Tractor Factory, the Turbine Factory, the Locomotive Factory and the Electromechanical Factory. In view of the conflict that later arose in Kharkov, it is interesting to note the emphasis placed by Ioffe on the connection between science and industry. This was precisely what the party leadership wanted to hear.

Soon thereafter, on 18 June 1928 the Scientific-Technical Administration of Vesenkha USSR passed a resolution in which it approved the organisation of a physico-technical laboratory in Kharkov in cooperation with the Scientific-Technical Administration of the Ukrainian Vesenkha, charged the Leningrad Physico-Technical Institute with handling the organisation of this laboratory for the first two years and laid down the funding for the laboratory.

On 30 October 1928 the Ukrainian Council of People's Commissars followed this up with a resolution on the foundation of the Ukrainian Physico-Technical Institute, whose main tasks would be: *"to carry out research in various fields of both theoretical and applied physics; to provide services to industry in respect of the organisation of all kinds of physics research, as well as experiments and consultation on these questions; to provide assistance to industry in organising factory and central group laboratories insofar as related to methods of physics research"* (Ref. [2], p. 133). It was also in line with the conclusions of the Scientific-Technical Administration of Vesenkha drawn up at its meeting of 16 May 1928, which stated to consider it necessary to organise an institute in Kharkov whereby *"the physico-technical institute should involve in its work the scientific-technical forces of Ukraine and establish a close connection with factory laboratories, and scientific-research institutions of Vesenkha and Narkompros"* and Abram Ioffe was to be asked *"to assume the function of chairman of the scientific-technical council of the institute"* (Ref. [4], p. 110 ff).

Later in 1928, following the resolution of Narkompros, Ivan Obreimov, the designated director of the new institute, Aleksandr Ilich Lejpunsky (1903–1972), who was to become deputy director, and Kirill Dmitrievich Sinelnikov (1901–1966) were charged with the organisation of the new institute in Kharkov. Sinelnikov was however almost immediately sent to Cambridge on Kapitsa's request,[4] with a grant from the International Education Board to replace July Khariton.[5] Sinelnikov did not finish his term of two years in Cambridge,[6] but in the

---

[4]It was one of his ways to help Russian/Soviet physics.

[5]July Borisovich Khariton (1904–1996); Khariton's father Boris Osipovich (1876–1941) was expelled from the Soviet Union in 1922 on the so-called Philosophy Steamer. He worked in Germany and Latvia, was arrested by the Red Army in Riga and died on route to the Gulag. In the twenties this did not turn out to be an impediment for the son July Borisovich to get permission to go to Cambridge, nor later for him to play an important role in the development of the Soviet atomic bomb.

[6]During his stay he had time though to marry an English woman, Edna Cooper, who went with him to Kharkov. Her letters to her sister give a vivid, although not a very informed, insight as to what was happening at UFTI [5].

spring of 1930 was ordered by Obreimov to wind up his work in England and return immediately to Kharkov as the construction of the institute was near completion. To his considerable annoyance Kapitsa was not informed of this and wrote to Obreimov to express his displeasure in no uncertain terms. Obreimov's action was all the more remarkable as Sinelnikov had only a month and a half to the end of his contract. Kapitsa also warned Obreimov that "*such steps as you are taking can very easily put you in personal difficulties with your collaborators and myself. A certain amount of certainty in the future and freedom is absolutely necessary for a researcher and he cannot be treated like a soldier. I think that for the successful development of your institute, the happiness and freedom of your collaborators is essential, as actual success in science is attained by persons and not by apparatus*" (Ref. [6], p. 561). In spite of this altercation Kapitsa and Obreimov remained on good terms, also in the more difficult times that lay ahead.

One of the basic goals of the institute was to create a world-class low-temperature laboratory with powerful liquefiers for air, hydrogen and helium. At the time there were only three laboratories in the world where helium could be liquefied: Leiden, where, as Shubnikov had experienced, the production of liquid helium had become a routine affair, Toronto and Berlin. In the early thirties commercial helium liquefiers came on the market and the Kharkov institute managed to produce liquid helium in 1932, before Great Britain where helium was liquefied for the first time in 1933 at the Clarendon laboratory in Oxford; in Cambridge Kapitsa developed a new liquefaction method and started to produce liquid helium from early 1934.[7]

In Leiden during 1927 and 1928 Obreimov had acquainted himself with everything needed for setting up a low-temperature laboratory (at first still planned in Leningrad as part of LFTI as is clear from Obreimov's correspondence during these years with Kapitsa). He travelled to Berlin to purchase a liquid hydrogen plant for the new institute and obtained training at the German Heylandt[8] factory, as he reported in a letter of 25 October 1927 to Kapitsa in Cambridge (Ref. [6], p. 551).

When UFTI started to function early in 1929 its staff consisted of sixteen people, among whom the three founders mentioned above, three Ukrainian Kharkov University professors including Abram Slutskin,[9] as well as two foreign consultants

---

[7]For a description of Kapitsa's method see M. and B. Ruhemann, *Low Temperature Physics* (Cambridge University Press, 1937), pp. 41–42. Within a few years the Russians would obtain two world-class cryogenic laboratories: the one in Kharkov and later Kapitsa's institute in Moscow. For historical details see R. de Bruyn Ouboter, Cryogenics at the end of the 19th and the first half of the 20th century (1880–1940), *J. Phys.: Condens. Matter* 21 (2009), 1–8.

[8]Paul Heylandt (1884–1947) was a pioneer in the use of liquid gasses, among other things for early rocket technology. After World War II he was taken to Moscow where he worked in industry and research until his death.

[9]Abram Aleksandrovich Slutskin (1891–1950) was a native of Ukraine, entered Kharkov University in 1910, studied in Dresden in Germany from 1928–1930. After his return he became professor at Kharkov University and was employed at UFTI where he headed the Laboratory of Electromagnetic Oscillations. He played an important role in shaping radio science in the Soviet Union.

**Fig. 6.1** The main building of the Institute in 2016. A bit rusty at places but still standing very solid (*picture taken by the author*)

Pëtr Kapitsa from Cambridge for experimental work and Paul Ehrenfest from Leiden for theory. Slutskin was the only one of the Kharkov professors who would remain there through the thirties and play an important role in the fateful events to come.

In a remarkably short period Obreimov managed to build from scratch the main building of the institute, living quarters for the staff, mechanical and glassblowing workshops, a scientific library, furnished with all physics journals that were then published, laboratory facilities with new equipment and first-class instruments (almost all from abroad). Most of the buildings are still standing (Figs. 6.1 and 6.2) and were built on the territory of the former Cheka building and prison in Kharkov.[10] The reinforcement steel for the laboratory walls came from the Russian battleship *Imperatritsa Mariya*, one of three dreadnoughts of the same type built for the Imperial Russian Navy. The ship was sunk in 1916 at anchor in Sevastopol and scrapped in 1926. This scrap metal was used in the UFTI-building. All four rooms of the cryogenic laboratory were equipped with a very light roof on rails, so that in the event of an explosion the roof would go upwards and slide down again into place (similar to the light-weight roofs used in Leiden). This indeed happened when the Germans during their withdrawal in 1943 wanted to blow up the building, but did not succeed as the shockwave of the explosion travelled through the building and just lifted the roof.

Obreimov displayed a great amount of activity in these early years. He even found the time to lay out a garden and plant flowers over the whole territory of the institute. Everything was ready when in May 1930, many months before the official opening on 7 November 1930, the first Leningrad colonists as they called

---

[10]Countless people were murdered and buried there under the leadership of the notorious Stepan Afanasevich Saenko (1886–1973), who was an active executor of the "Red Terror" and commander of the Kharkov Cheka (the first Soviet secret police) in the first half of 1919. He miraculously escaped the repression of the thirties which swallowed up most former Chekists. The very mention of the building and address continued to strike terror into the inhabitants of Kharkov when UFTI was up and running. Not only the site for this institute, but also the new UFTI built in the early 1960s on a site 15 km out of town turned out to have been a killing and burial ground of the Soviet secret police.

**Fig. 6.2** The building in the
early 1930s. Compared to the
picture in Fig. 6.1 the top
floor of the back part has
disappeared (*photo from* B.I.
Verkin et al. (1990))

themselves or the Varangians[11] as they were called by the Kharkov people, joined
the institute from Leningrad. Rudolf Peierls who visited the new institute in the
summer of 1930 on his way to Moscow from the conference in Odessa has the
following to say about it: *"The institute was new, and its members were very proud
to show a visitor around. The equipment seemed sound but not luxurious; the
rooms in which the members lived were adequate but very simply furnished,
showed poor workmanship, and most doors did not fit properly"* (Ref. [7], p. 65).
That conditions were actually quite tough at times can be gathered from a letter
Kirill Sinelnikov wrote to his former Cambridge mentor Pëtr Kapitsa early in 1931.
He complains about the lack of assistants and the extremely difficult financial
situation. They had to spend the entire winter without fuel due to the inefficiency of
the institute (temperatures in the laboratory did not exceed 5–6 °C and were even
lower in their apartment). The power supply was strictly limited and the gas supply
was cut off in the middle of the winter, so Kirill and his wife had moved into the
kitchen (Ref. [6], p. 562).

The total number of Leningraders amounted to 22, but not all of them arrived at
the same time. Most of them were still in their twenties. Landau only came to
Kharkov in 1932. At first Dmitry Ivanenko[12] acted as head of the theory section.
When he left in 1931, probably after a conflict with Obreimov, Lev Rozenkevich
took over for a few months, after which Landau became head of theory at UFTI.
The total staff at UFTI in 1929–1930 consisted of 71 people, including 19 scientific
collaborators and 18 post-graduate students. It grew rapidly thereafter and early in
1932 the total staff already numbered 186 people, of whom 73 were scientific
personnel. It had initially been proposed to staff UFTI mainly with scientists from
the Kharkov community. However, all of them, apart from Slutskin and some of his
collaborators were, after the arrival of the Leningraders, forced to leave or left
voluntarily. This undoubtedly contributed to the difficulties that would arise later. It
implies that UFTI was very much a creation of Leningrad physicists, but not only of

---

[11]This was the name given by Greeks and East Slavs to Vikings who, according to legend, were
invited in 862 CE by the local tribes of Novgorod to come and rule over them. Between the 9th
and 11th centuries their descendants ruled the medieval state of Kievan Rus' (*Wikipedia*).

[12]Dmitry Dmitrievich Ivanenko (1904–1994).

them. It was also an international institute, much more so than any other institute in the Soviet Union. Was that perhaps also a reason why it was doomed? In the Soviet Union there has never been a more internationally oriented institute than UFTI, which of course also had to do with the great international experience that many of the leading physicists at UFTI had. Landau, Obreimov, and Shubnikov had spent time abroad, made contacts there and invited physicists from the West to come and visit UFTI. It became very much a creation of European science, when apart from the Leningraders a considerable number of foreign scientists started to arrive from Germany, Austria and Great Britain (Ref. [2], pp. 137–138).

The first foreign physicist to join the institute in Kharkov was Walter Elsasser. Obreimov and Elsasser had met in Leiden, and when the former was in Berlin in 1929 he called Elsasser and invited him to come to Kharkov for a year as a 'technical specialist'. Elsasser, who was still recovering from the ordeal with Ehrenfest, accepted, partly also for the 'princely' salary that Obreimov offered. He left for Russia just before the summer of 1930 and arrived in Kharkov probably in early June.[13] So he came to Kharkov at a very early stage, before Shubnikov and just a month or so before the first Leningraders joined the institute. At about the same time in July 1930 Paul Dirac and Boris Podolsky[14] were visiting Kharkov, where they had to put up with quite some discomfort, as Sinelnikov's wife recounts (a bed without a mattress, standing in line for tap water, power failures) (Ref. [5], p. 46). Elsasser is quite positive, however, about his experience in Kharkov. His accommodation was adequate and, although at first he tried to avoid registering for the privilege of buying at the store reserved for foreign specialists, he did in the end as the food served in the institute's cafeteria and for sale in Russian stores was of poor quality and very limited quantity. He soon discovered that he "*had fallen into the middle of a vast famine; everyone starved, although it seemed accepted practice not to speak about it*". It was 1930 and the worst was still to come. He found it impossible to invite any of his colleagues to share the food he bought in the special shop; they politely but firmly refused. About his colleagues he observes that they were very young, about his own age, which was twenty-six. "*They were clearly the children of the workers and peasants who had made the revolution, a selected group with the talent and the stamina to finish the educational process successfully.*

---

[13]He is mentioned in a letter of 6 July 1930 of Sinelnikov's English wife to her sister, albeit that his name is spelled incorrectly (Ref. [5], p. 50). The Sinelnikovs themselves arrived in Kharkov on 2 June 1930.

[14]Podolsky (1896–1966) was born in Taganrog in the Russian Empire and moved to the US when he was 13. He worked at UFTI with Fock, Dirac and Landau, writing a paper with Dirac and Fock in which they produced a rather simple proof that Dirac's field theory is equivalent to Heisenberg and Pauli's theory and consistent with the special theory of relativity. He also worked with Einstein and is known for the Einstein-Podolsky-Rosen (EPR) paradox. In a recent book (John Earl Haynes, Harvey Klehr and Alexander Vassiliev, *Spies: The Rise and Fall of the KGB in America* (Yale University Press, 2009), pp. 73–75) he is unmasked as having been a spy (code name: Quantum) for the Soviet Union during World War II. In 1942 he approached the Soviet embassy and proposed to come to the Soviet Union and work on the Uranium problem, but they had no need for him there.

*There were only a few representatives of the old bourgeoisie. (...) The band of young scientists whose life I shared were clearly an elite. They had not been selected solely on technical grounds. They were proud; they had great dignity; they never complained in spite of conspicuous hardships. The revolution, 13 years in the past, was still sufficiently near for these young people to feel that they themselves were the new society. There was no authority above them, only some bureaucratic machinery*" (Ref. [8], pp. 107–109). Some of these observations may be accurate, but few of the early Russian scientists at UFTI indeed had a peasant or worker background, although that may actually have been different for the technical and administrative support staff. They would also soon learn that the authority above them was more than just some bureaucratic machinery. After about half a year, in October, Elsasser fell gravely ill with jaundice, then endemic in Ukraine. Since all hospitals were filled to capacity and he did not succeed in getting himself admitted to one, he took a leave of absence and returned to Germany. By the time he had recovered it was 1931 and he decided not to go back to Kharkov to serve out his contract there, but he nevertheless considered the half-year he spent in Russia and, in particular, in Kharkov as the '*most profound external experience of his life*' (Ref. [8], p. 114).

Elsasser was soon followed by the physicist-engineer Alexander Weissberg (1901–1964). Weissberg was born in Krakow (then part of the Austrian-Hungarian Empire, now in Poland) into an orthodox Jewish merchant family that moved to Vienna in 1907. He studied physics and mathematics in Vienna, obtaining his engineering diploma in 1929; was a member of the Austrian Communist Party and later in Berlin of the German Communist Party. He arrived in Kharkov in March 1931 and played an important role in the early history of the institute.[15] After some time "*he convinced the relevant Commissariat that UFTI, which already had a first class cryogenic lab directed by Shubnikov, should be supplemented by an applied cryogenic lab (OSGO). The guiding idea was that the metallurgical industry releases into the air industrial gases which could be captured, cooled and separated into constituents to create the basis of a chemical industry. Alex was commissioned to construct the physical plant to house OSGO*" [10].

In 1932 Martin Ruhemann (1903–1994) and his wife Barbara, also a physicist, came to Kharkov at the instigation of Alexander Weissberg. Martin Ruhemann was a cryogenic specialist, who had moved to Germany from England with his parents as a schoolboy, obtained German nationality, and studied in Munich. When Hitler came to power in 1933, while living in Russia, Ruhemann again switched to British nationality. Both he and his wife Barbara were firm believers in the Soviet cause.

---

[15]Weissberg wrote a well-known book about his experience in Stalin's prisons: A. Weissberg-Cybulski, *Hexensabbat* (Frankfurt am Main, 1951) of which an abridged version has been published as a Suhrkamp Taschenbuch, and which has been translated into English under the titles "*Conspiracy of Silence*" [9] (British edition) and "*The Accused*" (American edition). Weissberg's book is remarkable for the high analytic precision with which he describes the Stalin terror. He used the name Cybulski in the Polish resistance during World War II and added it to his name afterwards.

Martin and Barbara Ruhemann were destined to direct the new OSGO lab after its completion. While waiting for this to happen they wrote their successful book on low-temperature physics.

Another important participant in the events to come was the theoretical physicist Fritz Houtermans (1903–1966), who had already visited UFTI briefly in August 1930 when returning from the physics conference in Odessa. Houtermans was of Dutch-Austrian-German origin and a member of the German communist party. He studied with James Franck in Göttingen and with Gustav Hertz in Berlin, where he probably also met Weissberg and Elsasser. When it became increasingly difficult for him as a communist in Germany, he first went to England, where in 1935 Lejpunsky persuaded him to come to Kharkov, in spite of negative advice from Wolfgang Pauli, who knew about the experience of Victor Weisskopf there.

There were several others, such as the Hungarian physicist Laszlo Tisza (1907–2009), who came to Kharkov on the recommendation of Edward Teller. Also the German physicist Fritz Lange (1899–1987), who was a good friend of Houtermans, stayed with him in England, and like Houtermans was persuaded by Lejpunsky in 1934/35 to come to Kharkov. Most of these foreigners were dedicated to the cause of the Soviet Union and/or wanted to leave Germany as it had become impossible for them to work there after Hitler had come to power. Many physicists with a Jewish background who could not get a job in Germany considered going to the Soviet Union, and especially to this new institute in Kharkov. Among them was the Austrian physicist Victor Weisskopf (1908–2002), who after gaining his PhD in 1931 in Göttingen had difficulty finding a job and stayed for eight months at UFTI, before he went to Copenhagen with a Rockefeller Foundation fellowship. He was not a communist, but wanted to see with his own eyes how Stalinism in Russia worked. In 1936 he was offered a professorship in Kiev, which he declined as his earlier experiences had not been very positive [11].

Another physicist who later became well known and spent some time at UFTI was Georg Placzek (1905–1955). With Niels Bohr he made a contribution in explaining the role of uranium 235 in a nuclear chain reaction (Ref. [12], p. 282 ff). Placzek's visit was later used by the NKVD in accusing Weissberg and others of anti-revolutionary activity and Trotskyism (Ref. [9], p. 112 ff).

Soon after his return to Leningrad, Shubnikov accepted Obreimov's offer to go to UFTI and already on 15 August 1930 he was appointed senior physicist (scientific supervisor (*nauchnyj rukovoditel'*)). At that time liquid air had already been obtained, an installation for the liquefaction of hydrogen had been ordered and talks about a helium installation were being held. Elsasser says about work at the institute in this early period that "*everything was still very raw; the government had ordered a new and very fancy machine for producing low temperatures from Britain* (sic), *but this did not arrive for many months. Much of the local activity consisted of marking time, since in Russia, where there is no free market for technical implements, any technical undertaking of this kind can only begin to function after a certain 'critical mass' is achieved*" (Ref. [8], p. 108).

Shubnikov had expected that Obreimov would appoint him head of the cryogenic laboratory, but Obreimov was in no hurry to do so. According to Trapeznikova, the reason for Obreimov to hold back on this was that he had invited Kapitsa to come and head the cryogenic laboratory in Kharkov and was still awaiting the latter's reply (Ref. [1], pp. 278–279). This does however not agree with the letters of 6 January 1929 from Obreimov to Kapitsa and of 1 February 1929 from Kapitsa to Obreimov (Ref. [6], p. 553 and 554) which show beyond doubt that already in 1929 Kapitsa had made it perfectly clear that he was not prepared to come to Kharkov to take up the directorship of UFTI or part of it. He only agreed to a consultancy role for the new institute, which was duly accomplished with the help of Lev Kamenev, who personally wrote to Kapitsa with details of the consultancy contract offered.[16] In this correspondence with Kamenev Kapitsa had also made it clear that he was engaged in purely scientific research, and was in no way prepared to turn his activity into a direction that would be of interest to industry (Ref. [6], p. 554 and 551). The only plausible reason to explain why Shubnikov did not immediately get to work at Kharkov, but was idle for the greater part of a year, was that Obreimov first wanted to investigate spectra of molecular crystals at low temperatures and wanted to use the cryogenic facilities (e.g. the new liquid hydrogen plant) first for his own research. Shubnikov was in the first instance interested in low-temperature solid state physics and it was a great blow to him that he could not start with the organisation of a proper cryogenic laboratory. He actually found himself out of work, did not even have his own room at the institute and after some time no longer bothered to show up at the institute. It also agrees with the statements Shubnikov made after his arrest by the NKVD in 1937 and although these statements, made under duress to the NKVD, must be treated with caution there will be some truth in them: "*After having arrived in the Soviet Union, I went to work at UFTI with the aim of organising there a low-temperature laboratory. But the former director Obreimov opposed in every possible way that I would take up this matter, and in fact until the autumn of 1931 I was without work at the institute*" and further "*On the basis of an understanding with former UFTI director Obreimov, after my return to the USSR I should work at Kharkov on the foundation of a low-temperature laboratory. Obreimov did not allow me to do this work, as he wanted the huge imported equipment to be used for conducting his own petty topic* (nebol'shaja tema)" (Ref. [2], p. 236 and 245).[17] This shows that he did not value Obreimov's research very highly and that he still felt frustration when recalling that period, even while being interrogated by the NKVD. This is also apparent from his remark in the same statement that Obreimov "*deliberately delayed the formation of the low-temperature laboratory as an important establishment of technical, defence-related and scientific work.*"

Trapeznikova writes about this period: "*for whole days he lay on his bed and read literature, mainly Lermontov. It happened that when I came home in the evening Lev Vasilevich lay in bed and the lights were out. Darkness. I had never seen him in such a*

---

[16]At the time Kamenev's influence was in decline and he had already been expelled (for the first time) from the party. He was one of the first old Bolsheviks to become a victim of Stalin's purges after Kirov's murder in 1934, and was executed in 1936 after the first great show trial.

[17]See Appendix 3 for the full statements.

*state and even feared that he would kill himself*" (Ref. [1], pp. 278–279). The irony is that all this happened while Shubnikov was urgently needed in Leiden to carry out some more experiments at helium temperatures with the crystals he had prepared, which now were lying idle for a couple of years, as recalled in the previous chapter. In one of his statements to the NKVD in 1937 he actually declared that he applied for a *ko-mandirovka* to go abroad, but that this was refused to him (Ref. [2], p. 236; Appendix 3). Shubnikov of course knew that De Haas wanted him to continue the work he had done in Leiden at helium temperatures, and it is possible that he indeed applied for a new *komandirovka* to Leiden, but no documents have come to light about this. At another point in his statements to the NKVD he says: "*I started to make efforts to obtain a komandirovka and a passport for foreign travel, intending to stay abroad permanently in case my position at UFTI would not change. By that time, against the wish of the director, the party committee proposed that I get busy with the organisation of the laboratory.*" This suggests that the process of obtaining permission for a *komandirovka* was cut short by the party committee's proposal.

Trapeznikova herself arranged to work with Kirill Sinelnikov and Anton Valter.[18] She still knew Valter from Leningrad, but met Sinelnikov for the first time in Kharkov. The work was not interesting to her, she did not know the installation, was completely unfamiliar with the method of the experiment, and nothing was explained to her. Subsequently, Valter and Sinelnikov discussed something and brought out two papers. Apart from them and her, Igor Kurchatov and Lev Rozenkevich were also co-authors.

Only in the autumn of 1931 at the urging of deputy director Lejpunsky, after a hydrogen liquefier had arrived from the company Hoek[19] in the Netherlands which nobody could get to work, Shubnikov was placed in charge of the cryogenic laboratory. Lazarev tells the story that Obreimov had been warned about this by Meissner: "*You want to have a liquid hydrogen machine? The question in this respect is whether you will have a sufficiently educated mechanic in order to service it. I myself cannot obtain such a mechanic and service the installation myself. Others work on it, but I cater for them. German mechanics are of course better educated than Russian mechanics. If I cannot find sufficiently educated mechanics in Germany, then you will find them even less in Russia. Consequently your fate is the following: You will be the mechanic at the machine, and others will work with liquid hydrogen. However, I will be glad to help you*" (Ref. [13], p. 303). But no help from Meissner was forthcoming. Obreimov himself could indeed not operate the equipment and was forced to turn to Shubnikov. He was the only one who could get the machine to work and, according to his statement to the NKVD in 1937 (Ref. [2], p. 245), it was the party committee who proposed against the wish of the director that Shubnikov busy himself with the organisation of the laboratory. He was appointed head of the cryogenic laboratory and immediately came to life.

---

[18]Anton Karlovich Valter (1905–1965).

[19]W.A. Hoek's Machine- en Zuurstoffabriek NV in Schiedam; later it took over the company Loos en Co and became Hoek Loos; it still exists as part of the Linde Group. The helium machine used by Shubnikov and his group to liquefy helium in 1932 came from Linde in Germany.

The institute continued to grow very rapidly. The scientific staff was divided into brigades (laboratories), headed by a scientific supervisor (*nauchnyj rukovoditel'*). Each brigade had a brigadier (*brigadir*) who was the assistant of the scientific supervisor in organising the scientific work and who led the social life of the brigade. So in a sense the scientific supervisor was the 'spiritual' head and the brigadier the 'secular' head of the brigade, whereby the brigadier was subordinate to the scientific supervisor. On the staff of each brigade there were, in addition, senior engineers (*starshie inzhenery*), engineers (*inzhenery*), laboratory workers (*laboranty*) and laboratory assistants (*preparatory*), so a clear hierarchy was created. The physicists were called engineers to emphasize their connection with applied work and industry. Early in 1932 the institute consisted of five major laboratories: atomic/nuclear physics under Lejpunsky, Sinelnikov and Valter; low-temperature physics under Shubnikov and Martin Ruhemann; the physics of crystals under Obreimov; ultra-short waves under Abram Slutskin; and theoretical physics at first until 1932 under Ivanenko and later first under Lev Rozenkevich and then Lev Landau. Apart from Slutskin and Ruhemann all the directors (and many of the other physicists) came from LFTI. At the end of the year the number of brigades had already increased to fifteen, divided into two sectors of which one was led by Obreimov and the other by Lejpunsky. In addition there was a Scientific Council chaired by Ioffe, while Kapitsa, Fock, Ehrenfest, Gamow and the radio physicist D.A. Rozhansky[20] became consultants.

The management of the institute further included a technical director who was in charge of technical matters, such as the state and operation of the workshops and the scientific equipment; an assistant to the director for technical plans and personnel, who was also the head of the information and publishing office; and a deputy director for administrative matters responsible for the work of the secretariat, accounting and supply (Ref. [2], p. 139).

The special department for theoretical physics was at that time unique in the Soviet Union and Obreimov must be credited with having understood that theoretical physics was on the rise, borne out, of course, by the development of quantum mechanics in the twenties. In the late twenties Landau and Gamow, inspired especially by Niels Bohr's institute in Copenhagen, had tried to establish a separate institute for theoretical physics in Leningrad. They wanted to see theoretical physics recognised as a discipline in its own right, not subordinate to experimental physics (as the 'handmaiden of experiment'). Ioffe and Rozhdestvensky were, however, sharply opposed[21] to such an idea as they considered it harmful to separate and isolate

---

[20]Dmitry Appolinarievich Rozhansky (1882–1936) had worked at Kharkov University from 1911–1921. From 1923 he was a professor at the Polytechnic Institute in Leningrad, and also worked at LFTI, where he was head of the Electrical Oscillations group and one of Ioffe's earlier collaborators in building up Soviet physics.

[21]Isaak M. Khalatnikov (b. 1919), a well-known Soviet/Russian physicist, long-time collaborator of Landau and director of the Landau Institute for Theoretical Physics in Moscow from 1965 to 1992, claims in his at places rather superficial and uncritical autobiographical sketch *From the Atomic Bomb to the Landau Institute* (Springer Verlag, Heidelberg, 2012) p. 62) that Nikolaj Bukharin was actually in favour of the idea.

theoretical physics from experimental centres such as LFTI and the State Optical Institute. The fact is that Ioffe, Rozhdestvensky and other older physicists were not very impressed with theoretical physicists in general and considered them merely 'calculators' (*vychisliteli*) and 'not thinkers' (*ne mysliteli*), although this seems to be at odds with Ioffe's friendship and admiration for Ehrenfest. It is more probable that they were not very impressed by Landau and Gamow. In any case Ioffe had a very low opinion of Landau. According to Obreimov Ioffe had said: "*We already suffer from hypertrophy of theoretical physics. In LFTI we have Frenkel, at the university Krutkov and Bursian, and that is enough*" (Ref. [14], p. 45). So no separate institute for theoretical physics was formed, although the outcome was rather close. The affair was one of the factors that spoiled the relationship between Landau and Ioffe.

But their efforts had not been completely in vain as they inspired Obreimov, who recognized the significance of theoretical physics, at any rate more so than Ioffe, to set up a separate department for theoretical physics in Kharkov as part of UFTI, instead of dispersing them over the various experimental laboratories. Ioffe also found this important, as is clear from the letter he and Obreimov wrote to Ehrenfest in March 1929, imploring him to become a consultant for the new institute: "*one of the most important matters in the organisation of physics is the organisation of theoretical physics*" (Ref. [3], p. 285).[22]

In the early thirties, with its emphasis on industry and applied science, there was a general feeling among theoretical physicists that their position was rather uncertain and it was feared that the existing theory groups would be subdivided among the various experimental laboratories and/or assigned to industrial projects to provide support without the theorists being able to do their own research into topics unrelated to those projects (Refs. [15], pp. 90–91; [16], pp. 231–234).[23] That this fear was very real was shown by the order promulgated at LFTI in 1936

---

[22]This letter was a few years earlier than Ioffe's "hypertrophy" statements and it should be realized in this respect that Ioffe and Obreimov desperately wanted Ehrenfest's backing and advice and so were probably inclined to overstate matters. That they were bending the truth somewhat is also clear from what they further said in their letter: "the young theorists are not very keen to leave Leningrad and Moscow, as they would lose guidance". Landau and Gamow were still abroad, and Landau would in any case soon tire of guidance from Ioffe, for whom he had little respect. It would also soon emerge that Ioffe himself could do with some guidance from theorists when Landau pointed out that Ioffe and Kurchatov's calculations on the possibility of thin-film insulators were faulty.

[23]In the short-term view of party and state to keep tight control and plan scientific activity in line with the planning of the rest of the economy in order to move the country forward towards industrialization, it was not so unreasonable, I suppose, to see theoretical physics first and foremost as the 'handmaiden' of experiment and industry i.e. to carry out practical calculations. Landau and other scientists saw the order differently and considered the natural growth of scientific knowledge as the prime mover of technological development, but for this a certain technological basis must be present in a country which, in the Soviet Union at that time, was still lacking.

decreeing that theorists were obliged to: (a) attend the general institute seminars, (b) attend the theory seminar, (c) present a schedule of the hours they devoted to advice and to getting acquainted with the work of the specific laboratory they were attached to as a theorist. (Archive LFTI ANN SSSR cited in Ref. [17], p. 52.) It was also shown by Tamm's experience in Moscow where from 1934 he continuously had to defend the utility of his young theory group at the Physical Institute of the Academy of Sciences (FIAN) in Moscow against the ongoing pressure that "*theoretical research needs to be completely tied into the agendas of the experimental laboratories*". The trouble with Tamm's very small group (with very few (graduate) students) was that they all worked on different subjects. There was no 'collective attitude' (*kollektivnost'*) in the group. When Jury Rumer, who mainly worked with Landau, was arrested in 1938 the group became even smaller. Moreover, Rumer's exposure as an 'enemy of the people' neither helped and the group was temporarily dispersed among the various labs at FIAN.

Since Ehrenfest, who was Obreimov's first choice (Ref. [6], p. 553), had declined and Landau was still abroad, Obreimov invited Dmitry Ivanenko to become the first head of the UFTI theory brigade. The invitation was a surprise to Ivanenko as he hardly knew Obreimov from Leningrad. He did not really want to leave Leningrad, which at that time undoubtedly was the centre of physics in the Soviet Union, but his sister and father lived in Kharkov, the position offered was good with the promise of an apartment and trips abroad. The latter was perhaps most important for Ivanenko as he had so far missed out on such trips (contrary to his friends Landau and Gamow), but in Kharkov too they never materialized (Ref. [18], p. 56).

Ivanenko was only two years older than Landau and had been part of the theory group under Jakov Frenkel at LFTI. As a student at the university he belonged to the group around Gamow and Landau (the so-called '*jazz band*') and was one of its most important members (they called themselves the three musketeers and each considered himself to be a genius[24]). He was one of the most important first-generation Soviet theoretical physicists, published with Fock, Landau and Gamow in the twenties, made contributions in various fields (nuclear physics, gravitation and field theory), such as the first neutron-proton model of the nucleus [19]. Although he very much wanted to go abroad in the twenties, he was never allowed to. At Kharkov he also became a professor at the university as did many of his colleagues at UFTI. He was in Kharkov for only two years, until 1931, when he returned to Leningrad. He probably left as he had difficulties with Obreimov. Eddie Sinelnikova's letter of 14 August 1930 (Ref. [5], p. 66) says that "*Obreimov has quarrelled violently with one theorist here, and it is possible that Ivanenko will refuse to return*". Ivanenko was succeeded first for a short time by Lev

---

[24]And probably not each other, although until around 1929–1930 Ivanenko and Landau were very close friends and saw each other daily.

Rozenkevich and in 1932 by Landau, who was glad to leave Leningrad and get away from Ioffe's domineering influence. Until Landau's arrival the theory group was very small: in 1930 Walter Elsasser, Ivanenko, Rozenkevich and the Armenian physicist Viktor Ambartsumjan[25] were the only members, although Fock and Gamow, both consultants to the institute, were often visiting.

When Ivanenko arrived in Kharkov there was nothing there yet; the institute still had to be built, a suitable place had to be chosen and so forth. Ivanenko played a role in all this together with Obreimov and Lejpunsky and soon managed to make the existence of UFTI known to the world by organising already in May 1929, before the institute had officially opened, the first all-union conference in theoretical physics, in which all practicing theoretical physicists in the Soviet Union took part, as well as Pascual Jordan and Walter Heitler from Germany. The total number of participants (not all theorists) was about 60. This was actually real 'hypertrophy of theoretical physics', as Obreimov notes (Ref. [14], p. 53), and was followed by several other meetings and visits by foreign theoretical physicists.

In 1931 Ivanenko organised the second all-union conference on theoretical physics at which problems of the quantum theory of ferromagnetism, the quantum theory of the electric conductivity of metals and semiconductors were discussed (Ref. [2], p. 160). It was Ivanenko's last activity in Kharkov before returning to Leningrad and making way for Landau as head of the theory department.

—

Another noteworthy event in the early years of UFTI was the founding of a purely Russian journal publishing research papers in English, German and French. In the twenties and earlier many Russian physicists published the results of their research in German journals, in particular in the leading physics journal at that time, *Zeitschrift für Physik*, in which also almost all pioneering articles of, for example, Heisenberg, Jordan, Born and Schrödinger on quantum mechanics appeared. Shubnikov's first articles with Obreimov about the work he did in Leningrad before his departure to Leiden were also published in this journal. The number of articles by Russians in *Zeitschrift* was much larger than by French or English authors, and the Russians in turn hardly published in English or French journals, although there are some exceptions, such as Dmitry Ivanenko who, for some reason, also published in the French journal *Comptes Rendus de l'Académie des Sciences* and Kapitsa for whom, as he worked in Cambridge, it was natural to publish in English language journals, such as the *Proceedings of the Royal Society*.

The special bond with Germany, due to the many Russian scientists that had visited Germany and/or had studied there in the late 19th and early 20th centuries, had survived World War I, and its rebirth after the war was helped by the fact that both Russia/Soviet Union and Germany had a pariah status at that time in the international community. The close connection is also apparent from the appeal in

---

[25]Viktor Amazaspovich Ambartsumjan (1908–1996), who was invited by Ivanenko. He became a famous astrophysicist, a member of the Soviet Academy of Sciences and a fellow of the Royal Society.

*Zeitschrift für Physik* in 1921 (vol. 10, p. 352) to German physicists to send articles that had appeared since the beginning of the war to Professor Westphal in Berlin, who would then take care of their transport to the House of Scientists in Petrograd.

After 1930 the number of publications by Russians in *Zeitschrift für Physik* rapidly decreased from a peak of about 70 per year in 1926–1930 to just a dozen in 1935 and 1936. The decrease is connected among other things with a policy discouraging publication in foreign journals because of the build-up of socialism in one country, the corresponding isolation of the Soviet Union, and the rise of Nazism. But it is also due to the fact that in 1932 the Ukrainian Physico-Technical Institute in Kharkov started publishing its own journal, the *Physikalische Zeitschrift der Sowjetunion*, which published articles in English, French and German. (Ref. [15], p. 94.)

Alexander Weissberg, together with Dmitry Ivanenko, was instrumental in setting up this new journal. In his book Weissberg describes (Ref. [9], p. 180–181) how when arriving at UFTI in the summer of 1931 he was struck by the fact that there was no central organ for Russian physics in any world language. At the same time, Russian physicists were unwilling to have their work published only in Russian, because that would isolate them from world physics and cause their work not to be noticed, as very few Western physicists read Russian. So he proposed founding a journal for Russian physics to be published from Kharkov in German, English and/or French. Together with Lejpunsky he went to Moscow to discuss the idea with Bukharin, who approved and wrote a letter on their behalf to the Central Committee. A few weeks later the institute was charged with organising such a journal. The ubiquitous Abram Ioffe of course became chairman of the editorial board, which also included Obreimov and Lejpunsky from UFTI, and an equal number of Leningrad and Moscow physicists. Apart from Lejpunsky all were born in the 19th century; none of the younger UFTI physicists were included on the board. The editorial staff, responsible for the day-to-day work on the journal, consisted of Lejpunsky, Rozenkevich and Weissberg. Ivanenko, in his own words, had to do the 'donkey work'. He collaborated in the preparation of the first two volumes of the journal, but left Kharkov before the first volume came out in 1933. The journal was published from 1933 to 1938 by Narkomtjazhprom and stopped after Lejpunsky's arrest in 1938.[26]

In his preface to the first volume of the journal Ioffe thanks the foreign journals, in particular *Zeitschrift für Physik*, for the hospitality offered to Soviet authors, but since Soviet physics has now come of age it must publish its own journal. The new journal will inform the world in German, English and French about all significant work of Soviet experimental and theoretical physicists. Ioffe expresses the hope that through this journal the relations with foreign researchers will become closer [20]. He considered it an attempt to internationalise Soviet physics even more, while the

---

[26]It was followed up by the English-language journal *Journal of Physics USSR*, published from Moscow with Sergej Vavilov as chief editor. In 1947 all foreign-language journals in the USSR were closed under the pretext of the fight against 'cosmopolitanism'.

country as a whole had actually embarked on the road of becoming ever more isolated and inward looking.

In the issues of the new journal all the well-known names are encountered: Frenkel, Fock, Ioffe (still spelling his name in translation as Joffé, and apart from the first volume not publishing much in this journal), Sinelnikov, Kurchatov, Obreimov, Landau, Shubnikov, etc. Some new names are Jakov Borisovich Zeldovich[27] (1914–1987), working at Nikolaj Semënov's institute in Leningrad, Dimitry Ivanovich Blokhintsev (1907–1979), a student of Mandelshtam and Tamm in Moscow, various students of Landau, such as the brothers Evgenej Lifshits and Ilja Lifshits, Aleksandr Kompaneets, Laszlo Tisza from Hungary, and Isaak Pomeranchuk. There were also occasional contributions from foreign physicists, mainly from those who had been working for some time at UFTI, either as visitors or in a more permanent capacity. There is, for instance, the work of Boris Podolsky and Vladimir Fock and their work with Paul Dirac, as well as Dirac's classic paper on the *Lagrangian in Quantum Mechanics* [21], which inspired Feynman eight years later in formulating his path-integral approach to quantum mechanics (Ref. [22], p. 128–132).[28] Since the Soviet journal was not very well read or known abroad, there was also a tendency among Soviet physicists to publish a short communication in a foreign journal, e.g. in *Nature*, when they thought that an important discovery had been made, which was then followed up by a more extensive paper in the Soviet journal. Shubnikov, for instance, followed this practice several times. The theory group at UFTI published extensively in the *Physikalische Zeitschrift der Sowjetunion*, and Landau especially was a frequent contributor, publishing there, among other things, his theory of phase transitions and of superconductivity. In each volume of the journal 60–70 articles were published, so in the 12 volumes in total about 800 articles, letters, preliminary announcements and discussion notes appeared.

# References

1. O.N. Trapeznikova in B.I. Verkin et al. (1990), p. 256–291.
2. Ju.V. Pavlenko, Ju.N. Ranjuk and Ju.A. Khramov, *"Delo" UFTI 1935-1938* (The "UFTI" Case 1935–1938) (Feniks, Kiev, 1998).
3. N.Ja. Moskovchenko and V.Ja. Frenkel, *Ehrenfest-Ioffe Nauchnaja Perepiska* (Ehrenfest-Ioffe Scientific Correspondence) (Akademija Nauk SSSR, 1990).
4. N.M. Mitrjakov et al. (editors), *Nauchno-organizatsionnaja-dejatel'nost' akademika A.F. Ioffe–Sbornik dokumentov* (Scientific organisational activity of Academician A.F. Ioffe–Collection of documents) (Nauka, Leningrad, 1980).

---

[27]At the age of seventeen Zeldovich became a laboratory assistant at the Institute of Chemical Physics and remained at that institute for the rest of his life. He was a very prolific physicist making a number of important contributions in various fields.

[28]Gleick tells us that Feynman came across Dirac's paper only by accident, as *Physikalische Zeitschrift der Sowjetunion* was "not the best-read journal".

5. Lucie Street (ed.), *I married a Russian. Letters from Kharkov* (George Allen & Unwin, London, 1944).
6. P.E. Rubinin, P.L. Kapitza and Kharkov. Chronicle in letters and documents, *Low Temp. Phys.* 20 (1994) 550–578.
7. Rudolf Peierls, *Bird of Passage* (Princeton University Press, 1985).
8. Walter M. Elsasser, *Memoirs of a Physicist in the Atomic Age* (Science History Publications, Adam Hilger, Bristol, 1978).
9. A. Weissberg, *Conspiracy of Silence* (Hamish Hamilton, London, 1952).
10. Interview of Laszlo Tisza by Kostas Gavroglou on 1987 November 15, Niels Bohr Library & Archives, American Institute of Physics, College Park, MD USA, www.aip.org/history-programs/niels-bohr-library/oral-histories/4915-1.
11. H. Kreissler, *A Scientist's Odyssey: Interview with Victor Weisskopf* (1988), in the series "Conversations with History, (http://globetrotter.berkeley.edu/conversations/Weisskopf/weisskopf0.html)
12. R. Rhodes, *The Making of the Atomic Bomb* (Penguin Books, London, 1986).
13. B.G. Lazarev in B.I. Verkin et al. (1990).
14. I.V. Obreimov, in V.P. Zhuze (ed.), *Vospominanija ob A.F. Ioffe* (Remembering A.F. Ioffe) (Nauka, Leningrad, 1973).
15. A.B. Kojevnikov, *Stalin's Great Science* (Imperial College Press, London, 2004).
16. K. Hall, The Schooling of Lev Landau, The European Context of Postrevolutionary Soviet Theoretical Physics, *Osiris* 23 (2008) 230–259.
17. G.E. Gorelik and V.Ja. Frenkel, *Matvei Petrovich Bronstein and Soviet Theoretical Physics in the Thirties* (Birkhäuser, 1994).
18. G.A. Sardanashvili, *Dmitry Ivanenko – superzvezda sovetskoj fiziki. Nenapisannye memuary* (Dmitry Ivanenko – superstar of Soviet Physics. Unwritten memoirs) (URSS, Moscow, 2010).
19. D. Iwanenko, The neutron hypothesis, *Nature* 129 (1932) 798.
20. A. Joffé, Vorwort, *Phys. Z. Sow.* 1 (1933) 3.
21. P.A.M. Dirac, The Lagrangian in Quantum Mechanics, *Phys. Z. Sow.* 3 (1933) 64–72.
22. James Gleick, *Genius: Richard Feynman and modern physics* (Little, Brown and Company, London, 1992).

# Chapter 7
# History of UFTI in the Thirties

From the history of the first few years and the stormy development of the institute in Kharkov, from Shubnikov's role in building the low-temperature laboratory and Landau's flowering theory group, one might assume that everything was going well at UFTI. Until 1931 the institute was under the control of Vesenkha, the scientific department of the Soviet Council of Minsters (Sovnarkom). A resolution of its Scientific-Technical Administration had set the foundation of the institute in motion in 1928. Funding came from both Vesenkha and Glavnauka Ukraine, a department of the Ukrainian Commissariat for Education, and the task of the institute was to establish close relations with Ukrainian industry.

In 1932 Vesenkha was abolished and in the ensuing reorganisation UFTI came under control of Narkomtjazhprom (People's Commissariat of Heavy Industry), which from January 1932 until his suicide[1] in February 1937 was headed by Sergo Ordzhonikidze (1886–1937). Ordzhonikidze, but also Nikolaj Ivanovich Bukharin (1888–1938) and Georgy Leonidovich Pjatakov[2] (1890–1937), a leading Bolshevik revolutionary and member of the Left Opposition, executed in 1937, were highly placed protectors of, in particular, UFTI's deputy director Lejpunsky. As can be imagined, this change of control came to imply a change of direction for the scientific work at the institute, or at any rate a stronger emphasis on the connection with industry.

The scientific activity of the institute was of high quality, such that the institute started to vie with the Leningrad Institute for top position in the Soviet Union, at least that was Lejpunsky's opinion as he wrote in a letter to Kapitsa in Cambridge on 2 February 1932 (Ref. [3], p 564). In any case its size and budget began to rival

---

[1]Forced suicide or possibly even plain murder according to Ref. [1], p. 167 ff.
[2]Pjatakov was the brains and driving force behind the industrialization of the country; Ordzhonikidze depended entirely on his genius. The former created a major industrial base against all the obstacles that resulted from Stalin's system. His services to the Soviet government were extremely valuable, and still Stalin had him executed, although he had completely abandoned his former oppositionist stance. (Ref. [1], p. 140.)

© Springer International Publishing AG 2018
L. J. Reinders, *The Life, Science and Times of Lev Vasilevich Shubnikov*,
Springer Biographies, https://doi.org/10.1007/978-3-319-72098-2_7

the parent institute in Leningrad. As a sign of its increasing status in the world of physics a further international conference in theoretical physics was organised by the Kharkov institute from 19–23 May 1934, presided over by Landau and visited by a number of physicists from the West, including Niels Bohr and Leon Rosenfeld.[3] On this occasion Niels Bohr spoke before a packed audience in a lecture theatre at the Kharkov X-ray Institute, the chief centre for medical X-ray research in Ukraine. In the audience was also Vladimir Zatonsky (1888–1938), a physicist by training (graduate of Kiev University) and at that time the Ukrainian Commissar of Education. Bohr said about the institute: "*I welcome the opportunity to express my feelings of great delight and pleasure with which I looked at the beautiful new Physical-Technical Institute in Kharkov, where excellent conditions for experimental work in all areas of modern physics are used with great enthusiasm and success under outstanding leadership and in close cooperation with brilliant theoretical physicists*" (Ref. [4], p. 79).

Lejpunsky had so far served as deputy director of the institute and as party watchdog, after having become a party member in 1930. The purpose of his letter to Kapitsa was, however, a much graver matter, as he writes that further progress is being hampered by the fact that Obreimov can only handle a small collective of people and that "*under the changed conditions of work at the institute, some queer dealings by Ivan Vasilevich have placed him in such a position that he can no longer continue as the director of the institute*". What these 'queer dealings' were is not revealed, but that it was a serious matter becomes clear from what follows: "*He has in recent times completely exhausted the remainder of his reputation. (...) Outside the institute also, he does not enjoy any special respect*". Lejpunsky concluded that Obreimov had to be replaced and he continued with offering Kapitsa the directorship of the institute. He did not write this letter just on his own initiative as is clear when he says that "*these circumstances have forced all of us to raise the question of his replacement.*" This was obviously a delicate matter,[4] and Kapitsa chose not to reply to the letter. How the situation in Kharkov was resolved without Kapitsa's assistance is not clear, but the outcome was that Lejpunsky became director of the institute.

Obreimov was not completely side-lined, but became director of the laboratory for crystal physics and chairman of the physical-technical council of the institute and in 1933 was elected a corresponding member of the USSR Academy of Sciences, so there must still have been some people who had respect for him.

It also appears that relations between Lejpunsky and the other leading scientists were good and that he was respected by all (Ref. [5], pp. 289). Weissberg also paints a rather flattering picture of him in his book (Ref. [6], pp. 49–50). His style of leadership was devoid of any bureaucratic behaviour. He seldom used

---

[3]The conference, including its non-scientific program, has been described in some detail by Crowther who also attended the conference (Ref. [2], pp. 112–125).

[4]Which also explains why except for this letter, I have been unable to find any mention of it in the documents that discuss UFTI's history or Obreimov's life and work. The history of UFTI in this period (1932–1937) is anyway scantily dealt with in the literature.

administrative methods and he granted full freedom to people he trusted. He was easy to associate with and accessible to the collaborators of the institute. When needed, anybody could find the director in his laboratory and discuss his problems with him. In business matters he was consistent and principled, and enjoyed a great authority over the collaborators at UFTI, who trusted him and valued his opinion. Fig. 7.1 shows Lejpunsky and some of the other main actors of our story on the steps of the UFTI building in 1934.

Nevertheless a conflict had been brewing at the institute for some time as the UFTI collective was to some extent divided. On the one hand there was the director and the scientific heads of the various laboratories with their assistants, post-graduate students, talented engineers and all those for whom science was their life. These people were constantly busy with work, not from nine to six, but throughout the day. The other camp, if it can be called that, consisted of people of a rather different mould. They did physics more or less by accident, as a nine-to-five job, and had arrived at UFTI via *rabfaks* (workers' faculties)[5] which they considered the top of their learning. They attended theoretical seminars and practical classes without understanding much of the scientific problems discussed. This annoyed both sides: it angered the scientists that they had to spend time on instructing people who did not have elementary knowledge and did not want to learn. The others were upset by the fact that they were still compelled to learn, while also being burdened much by party, Komsomol, social and organisational matters. Gradually seminars and practical classes started to be held less and less often at UFTI, and stopped altogether with the departure of Lejpunsky on *komandirovka*.

The conflict started to spread when Obreimov (when still director) proposed the idea of forming a collective of UFTI collaborators of professional quality. He proposed that young collaborators work at the institute for no longer than one to two years, during which time they should obtain the necessary knowledge and practical skills. After this, they should go and work in industry or at schools of higher learning. Only the most talented, who considered physics their vocation, should remain at the institute. In this matter Obreimov was supported by the other leading scientists at UFTI, such as Landau and Shubnikov. Abram Chernets, in his statement in Shubnikov's rehabilitation procedure (Appendix 4) says in this respect that Shubnikov expressed the opinion that "*only people who have a special talent and undergo special training can engage in scientific work. At the time this was directed at a group of people—graduate students who came from the factory to the institute without sufficient training, and who had not obtained any special education. In one of his notices in the* (institute) *newspaper he even wrote that, after all, a hare can learn to strike a match, but that there is no sense to this.*"

---

[5]Such faculties had been organised with the laudable idea of giving young working class people who often lacked any proper school education a rapid preparation for a study at university. They reached their greatest importance at the end of the 1920s and early 1930s when there were more than 1000 *rabfaks* in Moscow, Leningrad and other important universities in the country. In the study year 1925/26 about 40 per cent of those enrolling at university were graduates of *rabfaks*.

The other side could in no way allow this plan to be realised, since, if it was, they should have to say good-bye to the prestigious work and privileges enjoyed at UFTI. They were also discontented by the fact that the salary of simple engineers, mechanics and laboratory assistants at UFTI, most of whom belonged to the second camp, was three to four times less than the salary of the leading scientists of the institute and that they had no access to the special shops. The conflict was worsened by the circumstance that first Obreimov, and subsequently also Lejpunsky invited foreign specialists to work at UFTI, who worked conscientiously and associated with the UFTI elite, speaking in German and English.

The confrontation further intensified after Landau's April Fool's day joke in 1934. Weissberg writes the following about this: *"One day someone high up in the Soviet counsels got the brilliant idea of establishing a hierarchy for scientists. A Government Commission was appointed with sub-commissions in the provinces, to give each scientist his proper status. The highest grade was Doctor of Science, and the lowest was 'Under Scientist'—there was also an 'Over Scientist.' As the leading scientific institute in the Ukraine we came under fire, of course, and a commission arrived and discussed the whole important matter for weeks on end with the Director. On 1 April an official prikaz (order) appeared on the Institute notice-board recording the result of their labours. The list was simply grotesque. Leading physicists were set down as 'Under Scientists' and modest assistants found themselves 'Doctors of Science'. Many of those who had been harshly treated stormed off to the office to protest, only to find out that the Director knew nothing at all about it. Inquiries revealed that Landau had brought in the list and told the Director's secretary that it was official, and the innocent girl had stamped the document and put it up on the notice-board. Everyone, except Landau, had forgotten that it was April Fools' Day!"* (Ref. [6], pp. 164–165). But Landau's joke also contained a malicious streak as he had assigned low ranks to those he did not value as scientists.

It should also be noted that still under Obreimov's directorship letters started to arrive at Narkomtjazhprom, at the Central Committee, at the District and Regional Committees with complaints about the UFTI leadership. With the departure of Lejpunsky abroad the stream of such letters increased. As one of the answers to these "statements of workers" a party purge of UFTI began, as will be discussed below (Ref. [7], pp. 172–173).

In April 1934 (the decision had already been taken in July 1933), it was considered necessary for Lejpunsky to go to England to work at the Cavendish laboratory with Rutherford and Kapitsa. He arrived in England at the end of March, where he worked at the patent office in London and with Rutherford at the Cavendish.[6] It was during this *komandirovka* that Lejpunsky met Houtermans and Lange in England and invited them to come to UFTI. In the meantime his deputy

---

[6]He returned to Russia in the early autumn of 1934 for a couple of weeks, travelling with the Kapitsa's in their car to Leningrad. It was Kapitsa's fateful trip after which he was not allowed to return to England.

**Fig. 7.1** A happy picture of some of the main actors of our story on the steps of the main UFTI building, taken on the occasion of Kapitsa's last visit from England in September 1934. First row, from left to right: Shubnikov, Lejpunsky, Landau, Kapitsa; second row from left to right: Finkelshtein, Trapeznikova, Sinelnikov, Rjabinin

Vladimir Gej, a physicist and party member, who also headed the party office at UFTI and the study circle for Marxism-Leninism and party history, was appointed acting director for Lejpunsky as he had been before when Lejpunsky was away on business. He belonged to the first cohort of physicists who moved from Leningrad to Kharkov in May 1930.

However on 1 December 1934, the day of the murder of the Leningrad party leader Sergej Kirov, while Lejpunsky was still abroad, rather suddenly and unexpectedly a new director was appointed in the guise of Semen Abramovich Davidóvich,[7] who nobody knew and had no scientific merits whatsoever, but was a Soviet style administrator. This very summary, negative qualification is the only one given in the Russian/Soviet literature about Davidovich, also in post-Soviet publications. According to some information he was a native of Leningrad and a friend of Gej, who "dragged" him over to UFTI for the role of director, but that seems rather unlikely. The leadership of Narkomtjazhprom must have had a hand in this, although they may of course have acted on Gej's advice. At the time of his appointment as director of UFTI Davidovich did not have a scientific degree, nor

---

[7]The stress is on the letter 'o' contrary to the patronymic Davídovich.

had he published a single scientific work (Ref. [7], p. 166). Davidovich always gets a negative press, and perhaps rightly so. In a passage that is also very revealing about the worsening atmosphere at the institute, Alexander Weissberg writes about him:

*"Davidovich came into the Institute with the firm intention of showing what he could do. Our scientific liberty he regarded as licence and he was determined to put an end to it—to Sovietize us, as he put it. As it was, he felt the director had far too little authority. Outside his office he put up a notice 'Interviews on Wednesdays and Fridays from 3 to 5.' At first we were inclined to grin. No one took advantage of the invitation. Almost every problem which arose could be settled without approaching the director at all. We therefore ignored him and got on with our work.*

*But one day he summoned Professor Shubnikov, the head of our low-temperature laboratory. Shubnikov appeared at the stated time and was left kicking his heels for an hour. He then left a message with the secretary saying that he had work to do and that if Davidovich wanted him he would be in the laboratory. Shubnikov's experience was only the first of a series of pompous and tactless actions which greatly irritated our scientific staff. Davidovich was a small-souled man with a narrow outlook but an overweening ambition. He insisted on his own importance as director and sought to introduce that atmosphere of subordination which had arisen in recent years in Soviet factories. But the men he had to deal with were scientists of note used to liberty of action, which was essential to their work. They were all loyal and devoted to their jobs, but they were not prepared to knuckle under to a stupid and narrow-minded bureaucrat who was not a scientist and knew nothing about scientific problems.*

*In his fight to Sovietize us and enhance his own position Davidovich even violated the sacred principles of Soviet management. The most important of Stalin's famous Six Points for the increase of labour productivity was one-man management. Everyone was to have only one immediate superior: for the workman it was his foreman (or brigadier as he was called); for the foreman it was his departmental manager and so on up the hierarchy. A factory director, for example, was not allowed to give orders direct to a workman and short-circuit the established channels. But Davidovich was anxious to set the assistants against the scientists and so he ignored this principle, and one day he went over the head of Professor Shubnikov and gave Rjabinin, Shubnikov's assistant, some special secret work to do for the Red Army. Now such work had absolute priority over all other, but our staff and our equipment were limited and a laboratory chief had to plan his work so that his resources were properly apportioned to the various tasks. Egged on by Davidovich, Rjabinin was constantly demanding this or that item of equipment for work about which his chief knew nothing. Professor Shubnikov then went to Davidovich and protested. It was exactly what Davidovich wanted. If he could manoeuvre the leading scientists of the institute into apparent opposition to the work for national defence his game was won. He did the same sort of thing in the other laboratories and before long the scientific work of the institute was almost paralysed.*

*Now in every Soviet institution, whether factory, office, or scientific institute, there is a secret sector with a double function. On the one hand it controls the political reliability of every member of the staff on behalf of the G.P.U.,*[8] *and on the other it supervises any work in the interests of national defence. For instance, mobilization orders are kept in its safe. Everyone who does any work of national importance must first be approved by the G.P.U. All our leading scientific men had been approved, and there was thus no reason for Davidovich to go over Shubnikov's head.*

*Almost every day he created some new cause for friction, and eventually my turn came. I was in charge of the building of our new experimental low-temperature station, and the work gave me a great deal of independence, though technically I was subordinate to the director of the institute, in this case Davidovich. In reality my instructions came direct from the Commissariat for Heavy Industry in Moscow. The former directors of the institute had always let me go my own way. I kept either Obreimov or Lejpunsky informed about the work in casual conversations; sometimes they gave me a friendly word of advice, and sometimes they even intervened with the Central Committee against this or that Soviet institution which had treated us unfairly.*

*When Davidovich was appointed I visited him as a matter of politeness and informed him about the purpose of the new scientific combination and the stage of the building and assembly work.*

*It was not long before a clash occurred. It was very difficult to obtain machine-tools, and I had fought a long and persistent battle in Moscow to obtain eight lathes and fraising machines. The administrative apparatus of the institute was over-loaded with incompetent officials who were incapable of obtaining the materials, etc. the institute required, but they cast envious eyes on my little machine park. Davidovich now stepped in and ordered me to hand over the machines to the institute proper. It meant that a few months later, when the workshops of the experimental station were ready to begin, I should have to go to Moscow and fight another battle for them, with much less chance of success. On the other hand, if the workshops were idle for lack of machines the responsibility was mine. Naturally I protested, but Davidovich insisted the national interests made his action necessary, and after that there was nothing more to be said.*

*He was rather afraid of me because he knew that I was often in Moscow and had the ear of the Commissariat, and so apparently he decided that I must go. There were plenty of grounds for friction. I had kept the purely administrative staff of the experimental station down to an absolute minimum, and I employed only eight people in my office whilst Davidovich employed about thirty for the same work in his. My technique was to pay my office workers as high a wage as the law would allow and to demand efficiency in return. No doubt Davidovich often had to listen to unfavourable comparisons between his administrative staff at the institute and mine at the experimental station.*

---

[8]State Political Directorate under the NKVD of the RSFSR (*Gosudarstevennoe politicheskoe upravlenie pri NKVD RSFSR*). The GPU was dissolved in 1923 and succeeded by the OGPU, which in its turn was dissolved in 1934 and succeeded by the Main Directorate of State Security under the NKVD.

*However, it was not actually Davidovich but the G.P.U. who raised the question of surveillance. This passion for guarding everything had developed into a cancer in the body of Soviet economy. The occupation which had the largest number of members was that of guard (watchmen in other countries), and I think I am right in saying that there were more of these guards than metalworkers and miners put together. Our institute had survived for years with nothing more formidable than a night porter to watch over it. That made the G.P.U. unhappy, and on a number of occasions they tried to force us to employ more guards. When Lejpunsky, and later on Gej, refused, the G.P.U. adopted a ruse. They sent some of their agents to break into the institute and steal some of the instruments. The next day they sent for Gej, who was Director at the time, and reproached him bitterly for his lack of watchfulness, and to clinch their case they showed him the stolen instruments. Then they demanded that a watch system should be introduced which would have cost the institute 88,000 roubles a year. To steal instruments worth that amount thieves would have had to drive up to the institute with a lorry. The guards were to cost more than their presence could possibly save. Gej tried to make this clear to the G.P.U., but then they brought up another argument: secret work was being done in the Institute. A spy might break in.*

*In fact the department of the institute in which secret work was carried on was guarded by steel doors and it was watched. And further, even if a spy had managed to break in he would have found out nothing. He would have seen nothing but a confusing jumble of glass piping and wires, a number of pumps and motors, and all the usual apparatus of a physics laboratory. What was really secret was the result of measurements, and that was in the heads of the physicists concerned. Even their notebooks would not have been of much use to an intruder, so that a spy—even if he were himself a trained physicist—would have gained no information of any importance by breaking into the institute. To find out anything he would have had to approach the scientists themselves, and they were not watched by guards.*

*But it was useless to reason with the G.P.U. All you got was slogans in return. The favourite one was 'revolutionary watchfulness'. Stalin had issued it and the G.P.U. applied it everywhere, whether there was anything worth watching or not. The country had to pay the costs of such indiscriminate watchfulness, and keep up a vast parasitic apparatus of guards.*

*One day Davidovich informed me that he considered that as a foreigner I ought not to hold such a responsible position; I would have to hand over control to a Russian Communist and continue my work under him.*

*I remember my discussion with him very well. He had received me in his flat late one evening. The flat had been furnished at the expense of the institute and he had chosen the furniture—and in very bad taste. In recent years it had become usual in Soviet factories for the director to have his flat furnished at the expense of the factory. It was just a straw in the wind. He was the first at our institute to introduce the disagreeable custom.*

*'Very well, Comrade Davidovich,' I replied, 'perhaps it's necessary for political reasons. I will gladly give up my present work and return to my normal work as a technical physicist. But I can't continue to do the same work in a subordinate position.'*

*'Does that mean that you propose to give up your work at the station?'*
*'Of course.'*
*'That sounds almost like a withdrawal of labour. You know what that means.'*
*I kept my temper with difficulty.*

*'Listen,' I replied, 'a bad organizer is better than two good organizers trying to do the same job. If I were to continue the work whilst someone else took the responsibility there would inevitably be conflicts and the work would suffer. But I'm quite prepared to carry on for, say, a month after the appointment of the new man in order to introduce him to the work.'*

*Davidovich didn't want to dismiss me because he knew very well how difficult it would be to find someone else to do the work efficiently. Commissariats and trusts fought tooth and nail for the services of good men, and if I had been a Russian I should have had a much more responsible job. Even so, the building of our experimental station was a complicated business and it required scientific knowledge in addition to organizing ability. He fought shy of taking the responsibility for getting rid of me. Construction work was difficult in the Soviet Union in those days. Everything was in short supply. If you had enough cement you were short of timber. If you had both you couldn't get the transformer you needed. If for the moment there were no supply difficulties you couldn't get workers. Or perhaps you had workers and your credits hadn't come through and you had to dismiss them because you couldn't pay their wages. And when the building was finished the assembly difficulties would begin. Davidovich himself was responsible for the building of our high-tension station, and the work had already been dragging on for five years, although it was not one-fifth of that involved in the experimental station. He knew perfectly well that if my dismissal were followed by an obvious decline in results the party would want to know the reason why. He therefore dropped the matter for the moment and intrigued through the party to make me give way. In the end I did"* (Ref. [6], pp. 156-161).

During his extensive visit of the Soviet Union and Kharkov in the winter of 1934–1935, so before the troubles at UFTI started, the British science journalist J.G. Crowther met and talked to Davidovich. In his book he gives a fairly extensive description of Davidovich without being prejudiced against him as all who write about him later clearly are. I will quote it here in its entirety as it brings this person who had such an enormous impact on UFTI's fate more to life and it is the only information available about him written by an outsider, not prejudiced by his own interests in the matter. Caution is required though, as all information presented by Crowther does probably originate from Davidovich himself: *"Leipunsky has recently been succeeded by Professor Davidovich,[9] and is now devoting himself entirely to research on atomic physics and the phenomena of high-tension electricity. Research physicists in the U.S.S.R. are not at all anxious to obtain directorships. These give them much administrative work, and distract them from the pursuit of physics.[10] (...) [T]he present director of the Physico-Technical Institute of the Ukraine, Professor Davidovich, is an administrator accustomed to scientific work, and not a physicist who continues to follow his own line of research besides conducting the administration. Like so many of the new directors and leaders, he is in his thirties, and was educated under Ioffe. The training of so many younger men in Ioffe's institute has made most of them known to each other.[11] This gives a peculiar homogeneity to the group of the younger Soviet physicists.*

*Davidovich studied at Leningrad at the same time as Leipunsky, Gej and other leaders of his departments. His acquaintance is particularly broad, as he began his physics studies three times. They were continually interrupted by the call of the Communist Party, of which he is an old member, to undertake some onerous special task. He had been an active Communist before the revolution and had been exiled from Russia on account of his political activities.[12] He spent part of his exile in Australia and there he acquired a knowledge of British institutions and the English language. Before he came to Kharkov he was engaged in the Leningrad Laboratory for Research in Road-making.*

*As Davidovich is primarily an administrator, the scientific lead in research will come from Leipunsky, who now devotes himself to pure research"* (Ref. [2], pp. 77–78).

The trouble with Obreimov and the change of director was happening against the background of the great purge that was being carried out in the party throughout the entire country since 1931. Within this framework a party commission was also investigating UFTI and although, as Weissberg puts it (Ref. [6], pp. 155), the

---

[9]Why he is called 'professor' here is not clear; he was certainly not teaching anything. Here it should also be noted that Lejpunsky was abroad when Crowther was at UFTI; Crowther gives the impression, which is odd, that he had already returned, but that was not the case; Lejpunsky actually returned to Russia for a short while in the autumn of 1934 while travelling with Pëtr Kapitsa and his family; before 12 October 1934 (Ref. [3], p. 567) he was back in Cambridge and only returned in September 1935, so Crowther is unlikely to have met him at UFTI; Gej was acting director and it was Gej who was replaced by Davidovich. However, Weissberg says in his testimony that he met Lejpunsky in April 1935 in Moscow when he was called from England to report to Ordzhonikidze, so it is not excluded that Lejpunsky was more often in the USSR in that year.

[10]Nothing is known about Lejpunsky having expressed any of these sentiments.

[11]This would mean that many if not all of the other Leningrad people at UFTI must have known him, while for instance Gorelik (Ref. [4], p. 77) says "nobody at the institute knew him and he had no relation to science", while Crowther says that he studied physics in Leningrad.

[12]If he is of the same age as Lejpunsky (born 1903) he must have been born around 1900 or even a few years later, and exiled from Russia when still a minor! Confirmation of this information would be very welcome.

institute may so far have been an island of freedom in the sea of Stalinist despotism, that may have been true for the scientific part of it, politically matters looked rather different.

In written memoranda and reports on the state of affairs at UFTI in the first half of the thirties composed by self-criticism teams from the office of the UFTI party organisation, it was noted that in this period communists did not occupy a leading position in the scientific work of the institute, and moreover there were too few of them. In 1931 only four of the 41 UFTI collaborators were party members, in 1932 six out of 43. Among the scientific workers party members were still fewer: 4.9% in 1931 and 7% in 1932 and 1933. In this respect it was also noted that *"the prospects of manning the scientific staff of the institute by party members were not favourable, since among physicists in the USSR there were almost no party members, nor were there suitable party members among those finishing physics higher schools in the near future"*. The main sphere of activity of the UFTI party organisation in the first half of the thirties was so-called sociopolitical and organisational work, including the issue of a wall newspaper, the creation and participation in various self-criticism commissions, the organisation and leadership of circles for the study of the classic works of Marxism, the organisation of sports competitions and 'socialist competition' brigades. The reports of the self-criticism commissions of this period note that *"at the institute there is a lack of organised shock-labour"*. As a consequence of this, *"emulation*[13] *and shock-labour are not yet the basic method of fulfilling the plan"*. Another shortcoming observed by the various self-criticism brigades was the fact that *"the work in the laboratories is not carried out during firmly established hours"*. However, since the working day of scientific workers far exceeded the established norm, this circumstance did not lead to absence from work and stoppages and the labour discipline among the scientific workers of the institute in the beginning of the thirties was considered satisfactory. In all reports of this period it was noted that the scientific workers worked very hard and fruitfully (Ref. [7], pp. 166–167).

At the end of 1934 a party purge was carried out at the institute along the line of the purges already ongoing in higher organs. A commission was appointed to carry out the purge of the primary party organisation at UFTI which started its work in the autumn of 1934. It was in fact a GPU field tribunal armed with denunciations. Pavlenko et al. (Ref. [7], pp. 167–169) have presented a short overview of its protocols. Its work at the institute started with the invitation to submit written and personal statements. In the final stage of the work of the commission, all-institute meetings were held. The party organisation was initially deemed ideologically healthy. Ideologically incorrect statements of some UFTI scientific workers on separate methodological questions (Lejpunsky, Landau), as well as a poor

---

[13]Socialist competition (or socialist emulation) was a form of competition between state enterprises and between individuals practiced in the Soviet Union and in other East-bloc countries. The word emulation is introduced to distinguish it from 'capitalist competition'. Implied was that 'capitalist competition' only profited those who won, while 'socialist emulation' benefited all involved. (*Wikipedia*)

attendance at meetings, close to 50%, were noted, but nothing very serious. However facts came soon to light which forced the commission to change its opinion on the party organisation into the opposite: the ideological state of the party organisation was now deemed unhealthy. What was the reason for this? Statements of some collaborators of the institute became known to the commission which it could not consider ideologically consistent, in which the party and even Stalin's regime were severely criticised. One party member, Stepan Shavlo,[14] who would later testify against Gorsky (see Chap. 10), had declared that he could not see any difference between the Hitler dictatorship and the Stalin dictatorship. Another that the government was engaging in speculation by buying bread for 1.30 rouble per pound and selling it for 1.20 per kilogram. And what was considered equally bad was that these statements were not opposed by those present at the meetings at which they were made. Early in December 1934, a few days after the appointment of Davidovich as director, the party commission presented its results at a general meeting of the institute. The party organisation at UFTI was declared 'ideologically unhealthy', since it lacked ideological consistency; there was no collective responsibility[15]; and the party organisation failed to exert any party influence on the work at UFTI. Shubnikov also spoke at this all-institute meeting. He was very critical of the work of the commission, did not agree with its conclusion and scolded it for not appreciating the enormous amount of excellent work done at the institute:

*"The conclusions on the work of the party organisation of the institute are intimately bound up with the conclusions concerning the work of the entire institute. These conclusions greatly resemble self-flagellation but not conclusions about the work of the party organisation.*

*In these conclusions we see solely the negative aspects of the work of the institute. These conclusions produce the impression of a death sentence for both the institute and the party organisation. In my opinion they have left out the most important fact of all, namely: starting from 1930, i.e. from the time when the construction of the institute began, and comparing it with the present day, we see that the institute is the largest, best technically equipped institute, the scientific level of the workers of our institute is higher than that in all other institutes, the work of our institute is faster than that of all institutes in the Soviet Union, and the work of*

---

[14]Stepan Trofimovich Shavlo (b. 1899) at the time a trainee in Gorsky's X-ray laboratory. He was one of the students sent by the party to higher education (the movement of the 'thousands' (*tysjachniki* or *parttysjachniki*): workers and young people with a proletarian background selected by the party and sent to universities and other higher education institutions at the end of the twenties to create the cadres needed to replace the old specialists (http://www.famhist.ru/famhist/landau/000ae7ee.htm). At Kharkov University they played a role in the student unrest that led to Landau's dismissal. Shavlo was still working as a physicist in the sixties.

[15]Collective responsibility (*krugavaja poruka*) also known as collective guilt is a concept in which individuals are responsible for other people's actions by tolerating, ignoring, or harbouring them, without actively collaborating in these actions.

*our institute is advancing so rapidly that in a year or two we shall be among the top in Europe.*

*This colossal work done in the institute is not seen at all in the conclusions of the Commission, nor is the role of the party organisation reflected in this work.*

*Unquestionably, the work of the party organisation has been weak, the number of party workers in our institute has been inadequate, but despite this it is also unquestionable that the party organisation does have a role in the achievements of our institute.*

*I will not dwell on the individual points of the conclusions with which I do not agree. All of them are separate shortcomings of the party organisation; they are correct; never can one find anyone or anything to be ideal, there are always flaws, but the question is how dangerous are these flaws. Of course, all of the flaws indicated in the conclusions are dangerous, but of course the party organisation at UFTI cannot be characterized as unhealthy on the basis of those flaws*" (Ref. [7], p. 168).

As a result of the work of the commission some people should have been expelled from the party, but in the end none were. The commission also noted positive points in the work of the party organisation. In particular, the party organisation was credited with the development and start of the programme of Landau's theoretical minimum. Shubnikov, Ruhemann, Trapeznikova, Rjabinin, Sinelnikov, Slutskin and others were singled out as the best shock-workers of 'socialist emulation'. All this was just a prelude to the more dreadful events that would fall upon UFTI in the following years.

Soon thereafter, in March 1935, the institute was charged with a number of defence-related orders (it is quite possible that Narkomtjazhprom had actually sent Davidovich to UFTI with the aim of introducing military topics in its research programme). It concerned the creation of powerful short-wave generators (work for the ultra-short wave department of Abram Slutskin who was keen to accept such orders), work on aviation engines running on liquid-hydrogen fuel and other tasks. It is now difficult to establish whether Davidovich, as the UFTI scientists claim, 'begged' for military topics in Moscow, or whether UFTI was charged with such topics by the Defence Council of Narkomtjazhprom. But the main problem must have been that the scientific heads of the various laboratories, i.e. the most prominent scientists of the institute who so far had determined scientific policy, had not been consulted and were from the very start excluded from participating in such defence-related work. Why this happened is not clear. They may themselves have refused to work on military topics as it would inevitably restrict their freedom of work.

The appearance at UFTI of work on defence-related topics had far-reaching consequences indeed. As soon as it became known that the institute would engage in such subjects, the Kharkov State Security Administration and the party District Committee started to work out measures ensuring a regime of secrecy. This was actually not an event particular to UFTI. According to Conquest (Ref. [1], p. 275) and also mentioned in the extensive quotation from Weissberg's book [6] above, in 1935 the NKVD suffered from a genuine security mania and all over the country a system with guards and watchmen was imposed on factories, research institutes and so forth. Conquest states that in a labour force of close to 80 million there were more than 2 million guards and watchmen (not counting NKVD militia), more than three times the number of miners and more than twice the number of railwaymen. In this atmosphere any failure or accident automatically became sabotage.

The defence orders implied, as seen in the quotation from Weissberg's book, that the freedom enjoyed at the institute was over, as were visits by foreign scientists. Laboratory work became secret, strict security measures were introduced, bureaucracy and 'red tape' increased (Ref. [7], p. 171).

The new measures and new type of work caused a conflict at UFTI that split the institute into two camps. One camp consisted of the scientific leadership of the institute, supported it seems by Narkomtjazhprom in the form of Pjatakov, Bukharin and Ordzhonikidze. On the other side of the barricade stood the director, the second camp of *rabfak* graduates who had already been disgruntled as mentioned above, and the scientists of the radiophysics division supported by the local NKVD and the regional party committee (Ref. [8], p. 55; Ref. [9], p. 60). The essence of the work concerned the development of a microwave generator and antennas for radar technology, which the Soviet Union did not yet have in those days, so work for Slutskin's department. A group of theorists and experimentalists led by Landau was against this kind of work with the argument that so much work of a technical applied nature would reduce the scientific standard of the institute. They demanded that the institute be split into two separate institutes (Ref. [9], pp. 57–58), one for fundamental research and one for applied research, i.e. that Slutskin and his department be removed from UFTI. It cannot be denied, in my view, that such a demand was rather strange as UFTI was set up with the main goal to establish contacts with and to support industry. It is obvious that such a split of the institute was completely at odds with the priorities set by party and state (science in the service of socialist construction), so it had to lead to a clash. Nor can it have been an accident that Slutskin's department at UFTI was the only one not manned by former Leningraders and foreigners, who apparently wanted to have the place all to themselves.

This tallies with what Sinelnikov's wife writes about this time to her sister in a letter dated 29 June 1935: "*And such trouble at the Institute for Kira* (i.e. Sinelnikov) ... *But the Institute is full of intrigues, firstly it was the scientists against the Administrative department, but now it seems to me wheels within wheels and some of the scientists are using dirty methods to obtain their own ends.*" And some time later in a letter of 19 September 1935: "*Kira has an awful lot of trouble and worry just now, I wish it would all finish. I guess the whole Institute just wants dynamiting and then a fresh start making. The silly thing is that it's chiefly the foreigners who are making the trouble, truly one or two Russians are with them, but it's all absurd, and anyhow a scientific institute is surely for work and not for scandals*" (Ref. [10], p. 231 and 242). It also agrees with statements by Leonid

Pjatigorsky,[16] reputedly Landau's first student and the fifth to pass the theoretical minimum exam, when interviewed by Ranjuk: "*A whole group of people, mostly 'guests' from Germany, took the view that Landau was doing important work for the organisation of Soviet theoretical physics. He should not be disturbed! UFTI should engage in fundamental problems of physics and Slutskin and his staff only disturb this and create difficulties for the organisation of Soviet physics. So Slutskin and his staff should be separated from UFTI so that they cannot interfere with this work*" (Ref. [11], p. 83).

And it also agrees with the genuinely sounding statement made by Abram Chernets, an engineer in Slutskin's laboratory, within the framework of the rehabilitation procedure of Shubnikov, Rozenkevich and Gorsky (Appendix 4; Ref. [7], p. 275): "*In a number of cases which I can now hardly remember [Shubnikov] was among a group of scientific workers who acted against the decision of the management and the party organisation. In general, at the time the institute consisted of two camps, one of which included the group of non-party scientists (some of the foreign specialists then working at the institute associated with this group), who behaved in a haughty manner to the other part of the scientific collective and, simply speaking, caused quarrels.*" So his view of the situation is not favourable towards Shubnikov and his friends, accusing them of behaving in a haughty manner and causing quarrels.

The situation at the institute became worse and worse. So far, many foreigners had visited the institute, but now visa applications were rejected, for example requests in 1936 from Eliza Wiersma from Leiden who had visited UFTI every year since 1932, as well as from De Haas and Francis Simon.[17] As has become clear

---

[16]Leonid Moiseevich Pjatigorsky (1909–1993) was a co-author of the first volume (on Mechanics) of Landau's famous course on theoretical physics. Pjatigorsky was a true Soviet child who lost both his parents and was himself heavily wounded (his right arm was later amputated) in 1919 in the anti-Soviet Grigorev rebellion in the Southern Ukraine, but was saved by the Soviets. He owed everything, down to his bare existence, to the Soviets and was therefore not surprisingly a faithful supporter of the Soviet government, which later brought him into difficulties with Landau, who suspected him, quite wrongly, of giving false testimony to the authorities, of having reported him to the NKVD as a German spy, accusations which were repeated later in biographies of Landau, but had to be rectified after a court case. From the NKVD documents the court established that Landau had never been accused of being a German spy. Weissberg (Ref. [6], p. 60) writes that Pjatigorsky was an informant of the NKVD. For more details about this tragic figure, see Ref. [9], p. 64 ff and Ref. [11].

[17]Sir Francis Simon (1893–1956) was a German and later British physical chemist and physicist who invented a method for liquefying helium in which helium is first cooled at high pressure by liquid or solid hydrogen and is then liquefied by a single adiabatic expansion. He also devised the method, and confirmed its feasibility, of separating the isotope Uranium-235. Among his other important achievements when working under Nernst in Berlin were the solidification of helium and of other gases by high pressures and the discovery of the specific heat anomaly in solid orthohydrogen. In 1936 he was able to produce the first liquid helium by using magnetic cooling at a laboratory at Bellevue near Paris. In the thirties he moved from Germany to Oxford. (N. Kurti, Franz Eugen Simon 1893–1956, *Biographical Memoirs of Fellows of the Royal Society* 4 (1958) 224–256).

from a letter dated 5 July 1935 from Lejpunsky in Cambridge to Landau, which reappeared in 2016,[18] in 1935 Landau and Shubnikov seriously considered moving to Moscow. Lejpunsky was not pleased with their attitude, urged them to continue the fight for the institute in which they had invested so much effort, and in any case to wait for his (Lejpunsky's) return. The letter also gives the impression that Lejpunsky did not really appreciate the seriousness of the situation. We again take up Weissberg's story here:

*"Things in the Institute went from bad to worse until finally our leading scientists decided they could stand it no longer. They got together and wrote a complaint to the Central Committee of the party and asked for the removal of Davidovich and the reappointment of Lejpunsky, who was then working in Cambridge at the Rutherford Laboratory. The scientific sector of the Party Control Commission was at that time under Modest Iosifovich Rubinshtejn,[19] a man who had travelled all over the world and spoke seven languages fluently. He was one of those Russian Marxists of the old school who happily combined the best traditions of Russian culture with those of the West. I knew him quite well and I explained to him what had happened at the Institute. He discussed the matter with a representative of the Central Committee and it was decided that a scientist should be director of the Institute. That decision automatically meant the removal of Davidovich. Unfortunately it was months before we heard about it, and in the meantime the struggle went on to the great discomfort of all of us and to the great detriment of the work. Davidovich launched a counter-attack; he informed the Kharkov G.P.U. that Weissberg and Landau were the leaders of a conspiracy to sabotage the work for the Red Army. The comrades in the party cell at the institute hesitated; they didn't yet know which side was going to win: the director, who probably had the backing of the G.P.U., or the scientists of the institute, who might be supported by the Central Committee of the party in Moscow. Even the Kharkov G.P.U. was cautious and it made no attempt to arrest either Landau or me.*

*Davidovich was quite convinced that I was the ringleader of what he called the conspiracy. The protest to the Central Committee had been very cautiously worded, and that made Davidovich suspect my hand behind it. He found it difficult to believe that anyone who was not a member of the party could have managed it so adroitly. Even for communications addressed to the highest party organs in the Soviet Union there was a settled ritual. For instance, it was ipso facto a counter-revolutionary act to draw up a collective document signed by eight or more people. This was regarded as the first step towards political organisation, and was counter-revolution.*

*In order to avoid this trap we made several appeals, each of which was signed by no more than three persons, which was quite in order. Collective documents*

---

[18]The letter was given by Korets's daughter Natasha Golfand-Korets to Mikhail Shifman at a conference in Jerusalem and published by Shifman in the Russian journal *Priroda* [12].

[19]Modest Iosifovich Rubinshtejn (1894–1969) was a Soviet economist. He worked at Gosplan from 1931 and was a member of the Party Control Commission from 1934.

were, of course, often drawn up, but only by the party; for instance when all the members of this or that collective farm signed a letter to the Beloved Leader and Father of the Peoples thanking him for the happy life he had given them.

Davidovich was anxious for arrests. He felt, quite rightly, that once someone was arrested that would start the ball rolling, but the Kharkov G.P.U. was not sure of itself; it was unwilling to risk the arrest of Landau, who was a scientist with a world reputation, and it was equally unwilling to arrest me, because I was a foreigner, and in those days they preferred not to arrest foreigners. In this situation the G.P.U. preferred smaller game and it arrested a young assistant named Korets who worked under Landau and had supported us in our struggle against Davidovich.[20]

First of all Davidovich had him expelled from the Komsomol. Allegedly he had made an attempt to conceal his social origin. He had drawn up two accounts of his past life, one, a very long and detailed affair, for the Komsomol, and the other, a briefer version, for the institute. In his longer account he had mentioned that during the civil war, and driven by force of circumstances, his mother had traded for a short while to keep the family—she had sold potatoes and apples on the market. In the shorter version given to the institute this detail had been omitted. However, the more detailed biography was at the disposal of the director and the party secretary of the institute at any time, and so Korets thought he had done everything which was required of him, as, indeed, he had. But Davidovich wanted to deal our group a blow somehow and so he brought the matter before the party and Korets was expelled from the Komsomol as a 'socially undesirable element'. A few weeks later he was arrested.

A few days after the arrest the usual party witch-burning was organized, and one member of the cell after the other rose to condemn Korets as a secret enemy of the people, a foreign spy, and so on—all in tones of the deepest conviction. At first I decided not to go to the meeting, but Komarov, who as an old Communist knew the atmosphere in the party very well, finally persuaded me to do so. 'Alex, if you don't go it will be regarded as a demonstration. Perhaps they'll arrest you too, or expel you from the country. This conflict won't be settled here in Kharkov, but in Moscow. The Central Committee will speak the final word. If you stage a demonstration, you will only be playing into the fellow's hands. The G.P.U. would regard it as a provocation.

It was perfectly true and I allowed myself to be persuaded. I did not speak against Korets, but when the resolution condemning him was put I raised my hand with the rest. I still experience a flush of shame and discomfort every time I think of it. When I look back on my whole party career, including the time I spent in the Soviet Union, I can find nothing I am more ashamed of.

The arrest of Korets drove our group underground. We were depressed and miserable and no one dared to say a word. Davidovich was openly triumphant. Our research work suffered and I wrote to Moscow. My letter was acknowledged, but I

---

[20]This affair will be discussed later in this chapter.

*was given no inkling of their intentions. Two months later Davidovich felt strong enough to dismiss me. At that time I had business in Moscow, but he refused to let me go unless I gave him my word of honour not to speak to Pjatakov or anyone else in Moscow about the situation. I refused and decided not to go. Instead I rang up Moscow on the phone and informed them of the situation. A week later Davidovich appointed the new director of construction and I continued my work under him. One afternoon—I had just come from the site and was lying down on the couch— my mother-in-law came in smiling happily.*

*'Alex, do you know who 's just come back? Korets!'*

*He had been released that morning on instructions from Moscow. The members of our cell began to feel ashamed of themselves and to wish they had never denounced him. Obviously, he couldn't be a spy or an enemy of the people, or anything of the sort if Moscow ordered his release. The general feeling now turned against Davidovich. The Central Committee in Moscow was stronger than the local G.P.U. in Kharkov.*

*It was still a few months before Davidovich was finally removed. Lejpunsky was recalled from Cambridge and appointed director in his place. That was in the autumn of 1935. At this triumph of good over evil, my courage rose again and I decided to stay on in the Soviet Union. I was to regret it bitterly"* (Ref. [6], pp. 161–163).

Moisej Korets was arrested on 28 November 1935, just one day after Lejpunsky was reinstalled as director. Korets was a collaborator of Landau from his Leningrad days, although they never published anything together. He was a physicist, but it is not clear whether he was a good one and why Landau needed him in Kharkov, where he acted as Landau's personal secretary at the institute and as his assistant at the university. There were quite a few much better physicists who would probably also have been better assistants. Leonid Pjatigorsky said about Korets, who as Landau's assistant had to do problem classes with the students, that Korets did not even understand the problems and generally caused laughter among the students (Ref. [11], p. 83). The only thing that can actually be said about Korets is that for some reason or other Landau liked him and wanted to have him around. According to a hand-written note from Landau from 1935, in which he gave a summary of the work that had been done in the theory group at UFTI in the second quarter of that year,[21] Korets was also preparing for the theoretical minimum and had actually considerably speeded up the work for it. He is, however, not known to have ever passed this exam, nor did he ever publish a research paper in physics. The physicist, and in his later years historian of science, Viktor Vorobev is in any case very negative about Korets: *"… twice expelled from the Komsomol and having a criminal record…"* and *"… he literally terrorized the students making it seem that he was carrying out Landau's instructions. … Korets did everything in order to incite the students against Landau, and Landau against the students. Given*

---

[21]A copy of this note was kindly given to the author by Professor V.I. Sokolenko during a visit at the Kharkov institute in June 2016.

*Landau's character, that was not hard to do. So a confrontation arose between the students and Landau"* and the students started to complain (Ref. [13], p. 96).

Due to their difference of opinion about splitting the institute enmity arose between Landau and Pjatigorsky. At one point Korets apparently came to Pjatigorsky[22] with the instruction from Landau to write an article for the institute's wall newspaper '*Impuls*' demanding the removal of Slutskin and his department from UFTI. Pjatigorsky refused to do this, made it clear that he did not at all agree with their action and actually wanted the theorists and other experimentalists to help Slutskin (Ref. [11], p. 84).[23] After Korets's arrest Pjatigorsky was heard by the court, while Shubnikov's wife and Korets's wife also attended the hearing. In his interview with Ranjuk, Pjatigorsky says about this: *"After a number of formal questions they asked me: is it true that M. Korets was against the work of Slutskin at UFTI on the radar problem? I answered: yes, that is true. Why did he think so, in your opinion? Why was he against Slutskin's work? I answered: out of stupidity, he did not understand anything. After this they said that I could go. I went to UFTI and into the library. (...)."*

He continues: *"Landau entered the library at UFTI and beckoned me with his finger. We went into his room and he asked me for the notebook with the list of physicists. I kept the book as Landau considered me his closest associate on the organisation of physics. He took the notebook, struck me from the lists of 'communists' and entered me in the list of 'fascists'.[24] What the distraught women— O. N. Trapeznikova and Korets's wife—told him, I never knew. Landau did not ask me a single question and told me not to come to the theory seminars"* (Ref. [11], p. 84). The documents of the Korets affair (*Delo Koretsa*) tell a slightly different story, namely to the question: *"What do you know about the counter-revolutionary activity of Korets and his associates?"* he answered: *"I know that in our institute there existed an anti-Soviet group consisting of Korets, Landau, Shubnikov, and the foreigners Weissberg, Martin and Barbara Ruhemann from Germany"* (Ref. [11], p. 85). So, in this testimony he unconditionally accuses Landau and Shubnikov of counter-revolutionary activity. From this testimony and that of others, the court came to the conclusion that Korets was guilty of disrupting defence-related work at the institute and on 26 February 1936 it sentenced him to a year and a half in prison. A very mild sentence compared to the ones that would be pronounced a year later. However, on 31 December 1935 Landau had already written a letter[25] on behalf of Korets to Vsevolod Balitsky, the Ukrainian People's Commissar of Internal Affairs,

---

[22]Others, such as Sinelnikov and A.K. Valter, were likewise put under pressure, and also refused to cooperate. This was the reason why Landau later never wanted to meet Sinelnikov (Ref. [9], p. 65). There was more friction between Landau and Sinelnikov as can be read in B.I. Verkin, S.A. Gredeskul, L.A. Pastur, Ju.A. Frejman and Ju.A. Khramov, Lev Vasilevich Shubnikov, *Priroda* 1 (1989) 89–97.

[23]Protocol of 5.12.1935 at http://www.ihst.ru/projects/sohist/document/ufti/piat1.htm.

[24]Landau had a passion for classifying, making lists of things and people (Ref. [11], p. 82).

[25]The letter has been reprinted (in Russian) in Ju.N. Ranjuk, "Delo UFTI" Historical comments to the book "The Accused" of Aleksandr Weissberg, http://www.sunround.com/club/22/ufti.htm.

in which he also complained about Davidovich, who for that matter had been sacked just a month earlier on 25 November 1935 by order of Ordzhonikidze and replaced by Lejpunsky. Landau's letter had an astonishing effect and resulted in a retrial at the Kiev regional court, which acquitted Korets. This happened in June 1936, so more than a year and a half after Kirov's murder. But it only amounted to less than two years respite since in April 1938 Korets was again arrested and this time he did not get off that lightly (Ref. [9], p. 88).

The final outcome was that Davidovich was fired from the institute and in 1937 was himself arrested and executed; Jury Rjabinin, a member of Shubnikov's group who had been carrying out defence-related work on direct orders from Davidovich without Shubnikov being consulted, was also forced to leave the institute, and Vladimir Gej returned to Leningrad.

It appears to me that this whole affair turned into a struggle between central party organs on the one hand and local party and NKVD departments on the other. So long as people like Ordzhonikidze, Pjatakov and Bukharin were alive and in favour, Lejpunsky, Landau and others at UFTI could feel safe and appeal to these people to overrule local authorities.

Let us now return to Kharkov University. In line with the party purge going on across the country, already before the murder of Kirov on 1 December 1934 in Leningrad, purges started among the cadres at Kharkov State University, which, according to a resolution of the party city committee, was full of anti-Soviet elements, in the form of great-power chauvinists, Trotskyists and nationalists. The purge process also had an ideological background, with as one of its manifestations the struggle against idealism in science. In this struggle Kharkov University clearly could not stand aside, and its so-called 'militant materialists' spared no effort in combating this idealism. The campaign was led by Semën Julevich Semkovsky (1882–1937), a doctor of philosophy and sociology and member of the Ukrainian Academy of Sciences. Landau attended Semkovsky's philosophy seminar, at which he reacted sharply to the latter's views on philosophical questions of physics. Landau had no difficulty in putting Semkovsky, who was virtually illiterate as far as physics is concerned, in his place and in the presence of 'militant materialists' he described materialism 'as a scholastic doctrine harmful to science,' which immediately earned him the label of 'idealist'. It was the time of the first great show trials in Moscow and also in Kharkov the 'enemies of the people' had to be exposed. The funny thing, if there is a funny side to this matter, was that Semkovsky, whose real name was Bronshtejn and who was a cousin of Leon Trotsky, was the first to be unmasked as the leader of a Trotskyist organisation (and in 1937 duly executed), but of course Landau with his idealism, who was moreover 'terrorizing' the students (see above) was just as dangerous. The complaints about Korets and the difficulties with the students were now used as a pretence by the university administration to try to get rid of Landau. The rector of the university, A.I.

Neforosnyj,[26] who had earlier invited both Landau and Shubnikov to come and teach at the university, now invited Landau to resign voluntarily, but when this did not work he was told that the administration of the university had to dismiss him for promoting idealism and showing disdain for the students. After having heard this, Landau immediately went to Shubnikov, who decided to hand in his resignation (his letter to that effect is dated 25 December 1936). Shubnikov mobilized the other UFTI scientists working at the university (Akhiezer, Pomeranchuk, Lifshits, Kikoin, Gorsky and Brilliantov), who likewise all handed in their resignation. This joint action was first labelled a 'strike' by the Academic Council and subsequently more ominously an 'anti-Soviet strike'. When attempts to reverse the situation did not work, all were fired for 'participation in an anti-Soviet strike'.[27] But Landau was not fired! The rector had no authority to do so and did not get permission from Kiev for Landau's dismissal. A few weeks later the rector himself was to be expelled from the party and dismissed from the university. The other 'strikers' now realized that they had made a gross and dangerous mistake. They wrote letters to the university administration admitting their political mistake, which in the end would actually contribute to Shubnikov and Gorsky losing their lives, and protesting the reason of their dismissal: 'participation in an anti-Soviet strike'. They asked the rector to let them keep their jobs. They were even called to Kiev where they had a meeting (without Shubnikov) with the Ukrainian People's Commissar of Education Vladimir Zatonsky. According to Akhiezer's reminiscences [14], the people's commissar started to reprimand them for having defended Landau, who was an idealist and did not accept the law of energy conservation. Pomeranchuk, with a downcast glance, noted that he would never have thought that a member of the Ukrainian politburo would entertain him on this question. Then Akhiezer, with great presence of mind, said: "*You know, comrade commissar, just yesterday Pomeranchuk and I had a talk with Landau on the problem of the scattering of light on light and somewhere at one point something had not been accurately calculated and he showed us a violation of the law of energy conservation. How he scolded us, I will remember it for a long time!*" With this it all ended, but it was too late, the wording was changed into 'voluntary dismissal', but the dismissals were upheld.[28] It was January 1937 and the NKVD was already closing in on UFTI for the final kill.

At precisely this time, on 23 and 24 January 1937, and, as if no dark clouds were gathering over UFTI, the physics group of the USSR Academy of Sciences had decided to hold its first-ever away meeting at UFTI in Kharkov in view of the

---

[26]Aleksej Ivanovich Neforosnyj (1897–1937). In January 1937 he was excluded from the party and relieved from his function as rector. He was accused of participation in a counter-revolutionary nationalistic organisation and executed. He was posthumously rehabilitated in 1957–1958.

[27]Ref. [9] (p. 97) because of "participation in an anti-Soviet strike" (*za uchastie v antisovetskoj zabastovke*). Ref. [13] (p. 99) uses a different formulation "for deliberate disruption of work" (*za soznatel'nyj sryv zanjatij*).

[28]Ref. [13], p. 97 ff; according to Gorobets (Ref. [9]), only Akhiezer managed to get the wording of the reason for his dismissal changed into this more acceptable version.

extraordinary results of the research carried out at UFTI, in particular at the low-temperature laboratory. This gathering came to be known as the First All-Union Conference on Cryogenics. Ioffe was chairing, of course, and a large number of scientists from Moscow, Leningrad, Sverdlovsk, Kiev, Odessa and Dnepropetrovsk were attending; in total about 200 people took part. Lejpunsky gave a general report on the work at UFTI. Low-temperature physics featured prominently on the agenda, especially its central problem, superconductivity. Both Shubnikov and Landau presented their work. The fruitful collaboration of theory and experiment in this remarkable work of the cryogenic laboratory was mentioned specifically.

In a resolution after Lejpunsky's lecture on the work at UFTI, the session noted that in the six years of its existence UFTI had become one of the leading physics institutes in the Soviet Union. It further noted the enormous scientific and technical value of the cryogenic laboratory at UFTI, which was on the same level as the best low-temperature laboratories in the world. In the field of low temperatures, as well as in the field of other crucial directions of work the institute had managed to establish a real connection with technology. The session also noted the importance of using the UFTI cryogenic laboratory by all physics institutes via *komandirovki* and the exchange of scientific workers. As a negative note the unsatisfactory participation of UFTI in training cadres in the Kharkov higher education institutes was mentioned, in spite of the successful work of individual workers in that direction.

It was probably a compulsory ritual that had to be performed at such occasions, but on a proposal of Ioffe the session also unanimously adopted a resolution expressing the anger and resentment of the Soviet scientists about the vile work of the Trotskyist bandits and demanded their destruction by the proletarian courts.[29] Nobody, I suppose, among the participants of the conference had at that moment any suspicion, not even the Trotskyists themselves, who like everybody else voted in favour of the resolution, that the most brilliant scientists applauded during the conference would soon be unmasked as inveterate Trotskyists and counter-revolutionaries.

It was at this meeting that Sergej Vavilov[30] estimated that more than one quarter of all physics research in the Soviet Union was carried out at UFTI. The NKVD, however, was not impressed. After all, the Beloved Leader of the country had said more than once that nobody is irreplaceable, and on 27 January 1937 it called

---

[29]Ref. [15]. It concerns a report of the conference reprinted from the Russian journal *Zhurnal tekhnicheskoj fiziki* 7 (1937) 884. Some information in this paragraph is from B.G. Lazarev, *Zhizn' v nauke, izbrannye trudy, vospominanija* (Life in Science, Selected Papers, Reminiscences) (Kharkov, 2003), p. 337.

[30]Sergej Ivanovich Vavilov (1891–1951) became one of the most important organisers and managers in Soviet Physics, vying with A.F. Ioffe for the top spot in this respect, culminating in his presidency of the Academy of Sciences from 1945. From 1932–1945 he was the scientific director of the State Optical Institute in Leningrad. He was the brother of the well-known biologist and geneticist Nikolaj Ivanovich Vavilov (1887–1943), who starved to death in prison after having become victim of Lysenko's persecution of genetics in the Soviet Union.

Alexander Weissberg to its office for questioning, which set in motion a train of events that would lead to the virtual destruction of UFTI in a little more than half a year. Landau understood that nothing good could come from this situation. He as yet handed in his resignation and left for Moscow to work at Kapitsa's newly founded Institute for Physical Problems.

On 30 January 1937 in Moscow, in the case of the so-called 'anti-Soviet Trotskyist Centre' Georgy Pjatakov, one of the most important patrons of UFTI, and other employees of Narkomtjazhprom were sentenced to death. The days of Nikolaj Bukharin were also numbered. All this implied that the support the leading UFTI scientists had enjoyed from highly placed party officials was rapidly eroding. This turn of events untied the hands of the Kharkov NKVD, which could now prepare itself for avenging its defeat in the Korets case.

# References

1. R. Conquest, The Great Terror: a Reassessment (Pimlico, London, 2008).
2. J.G. Crowther, *Soviet Science in Russia* (London, 1936).
3. P.E. Rubinin, P.L. Kapitza and Kharkov. Chronicle in letters and documents, *Low Temp. Phys.* 20 (1994) 550–578.
4. G.E. Gorelik, Lev Landau i Aleksandr Lejpunsky, *Priroda* 9 (2016) 77–83.
5. O.N. Trapeznikova in B.I. Verkin et al. (1990), p. 256–291.
6. A. Weissberg, *Conspiracy of Silence* (Hamish Hamilton, London, 1952).
7. Ju.V. Pavlenko, Ju.N. Ranjuk and Ju.A. Khramov, *"Delo" UFTI 1935–1938* (The "UFTI" Case 1935–1938) (Feniks, Kiev, 1998).
8. Ju.N. Ranjuk, M.A. Korets i L.D. Landau v kol'tse khar'kovskikh spetssluzhb, *Priroda* 1 (2008) 54–59.
9. B. Gorobets, *Krug Landau* (Letny Sad, Moscow, 2006).
10. Lucie Street (ed.), *I married a Russian. Letters from Kharkov* (George Allen & Unwin, London, 1944).
11. Ju.N. Ranjuk, L.D. Landau i L.M. Pjatigorsky, *VIET* 4 (1999) 79–91.
12. M.A. Shifman, Pis'ma A.I. Lejpunskogo L.D. Landau, *Priroda* 9 (2016) 74–76.
13. V.V. Vorobev, Lev Landau i "antisovetskaja zabastovka fizikov" (Lev Landau and the "anti-Soviet strike of physicists"), *VIET* 4 (1999) 92–101, edited by Ju.N. Ranjuk.
14. A.I. Akhiezer in B.I. Verkin et al. (1990), p. 337–339.
15. M. Dibilkovsky in B.I. Verkin et al. (1990), p. 51–52.

# Chapter 8
# Shubnikov's Scientific Work at UFTI

Shubnikov modelled his laboratory after the Leiden example, which implied among other things that he attached great importance to the workshop. In it he employed a large number of qualified craftsmen whom he acquired from the most unlikely places, such as Ivan Pavlovich Korolev who had earlier worked as a ship mechanic and turned out to be a very good head of the workshop. Shubnikov and the other physicists in his group were never seen at a workbench (contrary to Kapitsa for instance) and did not make their own apparatus as many physicists still did at the time. Figure 8.1 shows the first group of collaborators of Shubnikov's laboratory.

Already in the autumn of 1931 a large hydrogen liquefier from the Dutch firm Hoek was started up, yielding 12 l of liquid hydrogen per hour. This liquefier, which nobody apart from Shubnikov knew how to operate, was actually one of the reasons that after a long delay Shubnikov was finally appointed head of the cryogenic laboratory. At the end of 1932 he and the first co-workers of the laboratory, Jury Rjabinin[1] and the still very young, not yet seventeen year old, Aleksandr Sudovtsov,[2] who had just finished a factory-workshop training in repairing trams, obtained liquid helium with expansion liquefiers via the Simon method. This is a

---

[1]Jury Nikolaevich Rjabinin left UFTI after a conflict with Shubnikov and Landau, but he did so only in or after 1937, as is clear from Weissberg's book (Ref. [9], p. 43) where he also states that the conflict with Shubnikov had to do with the fact that Rjabinin played a very hostile role in the general conflict raging at the institute a few years earlier. It has also been noted in the lengthy citation from Weissberg's book in the preceding chapter that Rjabinin was used by Davidovich to sow discord among the scientists. Weissberg also says that he is not sure that Rjabinin is not an informer of the NKVD. Any evidence of wrong-doing by Rjabinin has not come to the author's attention. It is clear that Rjabinin was one of Shubnikov's most talented collaborators, and he was involved in much of the fundamental work carried out by the Shubnikov group between 1932–1937 without getting much credit. From UFTI he first went to the institute in Dnepropetrovsk and eventually to Moscow at Semënov's Institute of Chemical Physics (Ref. [10]).

[2]Aleksandr Iosifovich Sudovtsov (b. 1915) worked at the UFTI cryogenic laboratory from 1932 as a mechanic, and subsequently was a constructor (from 1937). He gained a candidate degree under B.G. Lazarev in 1954.

© Springer International Publishing AG 2018
L. J. Reinders, *The Life, Science and Times of Lev Vasilevich Shubnikov*,
Springer Biographies, https://doi.org/10.1007/978-3-319-72098-2_8

**Fig. 8.1** The group of first collaborators of the UFTI cryogenic laboratory. First row from the left: V.I. Bogatov?, N.S. Rudenko, N.M. Tsin, O.N. Trapeznikova, Ju.N. Rjabinin, A.I. Sudovtsov, Dogadin. Second row first two unknown, G.D. Shepelev (standing), Shubnikov, I.P. Korolev, V.I. Khotkevits, V. Sobolevsky, V.A. Maslov

method for liquefying helium whereby helium is first cooled at high pressure by liquid or solid hydrogen and is then liquefied by a single adiabatic expansion, invented by Francis Simon. The essence of this method is that if a gas is compressed, it cools down, and when it is subsequently expanded, part of it liquefies due to the Joule-Thomson effect (it is necessary to cool the gas to below the inversion temperature of the Joule-Thomson effect). In this way helium can be liquefied by hydrogen and hydrogen by nitrogen. The first cryogenic laboratory in the Soviet Union had come into being. In a short paper in a special edition of the *Physikalische Zeitschrift der Sowjetunion* in 1935 Shubnikov described the set-up and early years of the laboratory [1]. The staff of the laboratory at that time consisted of about 10 scientists and 25 technical personnel.

Also in other ways the laboratory profited very much from the one in Leiden, particularly due to the annual extended summer visits from 1932 to 1935 by Eliza Wiersma, the main assistant of De Haas in Leiden and from 1936 professor in Delft, who was very knowledgeable and informed of the latest developments in the field. He had been very close to the Shubnikovs in Leiden. At one time he even contemplated moving permanently to Kharkov, when suddenly in 1936 he was refused a visa.[3] Trapeznikova writes about this: "*Wiersma turned out to be of great help to the laboratory. Each year, up to 1935, he came to Kharkov and brought with him a great number of all kinds of things, without which we could not work. In Leiden he had learned about the new helium machine constructed by Francis Simon and immediately sent sketches of the machine made by Ehrenfest, who himself wanted to make such a machine. We had nothing for measuring low temperatures; for this special platinum thermometers were required. In order to prepare them one had to wind platinum wire on a porcelain cylinder, anneal all this at high temperatures and subsequently calibrate it. We did not have platinum of the necessary degree of purity, our porcelain was dirty which caused the platinum to become polluted in the annealing process. (...) From Leiden Wiersma brought us pure platinum wires and special porcelain cylinders, so that we could make our own thermometers. For keeping the liquids we needed Dewars. Metallic Dewars were soldered with tin. At low temperatures our tin cracked, and the Dewars became unfit for use. Wiersma brought us a large quantity of special solder, which did not crack at low temperatures. He brought everything we could not get in the Union. He brought revolution counters for winding transformers. We had poor weights for analytic scales; he brought weights. There was no adhesive plaster; he brought it. Everything that he could get and we did not have, he brought. Of course Wiersma did all this with the approval of De Haas. We obtained the help promised to us earlier in Leiden. Sometimes he sent something with Ehrenfest, who also visited Kharkov a few times. He helped a lot in the preparation of the cryogenic magnets, which were similar to those in Leiden: he drew sketches, counted windings. Wiersma very much helped in the build-up of the cryogenic lab, although there are few people who knew about this*" (Ref. [2], pp. 280–281).

That such help was offered and given was first of all due to the remarkable qualities of Shubnikov as a person and scientist, qualities which he for a great part had developed during his stay in Leiden. But it also showed the spirit of broad international cooperation which was characteristic of the science of that time. As C. P. Snow wrote: "*The scientific world of the Twenties was as near to being a full-fledged international community as we're likely to get ...The atmosphere of the Twenties in science was filled with an air of benevolence and magnanimity which transcended the people who lived in it...*" (Ref. [3], p. 258).[4] A letter by Ehrenfest to Shubnikov, written just after one of his visits to Kharkov, also bears witness to this (Ref. [4], pp. 13–14):

---

[3]Hollestelle in his biography of Ehrenfest (M.J. Hollestelle *Paul Ehrenfest—Worstelingen met de moderne wetenschap, 1912–1933* ((Leiden University Press, 2011), p. 259) suggests that Wiersma was in Kharkov for several years. That is incorrect and also unlikely given that Wiersma did not publish anything with Shubnikov, but did publish regularly in those years with De Haas on susceptibility measurements of various substances carried out in Leiden.

[4]Snow followed it up with the observation that "the discovery of atomic fission broke up the world of international physics".

Leiden, 16 February 1933

My dear Shubnikov!

... I very often think of you and of all my sweet Kharkov friends. Here I am all the time in a downcast state and in no way can I escape from this depression and absolute loneliness, although everybody is very sweet and kind to me. On my return journey I stayed for a few days in Berlin, it became there much better for me (perhaps, because I was not long at one and the same place), therefore last week I went to Germany, to Leipzig, to Debye, where a three-day conference on all possible magnetic issues was held (De Haas, Kapitsa and Kramers were also there). Of special interest was the lecture by Stern: by means of his own, cleverly improved, method of molecular beams (Stern-Gerlach method[5]) he had measured the magnetic moment of a hydrogen nucleus, by studying the deflections of molecular beams of para- and ortho-hydrogen in a magnetic field. Everything, it seemed, agreed well, although it was not noted that the contribution of the electron cloud (rotating with the molecule) had not been calculated correctly. It turned out that the electron cloud is, as it were, firmly connected to the molecule and rotates with it. But as a more meticulous calculation shows, due to a large "lagging" effect it falls behind the molecule. Then everything became immediately suspect. But Stern, comparing the results of the experiments on para- and ortho-hydrogen and almost without resorting to theory, showed that the magnetic moment of a hydrogen nucleus should be almost five times larger than known, if it were assumed (with reference to the Dirac equation) that the magnetic moment of an electron and proton should be inversely proportional to the mass of these particles.[6] With difficulty I managed to draw all those suffering from the failure, present here, in the discussion about Stern's results. But after half an hour I got the impression that there is hardly a way out. Bohr, among others, already succeeded in connection with this result to drop the very accurate remark APPROXIMATELY (?!) in the following sense. It is already known for three years that it is IN PRINCIPLE not possible to measure the spin of a freely moving electron or proton (due to the effect of the Lorentz force on a moving charged particle). But now Stern, by observing a neutral atom or molecule, wants to separately determine the magnetic moment of the nucleus and the electron cloud. Bohr, it seems, could more or less convincingly show that such separation is not possible inside an atom, as it concerns quantities of an order of magnitude less than the one characteristic for this situation. I could not find out anything about this remark of Bohr from those present (Bloch, Peierls, Muller etc. were here). But perhaps Landau can recover everything from these hints (greetings to him from me! I impatiently await his paper on the role of gravitation in the law of conservation of energy for the Amsterdam academy).

De Haas has invented a very clever machine for liquefying helium. In Leipzig he met Simon and there it became clear that Simon had recently already constructed a similar machine... I wanted to send you an as accurate as possible description of it, since I heard that it can be constructed in a few hours. But I was glad to hear recently that Wiersma had already sent you all the information about it. De Haas was excited about the sharpness of mind and inventiveness of Simon. Simon asked with great interest about the Ruhemanns...

Cordial greetings to you and all my dear friends,

Yours P. Ehrenfest

---

[5]The Stern–Gerlach experiment, carried out in 1922 by Otto Stern and Walther Gerlach, demonstrated that the spatial orientation of angular momentum is quantized. Later in this chapter Shubnikov's own measurement of the proton magnetic moment (together with Lazarev) will be discussed.

[6]This experiment carried out by Stern and Otto Frisch has been discussed in some detail in Isidor Rabi's biography (J.S. Rigden, *Rabi: Scientist and Citizen* (New York, 1987), pp. 99–102).

Appendix 1 to this book contains a memorandum that Shubnikov wrote in 1933 for the People's Commissariat of Heavy Industry in which he sets out his ideas and plans about the research to be carried out at the UFTI cryogenic laboratory. One of the main themes of investigation was superconductivity, which is not surprising in view of Shubnikov's earlier experience in Leiden, where research on this topic had been going on since Kamerlingh Onnes's discovery of the phenomenon in 1911. Although at that time he did not actively participate in research in superconductivity, it is difficult to imagine that a physicist of his standard would not show a keen interest in this remarkable phenomenon.

The scientific programme of the group was drawn up in close consultation with Landau. A list of research topics (Ref. [4], p. 14) compiled by Shubnikov including a number of published results, as well as work in progress has survived. The date of this list is unknown, but it has probably continually been revised over the years. Kikoin is mentioned in item 12, which suggests that this particular copy of the list is at the earliest from late 1935 as Kikoin was recruited by Shubnikov in the autumn of 1935 (Ref. [6], p. 321):

(1)  Magnetocaloric effect in salts of ferromagnetic metals;
(2)  Specific heat of superconducting alloys in a magnetic field;
(3)  Temperature below $1°K$;
(4)  Superconductivity $B = f(H)$;
(5)  X-ray studies of alloys;
(6)  Specific heat of salts of ferromagnetic metals and their magnetic susceptibility;
(7)  Dependence of the specific heat at the $\lambda$-point on the pressure. Methane;
      ...
(9)  Superconductivity $B = f(H)$ for uniaxial crystals of various orientations[7];
(10) Exact determination of $B$ in the interval $H = 0$, $H = H_c$. Penetration of the field. Paramagnetic effect.
(11) Paramagnetism of nuclear spins. Measurement of the susceptibility of $H_2$ at $T < 4.2$ K (Dorfman);
(12) Einstein-De Haas effect in superconductors (Kikoin);
(13) Improved Kapitsa machine;
(14) The impossibility of obtaining high magnetic fields by use of superconductors;
(15) Magnetic susceptibility of MnO;
      ...
(18) Obtaining ultrapure materials by use of a mass spectrograph;
(19) Resistance in a magnetic field at low temperatures;
(20) Generators for the liquefaction of air, hydrogen and helium;
      ....
(22) Joule-Thomson effect in helium;
(23) X-ray study of superconducting alloys;

---

[7]This item 9 is not in the list in Verkin et al., but is mentioned in Ref. [5], p. 59.

(24)  Breakdown field of liquid helium;
(25)  Influence of pressure on the λ point for $NH_4Cl$ and $NH_4Br$;
      ...
(29)  Destruction of superconductivity by a current for an alloy in a magnetic field;
      ...
(33)  Susceptibility of $O_2$ from 2 to 20 K;
      ...
(35)  Refrigeration apparatus for obtaining ultralow temperatures;
(36)  Double refraction in helium below the λ point, and also the Kerr effect;
(37)  The behaviour of the specific heat in liquid crystals;
(38)  Ferromagnetism in sulphides;
(39)  Optical study of the structure of the magnetized region in superconductors;
      ...
(41)  Verification of Ohm's law for CuAg for a current density of $5 \cdot 10^6$ $A/cm^2$ for T < 2.19 K.

The list shows the main topics of research in Shubnikov's laboratory. Superconductivity features prominently in about a quarter of the items; furthermore, there is a significant number related to magnetism with problems that were still very topical many years later, such as item 33, the study of the magnetic susceptibility of oxygen, which also had been, and still was being, extensively studied in Leiden; and item 41 that for the first time proposes the study of nonlinear phenomena in conductivity at high current densities. Systematic studies of the specific heat of liquid crystals (item 37) were only started after World War II and expanded into a broad field of research with a multitude of applications (Ref. [5], p. 60).

## Meissner Effect

One of the first scientific successes of the Kharkov laboratory was the independent discovery of the phenomenon now known as the Meissner effect, the phenomenon that magnetic field lines are forced out of a superconductor during its transition to the superconducting state. In a constant magnetic field Meissner and Ochsenfeld slowly cooled down a lead rod to below its transition point, that is the point at which the rod becomes superconducting. At the same time the magnetic flux was measured with the help of a coil placed beside the rod. Upon reaching the transition point a sharp increase of flux was registered, implying that on cooling a superconducting material through its transition point in a magnetic field the magnetic flux in the specimen does not remain constant, but the lines of force are driven out of the superconductor, increasing the flux in the neighbourhood. For studying this same problem Jury Rjabinin and Shubnikov had developed an experimental method by which the magnetization of long thin cylindrical samples placed in a homogeneous longitudinal field could be measured directly (Ref. [7], pp. 282–283). In the course of 1933–1934 they used this method for plotting the entire curve of the dependence of the magnetization as a function of the external field for semi-crystalline lead. They observed that in the superconducting state the magnetic induction is zero. Their experiment gave a direct experimental demonstration of this fact, while Meissner and Ochsenfeld had only established it indirectly on the basis of the change in the magnetic flux. The short communication of Rjabinin and Shubnikov was received by the editors of the *Physikalische Zeitschrift der Sowjetunion* on 21 April 1934 [11] (a slightly updated and extended paper was submitted to Nature on 3 July 1934 and published on 25 August 1934 [12]), while the note by Meissner and Ochsenfeld [13] was sent to the publishers on 16 October 1933 and published on 3 November 1933, so roughly half a year before the Shubnikov-Rjabinin paper, which actually contains a reference to the Meissner-Ochsenfeld paper, had even been sent to the journal. There is nevertheless reason to assume that the work was carried out independently.[8] This is borne out by the completely different methods and the comments in the paper of Rjabinin and Shubnikov that the penetration of the magnetic field into a superconducting material was already being investigated for a year from the moment liquid helium was obtained at the laboratory. Shubnikov understood the problem perfectly well and without any doubt arrived at the idea of the experiment independently of the work of Meissner and Ochsenfeld, although possibly not without being influenced by research carried out in Leiden (Ref. [4], p. 27). Moreover, there had been some delay in preparing the Rjabinin-Shubnikov paper for the printer both due to the difficulty of working with a small quantity of helium and because Landau at first rejected the results of the work, which went against the common knowledge of the time. According to the Soviet and Ukrainian experimental physicist Boris Georgievich Lazarev (1906–2001), who was one of Shubnikov's collaborators and in 1938 became his successor as head of the cryogenic laboratory, Rjabinin and Shubnikov's discovery was actually earlier then Meissner and Ochsenfeld's, and that this negative assessment by Landau, who is said to have called the results 'bullshit' (*chush' sobach'ja*), was the main reason that it was sent to the journal later. It shows that having a theorist close at hand is not always an advantage (Ref. [14], p. 304). Shubnikov was well aware of the importance of the work, but according to Trapeznikova he was not greatly upset when Meissner and Ochsenfeld beat them to it. When told he merely commented: "Well, then we are in good company!" (Ref. [2], p. 282).

An interesting detail in this respect is also that in 1932 Meissner and his mechanic actually visited UFTI to help get to work the helium liquefier that Shubnikov had purchased from the German firm Linde.[9] The machine was an exact copy of the one used by Meissner in the Physikalisch-Technische Reichsanstalt in

---

[8]At the time similar experiments were also independently being carried out in Oxford, Leiden and Toronto leading to a large number of publications shortly after the Meißner-Ochsenfeld publication (Ref. [7], p. 282).

[9]Linde was the company that marketed the helium liquefiers for which Meissner had designed important components. It was quite different from the one designed and constructed in Leiden by Kamerlingh Onnes, since the financing and personnel needed for this was "not even approximately available at the Reichsanstalt" (W. Meissner, Verflüssigung des Heliums in der Physikalisch-Technischen Reichsanstalt, *Die Naturwissenschaften* 13 (1925) 695–699).

Berlin but, in spite of this, Meissner allegedly couldn't get it to work and Shubnikov solved the problems himself (Ref. [15], p. 316).

The Meissner effect came as a complete surprise. It was still commonly thought at the time, in spite of some indications to the contrary, that the only difference between a superconductor below its transition point and the same material in the normal state was the value of the electrical resistance, namely that a superconductor was no more than a perfect conductor. This conviction was so well established that it led even Kamerlingh Onnes to a hasty interpretation of an experiment carried out in Leiden in 1924, whereby a hollow lead sphere was plunged into liquid helium in the presence of a magnetic field. The sphere became superconducting and more importantly the measurements indicated that the magnetic field inside the sphere, i.e. in the hollow, seemed not to have changed in the transition to the superconducting state. Then the external field was switched off and the field inside the sphere again seemed not to change. It looked as if the magnetic field had become 'frozen' inside the sphere. In view of earlier experiences with a ring instead of a sphere, this interpretation seemed clear and normal, and nobody took the trouble to repeat the experiment (Ref. [16], p. 50 ff). The idea of the frozen flux became accepted wisdom, which also explains Landau's reluctance to accept the results of the Shubnikov-Rjabinin experiment. The Meissner effect therefore implied a turning point in the history of superconductivity. The magnetic flux density inside the superconductor vanishes due to the effect of superconducting shielding currents induced close to the surface and flowing without electrical resistance. They generate a magnetic field that exactly cancels the magnetic field existing in the interior in the normal, non-superconducting state. These shielding currents have to flow without any losses, otherwise the superconducting state could not exist indefinitely in a magnetic field (Ref. [17], p. 74). It finally led to the discovery of the one property that is truly characteristic of a superconductor, not zero resistance, but perfect diamagnetism, i.e. the induction of a magnetic field in a direction opposite to the external applied field that exactly cancels out the external field. This is the effect discovered by Meissner. It does not necessarily follow from Maxwell's equations for a perfect conductor, but is not excluded by them either (Ref. [16], p. 57). This physical understanding of the results of Meissner and Ochensfeld was at first far from clear and simple. A number of theorists interpreted them as proof of the ideal diamagnetism of superconductors. In particular this formed the basis of the equations of the electrodynamics of superconductors proposed in 1935 by Fritz and Heinz London [18], who quote the papers of Rjabinin and Shubnikov in their work. At the same time many experimentalists disagreed with this interpretation. The experiments carried out in Berlin, Leiden, Toronto, Oxford and Cambridge, which on the basis of Meissner's method used hollow samples of various form and composition, oriented differently relative to the field, revealed a very whimsical picture of the induction. All this rather obscured and complicated the experimental picture of the phenomenon and did not at all contribute to the formation of a physically clear picture, to the point that in May 1935, at a discussion on low-temperature phenomena at the London Royal Society, Meissner himself said that the hypothesis of ideal diamagnetism put forward by London in no way explained all facts on the magnetization of superconductors.

Without belittling Meissner's priority, which is a historical fact, it should also be recognised that in their experiment Rjabinin and Shubnikov gave a soundly substantiated proof of the ideal diamagnetism of superconductors. They directly measured the dependence of the magnetic moment and the magnetic induction—the physically most interesting characteristics of the magnetic state—on the field in an experimentally 'pure' situation of long massive cylindrical samples in a uniform longitudinal field. From that time onwards these dependences became such a natural and integral element of practically every manual on superconductor physics that the initiators are practically never mentioned (Ref. [4], p. 27).

## Type-II Superconductivity

A second field of study, closely related to the first, was the study of the magnetic properties of superconducting alloys. In this field the greatest achievement by Shubnikov and collaborators was the discovery of what is now called type-II superconductivity (also called the Shubnikov phase). Superconductivity was discovered in 1911 in Leiden by Kamerlingh Onnes, who observed the resistance in a conductor drop to zero when cooled down to below a characteristic critical temperature. One of the other characteristics of superconductors is the vanishing of the magnetic induction when an external magnetic field is applied (Meissner effect). Kamerlingh Onnes had already observed that superconductivity breaks down when the applied magnetic field is increased above a certain critical field strength. Shubnikov's interest in studying the magnetic properties of superconducting alloys had already arisen in Leiden where from 1929 such studies were being carried out by De Haas and Jacob Voogd,[10] who among other things had noticed that for alloys the critical field is much higher than for pure metals (by a factor of several hundred). From a formula derived by Arend Rutgers[11] it followed that this should be accompanied by a large jump in the specific heat.[12] At the end of 1934, in virtually simultaneous papers by Keeley, Mendelssohn and Moore [19] from Oxford and by Shubnikov and Khotkevich [20] on calorimetric studies of superconducting alloys this jump was found, however, to be absent. While the English physicists restricted themselves just to the statement of this fact,

---

[10]Jacob (Jaap) Voogd (1904–1990) gained his PhD with De Haas in 1931; the title of his thesis was "Leiden studies of the superconducting state of metals 1927–1930".

[11]Arend Joan Rutgers (1903–1998) was a student of Ehrenfest and gained his PhD in 1930. In 1933 he became a lecturer at Ghent University in Belgium, was promoted to full professor in 1938, and remained in Ghent until his retirement.

[12]The *specific heat* is the amount of heat per unit mass required to raise the temperature by one degree Celsius.

**Fig. 8.2**   Vladimir Khotkevich (*from* B.I. Verkin et al. (1990))

**Fig. 8.3**   Jury Rjabinin in the laboratory (*from* B.I. Verkin et al. (1990))

**Fig. 8.4** Georgy Shepelev
(*from* Voprosy Atomnoj
Nauki i Tekhniki no. 1 (2006)
215)

Shubnikov and Vladimir Khotkevich (Fig. 8.2)[13] concluded for the first time that the idea of two phases, on which the derivation or Rutger's formula was based, did not correspond to the state of affairs in alloys.

This conclusion showed that Shubnikov had a deep understanding of the physical essence of the whole complex of magnetic properties of superconductors and that he had an extensive programme for their detailed investigation. He continued to work on this with Jury Rjabinin (Fig. 8.3) and in January 1935 they sent a paper [21] to the *Physikalische Zeitschrift der Sowjetunion* which has become a fundamental paper in the physics of superconducting alloys (a short version was published in Nature on 13 January 1935 [22]). The authors again used the method of plotting the curve of the magnetization as a function of the external field applied earlier for pure superconducting metals. The measurements were carried out on carefully homogenised and annealed samples of the lead-thallium alloy $PbTl_2$ and of a lead-bismuth alloy containing 65% Pb and 35% Bi, which had also been used in the work with Khotkevich. The fundamental discovery of Rjabinin and Shubnikov was that for alloys there are two critical fields, so that superconductors are divided into two types depending on how the superconductivity breaks down when the strength of the applied external field increases. In type-I superconductors the phase transition to zero magnetic induction occurs abruptly when the field exceeds the critical value. In type-II superconductors the following happens: up to the first critical value the magnetic induction remains nearly zero; when the external magnetic field exceeds this first critical value the induction gradually approaches the value characteristic for

---

[13]Vladimir Ignatevich Khotkevich (1913–1982), worked from 1932–1950 at UFTI after graduating from the Kharkov Mechanics-Machine Building Institute. He was the son of the Ukrainian author Gnat Khotkevich (1877–1938), who was shot in 1938 during the *Yezhovshchina* (Yezhov's reign of terror), accused of participation in a counter-revolutionary organisation. His mother was sent into exile. Vladimir continued to work at UFTI and from 1966–1976 was rector of Kharkov State University. The fact that he belonged to the family of an 'enemy of the people' and was surrounded by such enemies at his place of work does not seem to have hampered his career.

the non-superconducting metal; and the superconductivity disappears fully when the external magnetic field exceeds the second critical field.[14]

At the time the discovery was not appreciated, although the published results became instantly known, and were actually presented by Martin Ruhemann at the 7th International Congress of Refrigeration in The Hague in 1936. Experiments with alloys were often carried out with samples that contained lattice faults and in those days many people thought, a possibility also mentioned by Rjabinin and Shubnikov,[15] that such inhomogeneities caused the alloys to show deviating behaviour from ordinary superconductivity. However that possibility was ruled out in the classic paper published in 1936 by Shubnikov, Khotkevich, Shepelev (Fig. 8.4) and Rjabinin [24]. As he had shown in Leiden when making pure single crystals, Shubnikov was also a master in preparing pure alloys. In this respect Kurt Mendelsohn (1906–1980), a leading low-temperature physicist at Oxford University, noted (Ref. [25], p. 209): *"The real trouble here is that it is extremely difficult to make a homogeneous alloy, containing no lattice faults. Of the laboratories engaged in low temperature research in the thirties, Shubnikov's group in Kharkov had evidently the best metallurgical know-how"*. In this paper they gave the results of careful and comprehensive studies of a whole series of lead-thallium and mercury-cadmium alloys. For all of them qualitatively similar curves of the magnetic induction as a function of the external field were obtained with two critical fields and the dependence of these fields on temperature and impurity concentration was investigated. In particular, it was convincingly shown that when the impurity concentration increases the first of these fields decreases, and the second significantly increases. All these observations were made some twenty-five years before the experiment of Kunzler and collaborators [26] demonstrating type-II superconductivity in a niobium–tin ($Nb_3Sn$) alloy.

Vitaly Ginzburg (1916–2009) and Lev Landau, when constructing the Ginzburg-Landau theory of superconductivity in 1950, also ignored these results from Shubnikov's group, although they, especially Landau, were of course aware of Shubnikov's results,[16] but as Ginzburg states in his Nobel lecture: *"... an understanding of the situation was lacking, and Landau and I, like many others, believed that alloys were an 'unsavoury business' and did not take an interest in them"* (Ref. [27], p. 985). Later it turned out that the most spectacular application of the

---

[14]An extensive discussion can be found in [8, 23] (incidentally, the author A.G. Shepelev of the latter paper is the son of Georgy Dmitrievich Shepelev (1905–1942), one of the authors (his name spelled in various ways) of some of the original papers of Shubnikov and co-workers on this subject and Shubnikov's first PhD student, who defended his thesis in 1938. After Shubnikov's arrest and execution, for a short time Shepelev was head of the cryogenic laboratory; in the Second World War he volunteered and was killed in action at the age of 36 when defending Sevastopol).

[15]"if it appears that this unusual behaviour of alloys is caused by their inhomogeneity which may be due to the decomposition of the solid solution and the formation of a new very disperse phase" (Ref. [21], p. 125).

[16]In Ref. [8], p. 34 it is stated that at the time in 1936 Landau did not recognize the importance of Shubnikov's discovery and that "the publication of the paper was delayed by more than three months because Shubnikov failed to run it through Landau".

Ginzburg-Landau theory was actually to type-II superconductors. The first reference to the research by the Shubnikov group was made by Alexei Abrikosov in his fundamental 1957 work [28, 29] on the theory of type-II superconductivity, based on the Ginzburg-Landau theory. When comparing his calculations with Shubnikov's experimental results he found excellent agreement. For this work Abrikosov was awarded the 2003 Nobel Prize in physics, jointly with Vitaly Ginzburg and the American physicist Antony Leggett. Shubnikov's achievement in this field is now generally acknowledged. The French physicist and 1991 Nobel Prize winner Pierre de Gennes (1932–2007) was the first to introduce the name 'Shubnikov phase' for this type of superconductivity. Recognition for Shubnikov in the Soviet Union was rather late in coming. In Shpolsky's 1967 review article [30] Shubnikov is mentioned several times, but the discovery of type-II superconductivity, in spite of Abrikosov's work, is not. It is fitting in my view to close this subject here with the remark by John Bardeen (1908–1991), the only person who won the Nobel Prize for physics twice, who said that "*It should be noted that our theoretical understanding of type-II superconductors is due mainly to Landau, Ginzburg, Abrikosov, and Gorkov, and that the first definitive experiments were carried out as early as 1937 by Shubnikov*" [31].

In the work of the Shubnikov group described above critical currents, which break up the superconductivity, were also investigated. The values measured for thin wires of $PbTl_2$ were found to be unexpectedly small. Now it is known that appreciable critical currents in type-II superconductors can only exist at a rather high concentration of defects, providing the vortex pinning of Abrikosov,[17] while Shubnikov and co-workers did everything possible in order to have samples without defects and achieved great perfection in this. The characteristic value of the critical current of a wire of radius $r$ is equal to *(1/2)$H_c$r*. It produces a critical magnetic field on the surface of the wire. The hypothesis that this current coincides with the critical current was advanced in 1916 by the American physicist F.B. Silsbee (known as Silsbee's rule). It implied that the facts, already observed by Kamerlingh Onnes, on the breakdown of superconductivity by a magnetic field and by a current are two sides of one and the same phenomenon. Silsbee's rule is directly applicable to type-I superconductors, where the collapse of the superconducting state is sharp. For type-II superconductors the transition is gradual and the rule denotes only a point after which the mixed state develops. As shown by Shubnikov and co-workers, for type-II superconductors there are two critical values for the field: one after which a superconductive state begins reducing, and another one above which there is only a fully resistive state. The region between them is the mixed state, as indicated in Fig. 8.5.[18]

---

[17]The Abrikosov vortex is a vortex of supercurrent in a type-II superconductor theoretically predicted by Abrikosov in 1957. The supercurrent circulates around the normal (i.e. non-superconducting) core of the vortex.

[18]Copied from *Encyclopedia-Magnetica.com,* accessed on 28.12.2016.

**Fig. 8.5** Comparison of
superconducting behaviour
for type-I and type-II
superconductors

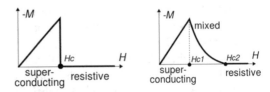

Silsbee's rule had the status of hypothesis until 1924 when De Haas and Willem
Tuyn conducted a direct experiment demonstrating its correctness. However a
precise verification of Silsbee's rule for pure superconductors was realized by
Shubnikov and Nikolaj Alekseevsky,[19] who obtained the transition curve of the
resistance of a tin wire as a function of the current [32, 33].[20] It makes use of the
huge thermal conductivity of helium that had just been discovered in Leiden by
Willem Keesom and his daughter Anne. This experiment illustrates how rapidly
Shubnikov took up and included new physical ideas and facts in his research (Ref.
[4], p. 30).

When the superconductivity in type-I superconductors is destroyed by an external
magnetic field rising above a critical value, an intermediate state[21] may emerge
consisting of a baroque pattern of regions of normal material carrying a magnetic
field mixed with regions of superconducting material containing no field. In 1937
Landau formulated a theory of this intermediate state in superconductors, according
to which a superconductor with a significant demagnetization coefficient under the
action of a sufficiently strong magnetic field takes on the structure of alternating
layers of these two phases arranged along the field. In the same year in an ingenious
experiment, whereby a spherical single crystal of tin was placed in a magnetic field at
a temperature below the critical temperature, Ilja Nakhutin[22] and Shubnikov
observed a strong anisotropy of the conductivity of such spheres [34, 35].[23]

In his reminiscences in the book by Verkin et al. Nakhutin writes about this (Ref.
[36], p. 319): "*A clever set-up of experiments invented by Lev Vasilevich made it
possible to obtain rather quickly results that confirmed the theory. A superconducting
tin sphere was included in a superconducting circuit of lead wire in which an induced
non-vanishing current was circulating. A magnetic field was imposed on the sphere. If
the current through the sphere flowed perpendicular to the magnetic field, it vanished
at values of the field equal to 2/3 of the critical value. However, if the current flowed
through the sphere parallel to the magnetic field, it vanished at a value of the field*

---

[19]Nikolaj Evgenevich Alekseevsky (1912–1993), graduated from the Leningrad Polytechnic
Institute in 1936; he worked at UFTI from 1936 to 1941.

[20]Ref. [33] has only Alekseevsky as an author as the paper was published when Shubnikov had
already been arrested.

[21]This was the name given to this state by Rudolf Peierls (R. Peierls, *Proc. Royal Soc.* A 155
(1936) 613).

[22]Ilja Evseevich Nakhutin (b. 1912) was a post-graduate student at UFTI in 1936. Later he became
the head of the laboratories of the All-Union Institute of Inorganic Materials in Moscow.

[23]Nakhutin was the sole author of Ref. [35] due to Shubnikov's arrest and execution in 1937.

*equal to the critical field. Thus, in the intermediate state, at fields from 2/3 $H_c$ to $H_c$, the sphere becomes non-superconducting in a direction perpendicular to the magnetic field, and superconducting in a direction parallel to the field. This nicely confirmed Landau's theory, according to which in the intermediate state the superconductor has a layered structure of alternating superconducting and non-superconducting layers, parallel to the direction of the magnetic field. When carrying out our experiments together with Lev Vasilevich, Landau came once or twice to the room where the installation stood, and showed an interest in the results."*

Nakhutin and Shubnikov's results implied an experimental verification of the intermediate state in agreement with Landau's theory. Almost at the same time Shubnikov and Alekseevsky observed the intermediate state at the destruction of superconductivity of metals and alloys by an electrical current [32, 33].

Work by Khotkevich and Shubnikov on the process of demagnetization of superconductors of circular form and of various composition under the simultaneous action of a field and current was related to this [37, 38].

Around 1936 research in superconductivity apparently slowed down. This question was addressed by Cor Gorter[24] in his talk at a conference in Oxford on the history of superconductivity before 1940. He said about this [40]: *"Many people have asked why the advance of our knowledge in this part of the science of superconductivity slowed down around 1936 so that in some respects quite a new start had to be made in recent years. I do not know a simple answer. One might, however, find a few more or less accidental and personal reasons besides the vague and more general causes. Among the first is the fact that the number of research workers in the field was small and that some of them almost simultaneously left it. Schubnikov disappeared, Mendelssohn concentrated a large part of his attention on the superfluid properties of helium II while—as already mentioned—I returned to magnetism. As to the properties of alloys, I feel that the lack of metallurgical facilities and experience also weighed heavily."* It is clear from Gorter's answers and from the exposé of Shubnikov's work given above that he indeed was sorely missed, as he was probably the only one at the time who would have been able to create the metallurgical facilities needed and moreover had the experience to do so. He would certainly have continued the work on superconductivity, especially on type-II superconductors which he had pioneered and perhaps no slowdown would have been witnessed, or in any case to a lesser extent.

---

[24]Cornelis Jacobus Gorter (1907–1980) was a Dutch experimental and theoretical physicist; he discovered paramagnetic relaxation and was a pioneer in low temperature physics. In 1948 he became director of the Kamerlingh Onnes Laboratory in Leiden as successor to Wander de Haas. Casimir regarded him as the most outstanding Dutch experimental physicist of his generation. In the thirties Casimir and Gorter did some work together on superconductivity. After Meissner and Ochsenfeld's discovery of the expulsion of the magnetic field from a superconductor, they formulated the thermodynamics of superconductivity, and somewhat later showed that many of the properties of superconductors could be interpreted in terms of a simple two-fluid model (Ref. [39], p. 176).

## Antiferromagnetism

A large cycle of work by Shubnikov and his co-workers was devoted to studying the magnetic properties of materials, in particular antiferromagnetism.[25] This work, too, was inspired by work carried out in Leiden by Herman Woltjer[26] in collaboration with Kamerlingh Onnes and later Wiersma [41–43]. They had investigated the temperature dependence of the magnetic susceptibility of various chlorides, in other words the degree of magnetization in response to an applied magnetic field. Landau provided a theoretical explanation for some of the phenomena [44] using a model of a layered antiferromagnet with a strong ferromagnetic coupling in each layer and weak antiferromagnetic coupling between the layers. Landau's paper was published earlier than Shubnikov's work, carried out together with his wife Olga Trapeznikova [45–47], and was referred to by them, so it might be thought that Landau's work actually initiated the set-up of the experiment. However, Verkin et al. (Ref. [4], p. 32) suggest that in view of the close contact between Landau and Shubnikov and Shubnikov's familiarity with the earlier Leiden measurements it is possible that Landau's paper was actually written in reaction to earlier Leiden data and preliminary data obtained by Shubnikov and Trapeznikova. A theory can be written down fairly quickly, certainly by someone like Landau, while the preparation of an experiment often requires more time. First of all sufficiently pure anhydrous salts were needed to be specially prepared at the Radium Institute in Leningrad. Chloride of bivalent iron was most pure and for this reason chosen as the first object of investigation. Next came $CoCl_2$ and then chlorides of nickel and chromium. The poor thermal conductivity of all salts investigated significantly increased the time needed for establishing thermal equilibrium. For measuring the specific heat a specially developed adiabatic calorimeter was used for the first time in the Soviet Union. Principal attention was paid to the thermometer which, when measuring the specific heat, is one of the main sources of error. The accuracy of about 4% obtained for the measurement of the specific heat was very high for that time. The results for ferrous chloride were unambiguous and impressive. However the other salts did not go that smoothly. In the conclusion to their article the authors note on the one hand good agreement of experiment with the prediction of Landau's theory, but on the other hand suggested the possibility of other reasons that might give rise to an anomaly of the low-temperature specific heat in other chlorides. Next

---

[25]The phenomenon that the magnetic moments of atoms in a material, formed by adding up the individual electron spins, align in a regular pattern with neighbouring spins pointing in opposite directions resulting in a vanishing total magnetization of the material.

[26]Herman Robert Woltjer (1887–1974), PhD with Pieter Zeeman in 1914; assistant to Kamerlingh Onnes from 1915, from 1927 he worked in Indonesia (at the time a Dutch colony) where he taught at the Bandung Technical University (*Technische Hoogeschool te Bandoeng*).

the heat capacity of other chlorides were measured (in which Miljutin[27] was also involved) [48–51]. The values of the Néel temperature[28] (in current terminology) which they found for the chlorides of iron, cobalt and nickel can still be found in the handbooks today. The absence of a specific heat anomaly in the chloride of manganese is now completely understood: its Néel temperature of 1.7 °K is significantly lower than the minimum temperature of 13 °K used by Shubnikov and his co-workers in their experiment.

In any experiment the lucky choice of the objects under investigation plays an important role. In this sense Shubnikov's choice was at the same time both lucky and unlucky. On the one hand, much was already known about the chlorides of the transition metals. On the other hand, because of their specific structure, these materials are unique antiferromagnets which show features of ferromagnetic high-temperature behaviour.[29] As a result, after having established, as is now clear, the existence of a phase transition to an anti-ferromagnetic state and having clarified the specific nature of the specific heat in this transition, Shubnikov called the transition temperature analogous to the ferromagnetic Curie point. For definitively establishing the magnetic nature of the observed phase transition it was necessary to observe the kink of the susceptibility predicted by Landau's theory at the same temperature at which the jump of the specific heat was observed. Therefore Shubnikov proceeded to investigate the low-temperature susceptibility of chlorides of transition metals. This work was carried out together with Simon Shalyt,[30] first as a doctoral student and later as a collaborator of the cryogenic laboratory. The first of their papers with results on the susceptibility of chlorides was published in May 1937 [52]. Detailed results followed a year later in a paper that has Shalyt as the only author since Shubnikov had by then already been arrested and executed [53]. In this case the Faraday method (measuring the force acting on the samples in an inhomogeneous magnetic field) was applied for measuring the susceptibility in the low-temperature phase. From a current viewpoint the main results of this paper are the observation of a sharp maximum (the kink according to Landau's theory) in the temperature curve of the susceptibility at the Néel temperature and a detailed investigation of its behaviour in a magnetic field. Unfortunately, the poly-crystalline

---

[27]Georgy Anatolevich Miljutin (1914–1990), assistant professor at Kharkov University. He worked from 1932 to 1940 at the UFTI cryogenic laboratory, did his degree work there (1936) while a student at the Kharkov Mechanics-Machine Building Institute.

[28]The Néel temperature or magnetic ordering temperature is the temperature above which an anti-ferromagnetic or ferromagnetic material becomes paramagnetic, i.e. the thermal energy becomes large enough to destroy the macroscopic magnetic ordering within the material.

[29]From his list of research items above it can be seen that Shubnikov planned to carry out research on the conductivity of manganese oxide (item 15) which as they showed in a further study is a classic anti-ferromagnet.

[30]Simon Solomonovich Shalyt (1911–1977) was a graduate of the Physico-Mechanical Faculty of the Leningrad Polytechnic Institute. He started his scientific career at UFTI under the guidance of Shubnikov and eventually became director of the Semiconductor Institute of the USSR Academy of Sciences in Leningrad.

structure of the sample did not allow the anisotropic susceptibility, predicted by Landau and so characteristic for antiferromagnetics, to be observed.

Thus, the results of the cycle of studies of the chlorides of transition metals was in essence the experimental discovery of anti-ferromagnetism (Ref. [4], pp. 31–34).


## Proton Magnetic Moment

Further work by Shubnikov, carried out in collaboration with Boris Lazarev, who came to Kharkov in 1934 for a couple of years from Ioffe's institute in Leningrad,[31] involved a novel measurement of the magnetic moment of the proton and was the first experimental detection of nuclear paramagnetism (that is, induced magnetism parallel and proportional to an applied external magnetic field). Nuclear magnetism is much weaker than atomic magnetism caused by the electron spin, due to the fact that the proton and neutron magnetic moments are a factor of thousand smaller than the electron magnetic moment. The proton magnetic moment had by then already been measured by two groups of researchers by deflecting a beam of molecular (in the second case atomic) hydrogen in a non-uniform electric field (see also the letter of Ehrenfest to Shubnikov quoted above) [54–56]. Following this, in 1934 Peter Debye had published a paper [57] devoted to a possible new method of measuring the nuclear magnetic moment. He drew attention to the fact that the paramagnetic susceptibility due to nuclear moments increases with decreasing temperature and in the absence of electron paramagnetism may become completely accessible to direct measurement. His estimate implied, however, that superlow temperatures of the order of $10^{-2}$–$10^{-3}$ °K were needed for this. A year later Jakov Dorfman came with a paper [58] in which more encouraging estimates were given. This is probably the reason why this problem appeared as item 11 on Shubnikov's list above. Lazarev, who worked in Dorfman's laboratory at LFTI, was undoubtedly aware of the latter's estimates and thought that measurements of the paramagnetic susceptibility must preferably be made on solid hydrogen in the region of helium temperatures, and not at superlow temperatures.

The experiment required the application of various subtle tricks and great skill. Special precision scales had been prepared in Leningrad and a method of preparing a sample for measuring via the Gouy method[32] was developed. The sample was a

---

[31]Officially Lazarev was attached to the new institute in Sverdlovsk (UralFTI), which was still under construction; therefore, for the time being the future employees of UralFTI stayed at LFTI (Ref [14], p. 305).

[32]Via the Gouy balance, invented by the French physicist Louis Georges Gouy (1854–1926) for measuring the magnetic susceptibility of a sample hung between the poles of an electromagnet. It measures the change in the mass of the sample as it is repelled or attracted by the high magnetic field between the poles. The change in weight is proportional to the susceptibility.

rather long cylinder of solid hydrogen (of normal ortho-para composition). The hydrogen used was thoroughly purified (mainly of oxygen impurities). A quartz ampoule of a special configuration and with a sufficiently large volume had to be blown. The composition of the hydrogen was controlled to an accuracy of 0.1% (only its ortho-component gives a contribution to the paramagnetic part of the susceptibility). In 1935 the set-up at UFTI was ready, and at the end of the year the first successful measurements were made, which were finished in the first half of 1936. With this nuclear paramagnetism had been directly measured. Lazarev and Shubnikov published a preliminary communication in the July 1936 issue of *Physikalische Zeitschrift der Sowjetunion* about the measurement of the magnetic moment of the proton [59], followed in April 1937 in the same journal with a detailed paper [60] outlining the method of the experiment and the results obtained. More than fifteen years later the American physicist and specialist in nuclear magnetic resonance Edward Mills Purcell (1912–1997; Nobel Prize 1952) wrote about this work that *"it was indeed an experimental triumph when, in 1937, Lasarew and Schubnikow succeeded in measuring the nuclear paramagnetic susceptibility of solid hydrogen"* (Ref. [61], p. 2). Another specialist in the field, Jack Powles (b. 1924) of Queen Mary College (London), wrote about the same time (Ref. [62], p. 451): *"By measuring the variation of the susceptibilities $\chi_n + \chi_e$ with temperature by the Gouy method, Lasarew and Schubnikow were able to deduce a value of the proton magnetic moment which was within 10 per cent of the present accepted value, merely by weighing. This must, I think, be regarded as one of the most beautiful experiments in physics."* In his 1966 book, Mendelssohn also credits Shubnikov and Lazarev with the discovery of nuclear paramagnetism when measuring the susceptibility of solid hydrogen (Ref. [25], p. 186). There can therefore be no doubt that this measurement of the magnetic susceptibility of solid hydrogen cooled down to 2 °K was indeed the great accomplishment it was called in a well-known review of Soviet physics in 1968 (Ref. [30], p. 717).

Apart from experimental skill, the work required a large amount of scientific courage and conviction of the correctness and scientific significance of one's physical ideas and obtained results. In 1935 Heitler, Willis and Teller had published a paper [63] in which they claimed that the magnetic relaxation time of solid hydrogen at helium temperatures is fantastically large—of the order of a million years. Clearly, if this estimate were correct, a sensible experimental set-up would be out of the question. In the work of Lazarev and Shubnikov it was shown that this relaxation time is actually less than a second. Practically simultaneously with their first communication a second paper [64] by Fröhlich and Heitler appeared, in which the previous theoretical estimate was corrected by 13 orders of magnitude, which fully agreed with the experimental result of Lazarev and Shubnikov (Ref. [4], p. 35).

## Nuclear Physics

In 1936 Shubnikov was also involved in nuclear physics research. His co-workers in this field included Igor Kurchatov,[33] who visited UFTI from Leningrad for considerable periods of time, Aleksandr Lejpunsky, the head of the nuclear laboratory at UFTI, the theoretical physicist Fritz Houtermans and several others. The research involved the investigation of the interaction of slow neutrons with matter, for example with liquid hydrogen, which was used here for the first time in a very large volume for that time, namely up to 50 l (Ref. [4], p. 36). As can be imagined, the interaction of neutrons (just discovered in 1932) with matter was a hot topic at the time. It was not yet known, however, that slow neutrons could be used to split nuclei. This research led to the publication of the four papers in Shubnikov's publication list in the back of this book with V. Fomin as the first author.

## Phase Transitions

A final topic that should be mentioned is Shubnikov's research into phase transitions: phase transitions in methane under pressure (with Trapeznikova and Miljutin) and the phase transition of helium I into helium II in collaboration with Abram Kikoin.[34] Helium I is a normal colourless liquid while helium II exhibits very unusual characteristics; it can for instance creep along surfaces. Helium shows such behaviour when it is cooled down below its so-called lambda point of 2.18 °K, where it becomes superfluid. The work with Kikoin [65] was the first publication of results of experiments with liquid helium in the Soviet Union and was, as far as Shubnikov was concerned, the initial stage of implementing extensive plans for investigating the condensed phases of helium. After having finished the optical experiments Shubnikov proposed to Kikoin to investigate the properties of a layer of helium II on the surface of a solid body, and also to measure the thermal conductivity of solid helium. In Shubnikov's list of research items, item 24 is the problem of the investigation of the breakdown field of liquid helium. Unfortunately, Shubnikov was not allowed to realize these plans. Some of the papers[35] resulting from this work were published in 1938 without Shubnikov, who had by then already been executed, as one of the authors.

This work of the UFTI cryogenic laboratory played an important role in the physics of phase transitions. It undoubtedly stimulated the interest of Landau in this

---

[33]Igor Vasilevich Kurchatov (1903–1960), who would later become the director of the Soviet atomic bomb project.

[34]Abram Konstantinovich Kikoin (1914–1999) was a student at the Leningrad Polytechnic Institute; in 1935 Shubnikov personally came to Leningrad to ask him to come to Kharkov, where he stayed until 1943. He later worked in Sverdlovsk (Yekaterinburg).

[35]See the publication list at the end of this book.

problem. However its significance goes much further. In the Soviet Union it laid the basis for the physics of cryocrystals and the physics of high pressure. The experience gained by Leonid Vereshchagin[36] in working with high pressures at UFTI enabled him to organise a laboratory of superhigh pressure at the Institute of Organic Chemistry of the USSR Academy of Sciences, which, in turn, became the basis for the Institute for high-pressure physics of the USSR Academy of Sciences, now known as the L.F. Vereshchagin Institute for High-pressure Physics (Ref. [4], p. 38).

# Applied Work

In addition to the fundamental scientific work mentioned above, the cryogenic laboratory also carried out work for industry (after all UFTI came at the time under Narkomtjazhprom, the Commissariat of Heavy Industry), the main customers being the commissariats of heavy industry and of chemical industry. Heavy industry was interested in oxygen for autogenous welding and cutting of metals, for instance, while the chemical industry had an interest in the properties of nitrogen, hydrogen and other gases. In the memo in Appendix 1 from 1933 for Narkomtjazhprom Shubnikov formulated the tasks of technical physics research in cryogenics of interest to industry. Because of the applied nature of such research it was often not published, but just written up in reports, with some exceptions, such as the paper by Trapeznikova and Shubnikov on an "Investigation of the equilibrium conditions of the gaseous and liquid phases of a mixture of oxygen and nitrogen" [66]. Other work of a purely technical nature, which Shubnikov did with Nikolaj Rudenko,[37] concerns the measurement of the viscosity coefficients of liquid $N_2$, $O_2$, CO, $CH_4$, Ar and $C_2H_4$ along the saturation line from the melting point to the boiling point at atmospheric pressure [67, 68]. The choice of substances is interesting as these are all substances that are important from an applied point of view.

In his memo, Shubnikov states that technological applications of low temperatures mainly involve the separation of gases. The main interest is the separation of air into its constituent parts: extracting oxygen from it, and separating coke-oven gases, mainly extracting hydrogen from it. For this task the low-temperature laboratory at UFTI proved too small and had no facilities for pilot plant work. Hence the idea was conceived (allegedly by Alexander Weissberg) to build a specialized physico-technical laboratory, the deep-cooling research station (*Opytnaja stantsija glubokogo okhlazhdenija* (OSGO)), in which small-scale operating plants could be installed. The construction of OSGO, just outside Kharkov, was directed by Weissberg. It was built in 1936 and contained well-equipped research laboratories,

---

[36]Leonid Fedorovich Vereshchagin (1909–1977).

[37]Nikolaj Semenovich Rudenko (b. 1907), completed his thesis work at the UFTI cryogenic laboratory, was a collaborator of the laboratory from 1933.

living quarters for staff and adequate factory space for erecting a plant for gas liquefaction and separation. Martin Ruhemann became the scientific head of this research station, which started to operate early in 1937. It employed about a dozen scientists and the total staff amounted to about 150. Ruhemann had to leave at the end of 1937 when he was told that his contract would not be renewed and he left the country soon afterwards. During World War II OSGO was destroyed by the Germans (Ref. [4], pp. 38–39; Ref. [69], p. 21).

As is clear from the above the scientific activity of the cryogenic laboratory in the six years (of which only four could actually be devoted to scientific work) that Shubnikov was its head resulted in fifty papers in a broad range of subjects in low-temperature physics. The work carried out in these few years is truly astonishing, not only as regards its quantity, but also the quality, which was on a par with the work performed at the best laboratories in the world. It established the Kharkov laboratory as the foremost laboratory of its kind in the Soviet Union and one of the leading laboratories in the world. From the experienced experimentalist he had become in Leiden, Shubnikov had grown into an outstanding scientist and a superb organizer.

# References

1. L.W. Schubnikow, Das Kältelaboratorium, *Phys. Z. Sow. Sondernummer* (1936) 1–5.
2. O.N. Trapeznikova in B.I. Verkin et al. (1990), p. 256–291.
3. C.P. Snow, The Moral Un-Neutrality of Science, *Science* 133 (1961) 256–261.
4. B.I. Verkin et al. (1990).
5. Ju. A. Khramov, Shubnikov Cryogenics School, *Sov. J. Low Temp.* 18 (1992) 57–63.
6. A.K. Kikoin in B.I. Verkin et al. (1990), 321–329.
7. M. and B. Ruhemann, *Low Temperature Physics* (Cambridge University Press, 1937).
8. A.G. Shepelev, The Discovery of Type-II Superconductors (Shubnikov Phase), in: *Superconductor*, edited by Adir Moysés Luiz (Sciyo, Croatia, 2010).
9. A. Weissberg, *Conspiracy of Silence* (Hamish Hamilton, London, 1952).
10. B. Gorobets, *Krug Landau* (Letny Sad, Moscow, 2006).
11. J.N. Rjabinin, L.W. Schubnikow, Verhalten eines Supraleiters im magnetische Feld, *Phys. Z. Sow.* 5 (1934) 641–643.
12. J.N. Rjabinin and L.W. Schubnikow, Dependence of magnetic induction on the magnetic field in superconducting lead, *Nature* 134 (1934), no. 3382, 286–287.
13. W. Meißner, R. Ochsenfeld, *Ein neuer Effekt bei Eintritt der Supraleitfähigkeit*, *Naturwissenschaften* 21, 1933, 787–788.
14. B.G. Lazarev in B.I. Verkin et al. (1990), p. 302–306.
15. A.I. Sudovtsov in B.I. Verkin et al. (1990), p. 315–318.
16. J. Matricon and G. Waysand, *The Cold Wars, A History of Superconductivity* (Rutgers University Press, 1994).
17. R.P. Huebner and H. Leubbig, *A Focus of Discovery* (World Scientific, 2008).
18. F. and H. London, The Electromagnetic Equations of the Supraconductor, *Proc. Roy. Society A: Mathematical, Physical and Engineering Sciences* 149 (1935) 71–88.
19. T. Keeley, K. Mendelssohn and J.R. Moore, Experiment of superconductors, *Nature* 134 (1934) 773–774.
20. L.V. Shubnikov, V.I. Khotkevich, Spezifische Wärme von supraleitenden Legierungen, *Phys. Z. Sow.* 6 (1934) 605–607.

21. J.N. Rjabinin, L.W. Schubnikow, Magnetic properties and critical currents of superconducting alloys, *Phys. Z. Sow.* 7 (1935) 122–125.

22. J. N. Rjabinin and L.W. Schubnikow, Magnetic properties and critical currents of supra-conducting alloys, *Nature* 135, no. 3415, (1935) 581–582.

23. Dimitri O. Ledenyov and Viktor O. Ledenyov, *The Nonlinearities in Microwave Superconductivity* (Brisbane, Kharkov, 2014).

24. Schubnikow L.W., Chotkewitsch W.I., Schepelew G.D., Rjabinin J.N., Magnetische Eigenschaften supraleitender Metalle und Legierungen, *Sondernummer Physikalische Zeitschrift der Sowjetunion, Arbeiten auf dem Gebiete tiefer Temperaturen* (1936) 39–66, later published as Shubnikov L.V., Khotkevich V.I., Shepelev G.D., Rjabinin Ju.N., Magnetic properties of superconducting metals and alloys, *Zh. Exper. Teor. Fiz.* (USSR) 7 (1937) 221–237.

25. K. Mendelssohn, *The Quest for Absolute Zero* (New York, 1966).

26. J.E. Kunzler, E. Buehler, F.S.L. Hsu, and J.H. Wernick, Superconductivity in $Nb_3Sn$ at high current density in a magnetic field of 88 kgauss, *Phys. Rev. Lett.* 6 (1961) 89–91.

27. V.L. Ginzburg, Nobel lecture: On superconductivity and superfluidity (what I have and have not managed to do) as well as on the "physical minimum" at the beginning of the XXI century, *Rev. Mod. Phys.* 76 (2004) 981–997.

28. A. A. Abrikosov, Nobel Lecture: Type-II superconductors and the vortex lattice, *Rev. Mod. Phys.* 76 (2004) 975–979.

29. A.A. Abrikosov, O magnitnykh svojstvakh sverkhprovodnikov vtoroj gruppy (On the magnetic properties of type-II superconductors), *Zhurn. Eksperim. i Teoret. Fiziki* 32 (1957) 1442–1452.

30. E.V. Shpolsky, Fifty Years of Soviet Physics, *Usp. Fiz. Nauk* 93 (1967) 197–376 [*Sov. Phys. Usp.* 10 (1968) 678–720].

31. J. Bardeen and R.W. Schmitt, International Conference on the science of superconductivity, *Rev. Mod. Phys.* 36 (1964) 1.

32. L.W. Schubnikow, N.E. Alexeyevski, Transition curve for the destruction of superconductivity by electrical current, *Nature* 138 (1936) 804; Krivaja perekhoda pri razrushenim sverkhprovodimosti elektricheskim tokom (Transition curve at the destruction of superconductivity by an electrical current), *Zhurn. Eksperim. i Teoret. Fiziki* 6 (1936) 1200–1201.

33. N.E. Alekseevsky, Zavisimost' kriticheskogo toka ot vneshnego polja v sverkhprovodjashchikh splavakh (Dependence of the critical current on the external field in superconducting alloys), *Zhurn. Eksperim. i Teoret. Fiziki* 8 (1938) 1098–1103.

34. L. Shubnikov and I. Nakhutin, Electrical conductivity of a superconducting sphere in the intermediate state, *Nature* 139 (1937) 589–590.

35. I.E. Nakhutin, Sverkhprovodimost' v promezhutochnom sostojanii (Superconductivity in the intermediate state), *Zhurn. Eksperim. i Teoret. Fiziki* 8 (1938) 713–716.

36. I.E. Nakhutin in B.I. Verkin et al. (1990), p. 319–320.

37. L. Shubnikov, V.I. Khotkevich, Kriticheskie znachenija polja i toka dlja sverkhprovodjashchego olova (Critical values of field and current for superconducting tin), *Zhurn. Tekhn. Fiziki* 6 (1936) 1937–1943.

38. V.I. Khotkevich, K voprosu o kriticheskikh znachenijakh polja i toka dlja sverkhprovodjashchego olova (On the question of the critical values of field and current for superconducting tin), *Zhurn. Eksperim. i Teoret. Fiziki* 8 (1938) 515–517.

39. H.B.G. Casimir, *Haphazard Reality: Half a Century of Science* (Harper & Row, New York, 1983).

40. C. Gorter, Superconductivity until 1940 in Leiden and as seen from there, *Rev. Mod. Phys.* 36 (1964) 3–7.

41. H.R. Woltjer, Magnetic researches XXVII. Magnetic properties of some paramagnetic chlorides at low temperatures, *Leiden Commun.* N173b (1926) 11–19.

42. H.R. Woltjer and H. Kamerlingh Onnes, Further experiments with liquid helium. Magnetic researches XXVIII. Magnetization of anhydrous $CrCl_3$, $CoCl_2$, and $NiCl_2$ at very low temperatures, *Leiden Commun.* N173c (1926) 23–29.

43. H.R. Woltjer and E.C. Wiersma, On anomalous magnetic properties at low temperatures of anhydrous ferrous chloride, *Leiden Commun.* N201a (1930) 3–7.

44. L.D. Landau, Eine mögliche Erklärung der Feldabhängigkeit der Suszeptibilität bei niedrigen Temperaturen, *Phys. Z. Sow.* 4 (1933) 675.

45. O.N. Trapeznikova and L.V. Shubnikov, Anomaly in the specific heat of ferrous chloride at the Curie point, *Nature* 134 (1934) 286–287.

46. O.N. Trapeznikova, L.V. Shubnikov, Über die Anomalie der spezifischen Wärme von wasserfreien Eisenchlorid, *Phys. Z. Sow.* 7 (1935) 66–81.

47. O.N. Trapeznikova, L.V. Shubnikov, Anomalija teploemkosti bezodnogo khloristogo zheleza, *Zhurn. Eksperim. i Teoret. Fiziki* 5 (1935) 281–291.

48. O. Trapeznikova, L. Schubnikow, G. Miljutin, Über die Anomalie der spezifischen Wärme von wasserfreiem $CrCl_3$, $CoCl_2$, $NiCl_2$, *Phys. Z. Sow.* 9 (1936) 237–253.

49. O.N. Trapeznikova, L.V. Shubnikov, G.A. Miljutin, Anomalija teploemkosti bezvodnykh $CrCl_3$, $CoCl_2$, $NiCl_2$ (Anomalous heat capacity of anhydrous $CrCl_3$, $CoCl_2$ and $NiCl_2$, *Zh. Exper. Teor. Fiz.* 6 (1936) 421–432.

50. O. Trapeznikova, L. Schubnikow, Anomale spezifische Wärmen der wasserfreien Salze $FeCl_2$, $CrCl_3$, $CoCl_2$ und $NiCl_2$, *Phys. Z. Sow. Sondernummer* (1936) 6–21.

51. O. Trapeznikova, L. Schubnikow, Anomale spezifische Wärmen der wasserfreien Salze $FeCl_2$, $CrCl_3$, $CoCl_2$ $NiCl_2$ und $MnCl_2$, *Sonderdruck aus den Berrichten des VII Kaltenkongresses* (1936) 1–14.

52. L.W. Schubnikow, S.S. Schalyt, Ferromagnetische Eigenschaften einiger paramagnetischer Salze, *Phys. Z. Sow.* 11 (1937) 566–570.

53. S.S. Shalit, Magnitnie svojstva nekotorykh paramagnetnykh solej (Magnetic properties of some paramagnetic salts), *Zhurn. Eksperim. i Teoret. Fiziki* 8 (1938) 518–530.

54. I. Estermann, O. Stern, Über die magnetische Ablenkung von Wasserstoffmolekülen und das magnetische Moment des Protons, II, *Z. Phys.* 85 (1933) 17–24.

55. I.I. Rabi, J.M.B. Kellog, and J.R. Zacharias, The magnetic moment of the proton, *Phys. Rev.* 46 (1934) 163–165.

56. J.M.B. Kellog, I.I. Rabi, and J.R. Zacharias, The gyromagnetic properties of the hydrogen, *Phys. Rev.* 50 (1936) 472–481.

57. P. Debye, Die magnetische Methode zur Erzeugung tiefster Temperaturen, *Phys. Z.* 35 (1934) 923–928.

58. J. Dorfman, Magnetic properties and nuclear magnetic moments, *Phys. Z. Sow.* 7 (1935) 126–127.

59. B.G. Lazarev, L.W. Schubnikow, Über das magnetische Moment des Protons, *Phys. Z. Sow.* 10 (1936) 117–118.

60. B.G. Lazarev, L.W. Schubnikow, Das magnetische Moment des Protons, *Phys. Z. Sow.* 11 (1937) 445–457.

61. E. M. Purcell, Nuclear Magnetism, *American Journal of Physics* 22 (1954) 1–8.

62. J.G. Powles, Nuclear magnetic resonance, *Science Progress*, 1956, v 44, no. 175, 449–471.

63. W. Heitler, H.H. Willis and E. Teller, Time effects in the magnetic cooling method I, *Proc. Roy. Soc. A* 155 (1935) 629–639.

64. H. Fröhlich, W. Heitler, Über die Einstellzeit von Kernspins in Magnetfeld, *Phys. Z. Sow.* 10 (1936) 847–848.

65. A.K. Kikoin and L.W. Schubnikov, Optical experiments on liquid helium II, *Nature* 138 (1936), no. 3493, 641.

66. O.N. Trapeznikova, L.V. Shubnikov, Issledovanie uslovy ravnovesija gazoobraznoj i zhidkoj fazy smesi kisloroda i azota (Investigation of the equilibrium conditions of the gaseous and liquid phases of a mixture of oxygen and nitrogen), *Zh. Tekhn. Fiz.* 4 (1934) 949–953.

67. N.S. Rudenko, L.W. Schubnikow, Die Viskosität von flüssigem Stickstoff, Kohlenoxyd, Argon und Sauerstoff in Abhängigkeit von der Temperatur, *Phys. Z. Sow.* 6 (1934) 470–477.

68. N.S. Rudenko, L.W. Schubnikow, Viskosität des flüssigen Methans und Äthylens in Abhängigkeit von der Temperatur, *Phys. Z. Sow.* 8 (1935) 179–184.

69. M. Ruhemann, Industrial Research and Development in the Soviet Union in J. Needham and J. Sykes Davies (eds.), *Science in Soviet Russia* (Watts & Co, London, 1942), p. 18–23.

# Chapter 9
# Repression at the Leningrad Physico-Technical Institute

The repression of physics and physicists at UFTI in 1937 was closely related to similar events in Leningrad. In the interrogations conducted by the NKVD interrogators with their victims, relations and connections with physicists still in Leningrad were time and again suggested. This of course had to do with the fact that the majority of the physicists at UFTI originated from the Leningrad Physico-Technical Institute (LFTI) and the NKVD was keen to reveal a grand conspiracy connecting LFTI and UFTI. Although such conspiratorial relations and connections existed almost exclusively in the nebulous minds of the NKVD officials in their unwavering zeal to uncover counter-revolutionary and Trotskyist conspiracies, there were close scientific and personal relations with LFTI and I will therefore first discuss briefly what LFTI had to go through in these years and before.

Immediately following the October Revolution, physicists and other scientists had been arrested as part of the repression of the intelligentsia who were considered to belong to the natural enemies of the revolution. In his article *The Economics of the Transition Period* Nikolaj Bukharin had identified nine strata of people hostile to the revolution, including the intelligentsia. He writes that they all had to be met with 'concentrated violence'. However, no special cases were constructed that only involved physicists. For instance, in connection with the fictitious Tagantsev[1] conspiracy (or the case of the Petrograd Military Organisation; *Delo PBO*) several scientists were arrested and executed. The case had been fabricated by the Cheka to

---

[1]Named after the geographer and academician Vladimir Nikolaevich Tagantsev (1889–1921). Tagantsev was tortured and forced to give the names of hundreds of people who were critical of the Bolshevik regime, out of which a conspiracy, the existence of a military organisation, was created.

© Springer International Publishing AG 2018
L. J. Reinders, *The Life, Science and Times of Lev Vasilevich Shubnikov*,
Springer Biographies, https://doi.org/10.1007/978-3-319-72098-2_9

terrorize intellectuals who might be in opposition to the Bolsheviks and to force-
fully bring home to them that the Bolsheviks would not hesitate to make short shrift
of allegedly disloyal members of the intelligentsia (Ref. [1], p. 161).[2]

After a relatively quiet period in the twenties during the NEP, in which the
struggle for power within the party was decided in Stalin's favour, he started his
cultural revolution in 1928, which also included an assault on 'bourgeois science'.
In May 1928 in a speech at the Eighth Komsomol Congress Stalin singled out
science: "*A fortress stands before us. This fortress is called science, with its
numerous fields of knowledge. We must seize this fortress at any cost. Young people
must seize this fortress, if they want to be builders of a new life, if they want truly to
replace the old guard... A mass attack of the revolutionary youth on science is what
we need now, comrades.*" [2] This was followed by an article published on 7
November 1929 in *Pravda* in which 1929 was singled out as 'the year of the Great
Break' (*Veliky Perelom*), a year of shattering transformation on all fronts of socialist
construction (Ref. [3], p. 233). An assault on the universities and research institutions followed [4]. In 1928 a campaign was started against 'wreckers' among
'bourgeois-specialists', in which engineers and technicians were accused of sabotage and conspiracy to overthrow the Soviet regime. The first and best known of
such cases is the Shakhty trial (*Shakhtinskoe delo*) in 1928, whereby a group of
engineers was arrested in the North Caucasus town of Shakhty and accused of
conspiring to sabotage the Soviet economy. A search for 'wreckers' was carried out
all over the country, filling up jails with representatives of the technical and
humanitarian intelligentsia. LFTI did not remain unaffected, when Dmitry
Rozhansky, one of the consultants at UFTI, was arrested. The reason was that at one
of the obligatory meetings at LFTI, organized to condemn the wreckers, he refused
to vote in favour of the death penalty for the accused in the Industrial Party Trial, as
he was of the opinion that the court should establish the guilt of the accused and
also since he was against the death penalty.[3] An article appeared about this in a
newspaper with the heading "*There is no place for Rozhanskys in the family of
Soviet scientists*".[4] Rozhansky was in prison for almost a year until July 1931 when
the case was dismissed for lack of evidence. Mikhail Viktorovich Kirpichëv (1879–
1955), the head of the Laboratory of Heat Engineering at LFTI, and Mikhail
Aleksandrovich Mikheev (1902–1970), two other friends and colleagues of Ioffe,
were also arrested in October 1930. There does not seem to have been any special
reason for their arrests, apart from the fact that they belonged to the
bourgeois-intelligentsia which, however, was also true for many others who were

---

[2]Apart from a 'chronicle of the terror' in the first years after the revolution, Ref. [1] contains letters
to Lenin personally and/or to the Cheka commission in charge of the case from various people,
including renowned scientists, written in support of those arrested; even a letter written by Maxim
Gorky in support of the poet Nikolaj Gumilev, husband of Anna Akhmatova, was sent to no avail;
98 people were executed, including Tagantsev and his wife; in total, criminal charges were brought
against 833 people. The victims were only rehabilitated in 1992.

[3]His friend and colleague Abram Ioffe had no such qualms.

[4]Leningradskaja Pravda no later than 5 October 1930); http://www.ihst.ru/projects/sohist/.

left in peace. Perhaps something had gone wrong at their laboratories which was interpreted by some zealous officials as sabotage. Kirpichëv was sentenced to ten years of hard labour, of which he served only half a year, after which he returned to his scientific activity. Just like Rozhansky, Mikheev was released after a year without charge (Ref. [5], pp. 110–111). In the *Gulag Archipelago* Aleksander Solzhenitsyn explains this turnaround (Ref. [6], p. 48). At the beginning of 1931 Stalin had proclaimed: "*We must move from a policy of destruction of the old technical intelligentsia to a policy of concern for them, of making use of them*".[5] Stalin's proclamation was the result of ever increasing complaints that the attempts to discover wrecking had gone too far and were harmful to the industrial enterprise. Every failure was attributed to wrecking; inquiries at industrial installations and research institutes were happening all the time and were consuming and virtually wasting large amounts of management time; engineers of crucial importance for the industry in question were put in prison, and so forth. So Rozhansky and many others were released, including Lev Landau's father, who had been arrested in 1930 (while Landau was abroad) and sentenced to ten years for sabotage in the oil industry.

## Arrests in Leningrad after Kirov's Murder

The first big wave of repression started immediately after the murder of Sergej Kirov (1886–1934), the first secretary of the party in Leningrad, on 1 December 1934. Already a few days after this event large numbers of people were arrested, with the NKVD taking special aim at Leningrad's elite and intelligentsia (Ref. [7], p. 37 ff). According to Frish (Ref. [8], p. 113), apart from resulting from the usual denunciations, the arrests were made according to lists compiled in advance on the basis of fairly random characteristics: people who had relatives abroad or knew foreigners, who were close to the aristocracy in Tsarist times, or who were formerly officials in the Tsarist government. The random element was reflected in the fact that some were arrested, while others with the same characteristics (and even close relatives) were left in peace. Most were released after a few days, but some were exiled and ordered to leave Leningrad within three or four days. According to Kosarev (Ref. [5], p. 115) in total almost 100,000 people were sent away from Leningrad without trial, which seems a rather high estimate. Frish says that it is difficult to say how many people were affected, while Ivanenko (Ref. [9], p. 180) gives a number of more than 40,000, but it is clear that it concerned at least many thousands. No official statements were made about the arrests, simply that a number of 'former people' had been sent away. This wave of arrests died down as quickly as it had come up.

---

[5]See also Ref. [4] about policy changes in the early 1930s.

In February and March 1935 a number of leading scientists at Leningrad State University and the State Optical Institute were arrested, including Vladimir Fock, Dmitry Ivanenko and various experimental physicists. Most of the specialists arrested, including Fock, were released without charge after a few days. Dmitry Ivanenko had been the first head of the theory department at UFTI and returned to Leningrad in 1932 when Landau arrived in Kharkov. Had he stayed in Kharkov probably nothing would have happened to him. Now he was also judged to belong to the 'former people' since, according to a copy of the court judgement (Ref. [9], p. 78) of 25 February 1935, he was the son of a non-hereditary nobleman,[6] who moreover was disenfranchised in 1928 and 1929. The same document also states that Ivanenko had connections with relatives abroad and, most importantly, that he was a member of a counter-revolutionary group led by Gamow[7] and organised illegal meetings in his apartment under the guise of scientific meetings. As a "socially dangerous element" (*sotsial'no opasnyj element*), which according to the criminal code of the Russian Federation is someone who has committed criminal activities or has connections with the criminal milieu, he was sentenced to three years in a corrective labour camp and sent to a camp in Kazakhstan. His wife and daughter were also exiled from Leningrad for three years. Abram Ioffe, with the help of Sergej Vavilov, sent an official complaint to the Central Committee, and others at LFTI collected signatures on a petition. It helped, for Ivanenko only stayed in the camp until shortly after 1 January 1936 when he was sent into exile to Tomsk. There he worked at the Siberian Physico-Technical Institute (one of LFTI's daughter institutes founded in 1932) and taught at the university until the end of his sentence in 1938, after which he went to Sverdlovsk. Apart from the nine-month stay in the camp Ivanenko could continue his work as a physicist, correspond with collaborators in other towns, and even make visits to Moscow, Leningrad and Kiev during this period.

A few times during his stay in Tomsk Ivanenko was threatened with renewed arrest; he had a particularly narrow escape in October 1937 when he was dismissed from both the institute and the university, but managed to 'clear the matter up' with the NKVD, and again in 1938 in connection with the UFTI affair as he was mentioned by Rozenkevich and Shubnikov in their interrogations by the NKVD as a member of the counter-revolutionary group led in Leningrad by Landau. He was

---

[6]*Lichnyj dvorjanin:* Personal nobility could for instance be acquired by admission to orders of knighthood of the Russian Empire. It was transferable only to the wife, not to children.

[7]It is a surprise to read that as early as February 1935 Gamow, who was still officially on *komandirovka*, was considered a counter-revolutionary. He was also still a corresponding member of the Academy of Sciences, an honour that was only taken away from him in 1938 (Ref. [5], p. 114).

also mentioned by Landau, as is clear from the protocol of Landau's interrogation by the NKVD of 3 August 1938 published in 1991 (Ref. [10], p. 139).[8]

## The 1937 Onslaught at LFTI

Stalin's concern about the harm done to the industrial enterprise by the arrests of bourgeois specialists did not last long and by 1937 the repression assumed a massive and much more deadly character. The first to be arrested at LFTI, already on 15 October 1936, was Viktor Bursian, who also headed the physics institute at the university, followed a few days later by Vsevolod Frederiks[9] and Jury Krutkov.[10] After the arrests the fantastic announcement was made that Bursian and Frederiks were terrorists who were plotting to assassinate a prominent member of the party and were involved in such a plot with their colleagues (Ref. [11], p. 78). A few months later, in February 1937, Fock was arrested as an 'accomplice'. Kapitsa, who was very shocked by this arrest (after all, it was the second time in two years that he was arrested), immediately wrote letters (Ref. [12], pp. 337–339) to Stalin and Mezhlauk, at the time the vice-chairman of Sovnarkom, and managed to convince them that Fock's involvement in a terrorist plot was extremely unlikely. Within a few days Fock was transferred to Moscow, where he had an interview with Yezhov, the dreaded head of the NKVD (only hearing at the end of the interview that the 'little man with a narrow pale face in a military uniform' (Ref. [8], p. 119) who had been speaking to him about the multitude of enemies surrounding the Soviet Union was Yezhov himself) and was released. Frish also relates how the colleagues of the arrested physicists viewed such arrests and that they did not doubt that indeed some crime had been committed: *"Neither I, nor any one of my close friends ever doubted the innocence of Bursian, Frederiks, and Krutkov. But we would be very surprised if someone had told us that all the defendants—every single one of them—would later be acquitted, that there was no case at all, that it was a fabrication from beginning to end. For us there was an unknown, but unquestionable criminal hiding behind each court case. But we tried to look for this criminal further away from us. So, a version of the arrest of our professors arose,*

---

[8]Ref. [10] is a publication of the Central Committee of the Communist Party, and it must have been one of the last publications of this august body. Only a selection from the file seems to have been published as there are no documents dated between 28 April and 3 August 1938. Landau must have been interrogated much earlier. The protocol could very well be an NKVD fabrication, as it contains some odd statements about Bohr's institute and Landau's 'idealism'.

[9]Vsevolod Konstantinovich Frederiks (1885–1944) who wrote widely on the theory of general relativity with Aleksandr Aleksandrovich Fridman (1888–1925). He died in 1944 of pneumonia when on his way to another *sharazhka* or according to another version on his way home after his release.

[10]Jury Aleksandrovich Krutkov (1890–1952), who was Ehrenfest's assistant in Leiden before World War I.

*whereby it was assumed that the 'reason' could be found outside the university walls. In the summer Sokolov,[11] Bursian, and Frederiks[12] had taken part in a geological expedition, fitted out to test a new method of mineral exploration. This made it possible to create a "geological version", according to which it had all started with a mythical wrecker-geologist, who had subsequently spread some slander in order to complicate matters for the other members of the expedition. We were all convinced of his real existence and speculated about his wrecking activity"* (Ref. [8], p. 118). This, however, was not as mythical as Frish wanted to believe, for the case indeed started with the geophysicist Ju.N. Lepeshinsky (1891–1937), working in Leningrad at the Central Research Institute of Geological Prospecting (CNIGRI) and a professor at Leningrad University. His arrest in the summer of 1936 in Zyrjanovsk in the Altai (now in Kazakhstan), where the geological expedition mentioned by Frish had taken place, started the 'geological branch' of the Pulkovo affair. When such an expedition did not yield the expected results or something went wrong, as it often would, the charge of 'wrecking' was easily made in the atmosphere of the mid-1930s. Since, as far as I know, the documents of these NKVD inquiries have not been made public, the actual course of affairs is difficult to establish, but not difficult to imagine. When interrogated Lepeshinsky will have made various statements, perhaps voluntarily, but most probably under coercion (beatings, sleep deprivation, continuous interrogation), denouncing other people, from which the NKVD started to spin its elaborate plot, catching more and more people in an ever widening net. As Beck and Godin [15] describe in their remarkable[13] little book written in 1951: *"The method of interrogation (...) consisted of making it the arrested man's primary task to build up the whole case against himself, more or less of his own free will, (...) to make it plausible in every detail, relating it to actual events or giving these the desired twist"* (Ref. [15],

---

[11]Most probably Pavel Timofeevich Sokolov (1900–1937); arrested on 10 October 1936, accused of espionage and terrorism and shot on 23 May 1937. Sokolov was professor of theoretical physics at Leningrad University (Central Research Institute of Geological Prospecting (CNIGRI)) and a geophysicist (seismic specialist). Not much is known about his work. The data here are from [13].

[12]Bursian was also interested in geophysics. Frederiks' main interests were general relativity and later liquid crystals, but he was also a consultant of the Geological Committee (*Geolkom*) at CNIGRI on electric prospecting and in this capacity participated in expeditions. In Sonin and Frenkel's biography of Frederiks [14] it is stated that Frederiks was in the south in the summer of 1936, but nothing is said about a geological expedition. There is still another geologist with the name Frederiks: Georgy Nikolaevich (1889–1938), who was arrested for the first time in 1935, exiled, but released after eight months, arrested several more times, for the last time in June 1937 and shot in February 1938 for conspiring to murder Soviet leaders. The two were relatives and belonged to an important noble family.

[13]This book is remarkable as it was written under a pseudonym by two former prisoners of the Stalinist regime in the thirties, namely the physicist Fritz Houtermans (who worked at UFTI) and Konstantin Feodosevich Shteppa (1896–1958), a very controversial Ukrainian historian who was arrested in 1938 and for some time shared a cell with Houtermans; he is alleged to have worked for the NKVD, the Nazis and later also for the CIA (see Shteppa's Wikipedia page http://en.wikipedia.org/wiki/Konstantin_Shteppa; Ref. [7], p. 292, also gives information on Shteppa). After the war Houtermans and Shteppa met again and wrote this book.

p. 45). *"Everyone was required to denounce at least one other person who had 'recruited' him, i.e. had persuaded him to engage in his counter-revolutionary activity and had thereafter directed it; everyone was also required to denounce as many other people as possible whom he had himself 'recruited' and induced to commit political crimes..."* (Ref. [15], p. 47). An example of this is Frederiks who mentioned the names of several of his arrested colleagues as participants in the counter-revolutionary organisation and is himself mentioned in the protocols of the interrogations of no less than 21 other people (Ref. [16], p. 413).

So Lepeshinsky's arrest was followed later in the year by the arrest of other geologists and of Bursian. From the testimony of others among the arrested, it was 'established' that for a long time Bursian had kept a revolver in his desk in his office, a revolver that had disappeared in 1935. Bursian himself vigorously denied this 'fact' and it most probably was the figment of someone's (artificially stimulated) imagination, but who cares and in one of the scenarios elaborated by the NKVD this 'imaginary' revolver was even supposed to have been used in an assassination attempt on Stalin (Ref. [16], pp. 411–418). Accused of participation in a 'counter-revolutionary fascistic' organisation Bursian was finally convicted to ten years in prison, where he died in 1945. The arrest of Frederiks can also be linked to this geological affair. He was convicted to ten years' labour camp. In the final years of his captivity he engaged in technical work in the oil fields. He also died while still in captivity.[14]

Some time later, Pëtr Lukirsky (April 1938) and Matvej Bronshtejn[15] (August 1937) were also arrested, leaving Frenkel the only theoretical physicist of importance in Leningrad who was not directly affected by the purges. Let me start with Lukirsky as his arrest also throws some light on the fate of Bronshtejn. At the moment of his arrest Lukirsky, one of the first generation of Soviet physicists, was a professor of physics at Leningrad University and a staff member at the Physico-Technical Institute. At the latter institute he was the leader of the group studying the scattering of X-rays and electrons (Ref. [17], p. 160). He was Olga Trapeznikova's advisor in her post-graduate work. Since 1933 he had been a corresponding member of the Academy of Sciences. Lukirsky's case has been studied in detail by Viktor Frenkel, who had gained access to the corresponding KGB/NKVD files [18]. Without revealing any names Frenkel quotes from an interrogation protocol kept in Lukirsky's file, dated 19 March 1935, of a staff member of LFTI arrested early in 1935. It emerges that already at that time Bronshtejn, Ivanenko and Lukirsky had been denounced as *"most openly expressing their counter-revolutionary views"* at LFTI. No explanation is given why these three physicists were singled out in this statement. It also cannot be maintained with any certainty that their arrest had anything to do with this

---

[14]Lepeshinsky was executed in May 1937. The suppression among geologists was extremely severe. Ref. [15] lists 970 repressed geologists, an astonishing number.

[15]Matvej Petrovich Bronshtejn (1906–1938). He was married to the writer Lidija Chukovskaja (1907–1996).

denunciation, in any case not as regards Ivanenko, who had already been arrested and convicted in February 1935, so before the date of this protocol. Moreover the protocol continues with naming others, among whom Anatoly Petrovich Aleksandrov (1903–1994) and Igor Kurchatov, who would later become very prominent Russian nuclear physicists, with Kurchatov the leader of the Soviet atomic bomb project and Aleksandrov from 1975 to 1986 President of the Academy of Sciences. The denunciation recalled here does not seem to have adversely affected their careers. If they had been sent to the camps, the history of Soviet physics would have been quite different. The astonishing aspect is that the NKVD investigator came with statements of three other people, all former members of LFTI or Leningrad University, who accused Lukirsky of being a member of a counter-revolutionary organisation. According to Frenkel it is remarkable that copies of the interrogations of all these other people have been kept in Lukirsky's file. On the basis of these statements, without himself having confessed or having been confronted with the accusers, in September 1938 Lukirsky was sentenced to five years in a labour camp for being a member of a 'fascist organisation' and for carrying out as such all kinds of anti-Soviet activity. He was sent to a camp in Solikamsk in East Russia.

In February 1939 at the request of Lukirsky's wife a whole group of scientists (including Sergej Vavilov, Abram Ioffe, Pëtr Kapitsa, Aleksej Krylov,[16] and Vladimir Fock) wrote a letter to Lavrenty Beria, the new head of the NKVD after Yezhov's fall from power in 1938, with the request to reconsider the cases of Lukirsky, Krutkov and Frederiks. In the letter they stressed that "*Lukirsky was a leading scientist in the field of photo-elements and the founder of an entire scientific school. His work was included in all textbooks*". Early in 1940 the Lukirsky case was indeed reviewed and it was established that he should be freed. He was, however, kept in prison in rather harsh conditions, as the decision had to be confirmed by various other bodies, whereupon Kapitsa asked the vice-president of the Academy of Sciences to appeal directly to the NKVD to speed up Lukirsky's case in view of his exceptional value for theoretical research on electronic phenomena. It wasn't until August 1942 that all the paperwork was completed and the review was finalized by a decision of the Special Council of the USSR NKVD (*Osoboe Soveshchanie pri NKVD SSSR, OSO*).[17] Lukirsky was freed in October of the same year and, although this intervention seems quite a success, it must be realized that at that time he had already served most of his five-year term.

Bronshtejn's arrest, like Krutkov's before him, is connected with the Pulkovo affair, as he had many astronomer friends, but also with the trouble at UFTI and perhaps with Lukirsky's arrest. Moreover Bronshtejn had already once been

---

[16]Aleksej Nikolaevich Krylov (1863–1945), academician, mathematician and naval engineer, also Kapitsa's father-in-law.

[17]This odious council, created by the same decree that introduced the NKVD itself, had the power to apply punishments by administrative means, i.e. without a trial. The punishments could be banishment (from the place of residence), exile, corrective labour camps for up to five years and deportation from the USSR.

censured for anti-Soviet behaviour ("*open revolt against the principles of dialectical materialism and Marxist ideology*") in his prank with Landau and Gamow about Gessen's entry about the ether in the *Great Soviet Encyclopaedia*.[18] The fact that his name was the same as Trotsky's real name was not the reason for his arrest, although it will not have helped. Bronshtejn was arrested at his parents' flat in Kiev for "*active involvement in a Leningrad counter-revolutionary organisation of intelligentsia who wanted to bring down Soviet power and set up a new political order that would allow intelligentsia to take part in state administration together with other social groups according to the Western pattern*" and work "*to create a basically fascist state that would be able to resist communism*" (Ref. [19], p. 144) (note that it was quite an elaborate plot the NKVD constructed against Bronshtejn). According to Uspenskaja (Ref. [20], p. 96) his arrest was triggered by a statement beaten out of the young Pulkovo astronomer Nikolaj Aleksandrovich Kozyrev (1908–1983) during an interrogation by the NKVD. He was a close friend of Bronshtejn and was heavily involved in the Pulkovo affair within which framework he had already been arrested in the autumn of 1936.[19] Bronshtejn, hinting at the similarity between his name and that of Trotsky, was claimed to have said at one time: "*Trotsky is gone, but he will still prove his worth. If Trotsky returns to power, I will call myself his nephew*" (Ref. [20], p. 96). When the NKVD questioned Bronshtejn, apparently using rather rough methods, he 'confessed' that Frenkel had recruited him into a fascist-terrorist organisation and that Ambartsumjan, Fock, Lukirsky, Landau, Bursian, Frederiks, Krutkov and others were also part of this organisation. Matvej Bronshtejn was executed by firing squad on 18 February 1938. Frenkel and Ambartsumjan were the only ones to be left untouched.

One of the main points here is to show that all of these affairs were plausibly related. People were mentioned in NKVD interrogations in the way described above in the citation from Beck and Godin's book and illustrated in the Lukirsky and Bronshtejn cases, and in this way were sucked into the NKVD whirlpool of intertwined plots. In my view there was not much randomness in all this, although being mentioned or denounced did not necessarily mean that arrest would definitely follow. Frenkel was also mentioned in several 'confessions' as a member, even the

---

[18]Boris Mikhailovich Gessen (1893–1936) had attacked Einstein's theory and declared that the ether, the medium postulated in the 19th century for the propagation of light and which Michelson and Morley in their experiment in 1887 and Einstein in his theory of relativity had done away with, must exist and that Soviet physicists should prove its existence and determine its true mechanical properties. Landau, Gamow and Bronshtejn had a good laugh about this and as a joke sent him a telegram stating that they were inspired by Gessen's article on the ether, that they were looking forward to his leadership in the search for phlogiston, and that old Einstein is an idealistic idiot. The telegram had far-reaching consequences. Landau and Bronshtejn were found guilty of anti-revolutionary activity and dismissed from their teaching jobs at the Polytechnic Institute, and two graduate students who had also signed the telegram lost their stipend and had to leave town. Gamow got off scot-free as he was working for the Academy of Sciences, which did not take measures.

[19]For his role in this affair see [20–22]. He also figures in Aleksander Solzhenitsyn's *The Gulag Archipelago*. Kozyrev was convicted to ten years in prison and freed in 1946 at the request of his fellow astronomers as a talented scientist.

leader, of the most fantastic terrorist organisations, but he was never arrested, although he certainly possessed some of the objective characteristics[20] that would 'warrant' arrest, even more so than many others, as he had spent considerable time abroad. As late as 1930, for example, he was a visiting professor at the University of Minnesota in that 'most despicable' of all countries, the United States of America. He had also played an important role in the discussion with the 'philosophers' about whether modern physics (relativity theory and quantum mechanics) could be reconciled with dialectical materialism. In this discussion the Czech-born 'philosopher' Ėrnest Kolman[21] had accused him in 1931 of Trotskyism and 'rotten liberalism'. Of all the physicists taking part in the 'debate' Frenkel was the most outspoken and courageous in his statements. He did not see much use of dialectical materialism for science. At a meeting with members of the party in 1931 he said: *"The dialectical method has no right to claim a leading role in science. Our policy of imposing the views of dialectical materialism on scientists and young people has been carried much too far. ( ...) What I have read in Engels and Lenin does not delight me at all. Neither Lenin, nor Engels is an authority for physicists. Lenin's book is a model of acute analysis, but it amounts to little more than the assertion of truisms, over which it is not worth breaking a lance. ( ...) Your philosophy is reactionary, and I hope that the party will soon be convinced of this. ( ...) As a Soviet person I personally cannot sympathize with an opinion that is harmful to science. There cannot be proletarian mathematics, proletarian physics, etc."*[22] It may be seen as surprising that these statements, which Frenkel made while being questioned by party members, had no further consequences for him. His son Viktor Frenkel also asked himself the question, why was Jakov Frenkel, who figures in practically all cases he has investigated, left untouched? The same applies more or less to Abram Ioffe, Igor Tamm and Nikolaj Semënov. Frenkel suggests that they were kept 'in reserve' (Ref. [18], p. 188). And who knows, if Yezhov's reign of terror, the Yezhovshchina, had lasted one or two years longer, their time may also have come.

---

[20]About these objective characteristics see Ref. [15], p. 89 ff. On the other hand Frenkel's father Ilja Abramovich Frenkel (b. 1863) was a member of *Narodnaja Volja* (People's Will), an organization which at the end of the 19th century propagated mass terror in the struggle against the autocratic regime. He spent six years in exile in Siberia and after his return to European Russia was kept permanently under surveillance by the tsarist police, so his son Jakov may have had the right pedigree to be respected by the masters of terror (Ref. [23], p. 1).

[21]Ėrnest Kolman (1892–1979), a loathsome character who was especially keen on 'wrecking in science' which he spotted everywhere, often with serious consequences for the persons he 'exposed'; he initiated the Luzin affair with the publication of the anonymous article "On enemies with a Soviet mask" (*O vragakh v sovetskoj maske*) in *Pravda* of 3 July 1936 and also accused other scientists, such as Landau, Vernadsky, Tamm, and Vavilov, of 'wrecking'.

[22]Quoted from the minutes of the meeting on 14 November 1931 in Ref. [24], p. 141.

# References

1. V. Goncharov and V. Nekhotin, *Prosim osvodit' iz tjuremnogo zakljuchenija* (Please release from prison) (Sovremenyj Pisatel, Moscow, 1998).
2. J. Stalin, *Sobranie Sochinenii* (Collected Works) (Partizdat, Moscow, 1947), vol. 11, p. 77; quoted in N. Krementsov, *Stalinist Science* (Princeton University Press, 1997).
3. David Joravsky, *Soviet Marxism and Natural Science* (Routledge and Kegan Paul, London, 1961).
4. M. David-Fox, *The Assault on the Universities and the Dynamics of Stalin's "Great Break," 1928–1932*, in: M. David-Fox and György Péteri (eds.), *Academia in Upheaval: Origins, Transfers, and Transformations of the Communist Academic Regime in Russia and East Central Europe* (Bergin & Garvey, London, 2000), p. 73–103.
5. V.V. Kosarev, Fiztekh, Gulag i obratno (belye pjatna iz istorii leningradskogo Fiztekha) (*Phystech, Gulag and back (white spots from the history of the Leningrad Physico-Technical Institute*) in: *Chtenija pamjati A.F. Ioffe* (Readings in memory of A.F. Ioffe) (Nauka, St. Petersburg, 1993), 105–177.
6. Alexander Solzhenitsyn, *The Gulag Archipelago* (Collins/Fontana, 1975), Vol. 1.
7. Robert Conquest, *The Great Terror: A Reassessment* (Pimlico, London, 2008).
8. S.E. Frish, *Skvoz' prizmu vremeni* (Through the prism of time) (Solo, St Petersburg, 2009).
9. G.A. Sardanashvili, *Dmitry Ivanenko–superzvezda sovetskoj fiziki. Nenapisannye memuary* (Dmitry Ivanenko–superstar of Soviet Physics. Unwritten memoirs) (URSS, Moscow, 2010).
10. V. Vinogradov and N. Mikhajlov (eds.), Lev Landau: god v tjur'me (*Lev Landau: a year in prison*), *Izvestija TsK KPSS* 3 (1991) 134–157.
11. Ju.S. Vladimirov, *Mezhdu Fizikoj i Meta-Fizikoj* (Between Physics and Metaphysics) (Librokom, Moscow, 2010).
12. J.W. Boag et al., *Kapitza in Cambridge and Moscow* (North-Holland, 1990).
13. V.P. Orlov (ed.), *Repressirovanye Geologi* (Moscow, 1999).
14. A.S. Sonin and V.Ja. Frenkel, *Vsevolod Konstantinovich Frederiks* (Nauka, Moscow, 1995).
15. F. Beck and W. Godin, *Russian Purge and the Extraction of Confession* (Hurst & Blackett, London, 1951).
16. V.Ju. Zhukov, *Pulkovskoe Delo* (The Pulkovo Affair), in V.P. Orlov (ed.), *Repressirovanye Geologi* (Moscow, 1999).
17. P.R. Josephson, *Physics and Politics in Revolutionary Russia* (University of California Press, 1991).
18. V.Ja. Frenkel, Trudnye gody Petra Ivanovicha Lukirskogo (*The Difficult years of Petr Ivanovich Lukirsky*), *Zvezda* 10 (1996) 179–197.
19. G.E. Gorelik and V.Ja. Frenkel, *Matvei Petrovich Bronstein and Soviet Theoretical Physics in the Thirties* (Birkhäuser, 1994).
20. N.V. Uspenskaja, Vreditel'stvo…v dele izuchenija solnechnogo zatmenija (*Sabotage …. in the study of the solar eclipse*), *Priroda* 8 (1989) 86–98.
21. R.A. McCutcheon, The 1936–1937 Purge of Soviet Astronomers, *Slavic Review* 50 (1991) 100–117.
22. A.I. Eremeeva, Political repression and personality: The history of political repression against Soviet astronomers, *Journal for the History of Astronomy* 26 (1995) 297–324.
23. V.Ya. Frenkel, *Yakov Ilich Frenkel: His work, life and letters* (Birkhäuser, Berlin, 1996).
24. V.Ja. Frenkel, Zhar nad peplom (*Heat over ash*), *Zvezda* 9 (1991) 129–148.

# Chapter 10
# The UFTI Affair: The Case of Weissberg and Weisselberg

## Introduction

Much of the information in this and the following chapters comes from the book *"Delo" UFTI 1935–1938* by Jury Pavlenko, Jury Khramov and Jury Ranjuk, published in 1998 in Kiev (Ref. [1]). Early in the nineties, just before or after Ukraine became an independent country, Ranjuk was asked by the then director of UFTI to investigate the possibility of writing a history of UFTI in Ukrainian. I do not know whether such a history ever appeared, but in the research for it he, of course, also came across the events in the thirties. He requested and obtained access to the archives of the KGB in Ukraine which contained the files of the UFTI affair (*Delo UFTI*). Through his efforts a great part of these files was declassified and could eventually be published in his book co-authored with Pavlenko and Khramov and on the Internet (both in Russian). Appendices 2, 3 and 4 to this book contain translations of a great part of the documents in these files, while here in the text only the most salient passages are quoted.

The UFTI affair consists of a number of affairs: the Korets affair already described in Chap. 7, which had just been a prelude and did not have a grim ending; the case of Weissberg and Weisselberg, which started the NKVD onslaught on UFTI; the Katod-Kredo[1] affair, which in particular involved Shubnikov, Rozenkevich and Gorsky; and the arrests of various other physicists at UFTI, such

---

[1] The name Katod-Kredo was used in the telegram from Moscow, dated 24 July 1937, ordering the arrest of Shubnikov, Rozenkevich and Pëtr Komarov (Pëtr Frolovich Komarov was a party member and the UFTI party organiser; after Weissberg's arrest in 1937 he led the construction of OSGO; was arrested and died in prison (Ref. [3], p. 136)). Jury Ranjuk did not succeed in finding a personal case on Komarov in the UFTI archives, nor a file at the NKVD (Ref. [1], p. 226). For details on Komarov, see also Ref. [2], Chap. 2 and p. 76, and Chap. 3, p. 118 and p. 126ff, which describes a confrontation with Komarov while in prison.

© Springer International Publishing AG 2018
L. J. Reinders, *The Life, Science and Times of Lev Vasilevich Shubnikov*,
Springer Biographies, https://doi.org/10.1007/978-3-319-72098-2_10

as Obreimov and Lejpunsky. Closely connected with the UFTI affair is the case of Landau, Korets and Rumer, which unfolded in Moscow as Landau had fled UFTI early in 1937 to seek refuge at Kapitsa's new Institute for Physical Problems, where the NKVD caught up with him in 1938.

Early in 1937 the mighty protectors of the 'anti-Soviet elements' at UFTI had all quit the scene in one way or another and the time had come for clearing out this nest of anti-Soviet Leningrad elements.

## The Case of Weissberg and Weisselberg

Chapter 7 ended with Alexander Weissberg (Fig. 10.1), the German physicist-engineer, who was in charge of the construction of OSGO, being called in for questioning by the NKVD early in 1937. This questioning continued for a couple of weeks, during which time he was repeatedly sent home and then called back to the NKVD office, until, on 1 March 1937, he was finally arrested and the UFTI affair was thrown into gear. Weissberg's arrest must have been part of a carefully laid out plan; it was the first test of who was actually on top. There was nothing random in this act of state terror. Three days after Weissberg's arrest, on 4 March, another foreigner was arrested, the chemist Konrad Weisselberg (Fig. 10.2).[2] At first one joint case was opened for the two of them, but subsequently their cases were separated and their fates diverged.[3]

The first document in Ref. [1], dated 23 May 1937, serves as an illustration of the seemingly absurd accusations against Weissberg and Weisselberg, but especially also to what extraordinary lengths the NKVD went to keep up a semblance of legality and due procedure. It concerns the application for a 'permit' to extend their custody, as if there indeed existed a proper judicial organ that looked after the rights of those arrested. The submission of such an application was a routine matter, but still required a lot of paperwork. The relevant part of the document reads:

---

[2]Konrad Weisselberg (1905–1937) was born into a Jewish family in Bârlad (Bariach) in present-day Romania, not far from the border with Ukraine. At some point the family moved to Vienna, where Weisselberg studied chemistry at the university, earning a PhD in 1930. Like Weissberg, Weisselberg was first a member of the Social Democratic Party of Austria and from 1933 of the Austrian Communist Party (Ref. [4], pp. 546–547).

[3]According to the authors of Ref. [1], p. 197, who have not seen Weissberg's file, which remains hidden in the KGB archives, people who have seen the file claim that it is in agreement with Weissberg's book *Conspiracy of Silence* (Ref. [2]). From Weissberg's book it is not apparent that at first a single joint case was opened for the two of them.

**Fig. 10.1**  Alex Weissberg

(...) Weissberg and Weisselberg were arrested on the accusation of having arrived in the USSR from abroad for carrying out counter-revolutionary, Trotskyist activity, and that, when being in the USSR, they were connected to the Trotskyist Placzek who had arrived from Denmark.

1. Additional inquiry and information has established that Weissberg, being a rightist, while abroad was connected with the prominent rightists who carried out counter-revolutionary and sabotage work in the USSR – Karl Frank[4] and Willie Schlamm.[5]
2. That at UFTI there exist two counter-revolutionary sabotage groups (the Landau group – Trotskyist and the Weissberg group – rightist), which combined in 1935 in the interests of the disruption of defence-related tasks set for UFTI by party and government.
3. That in these counter-revolutionary-sabotage groups a leading role is played by the prominent rightist Lejpunsky, who is connected with Bukharin and Orlov, has connections among the liquidated terrorist group of Nikolaev-Katalymov, and is subject to arrest together with other participants of the counter-revolutionary group (Shubnikov, Rozenkevich, Komarov and others),[6] that in connection with this the inquiry requires more work and cannot be completed in the appointed time and that the period of custody of Weissberg expires on 1 April 1937 and of Weisselberg on 4 April 1937.

### IT IS DECIDED:

To submit a petition to the Central Executive Committee[7] of the USSR for an extension of the period of custody for one month, i.e. until 1 June 1937, for the accused Weissberg, Alexander Semënovich, and Weisselberg, Konrad Bernardovich (Ref. [1], pp. 197–198).

---

[4]Meant here is Karl Borromäus Frank (1893–1969), who was born in Vienna and was a member of the Austrian Communist Party. Weissberg knew him from 1918 when he was still a schoolboy (Ref. [2], p. 152). Frank emigrated to Germany in 1920 and was a member of the German Communist Party from 1920. From 1929 he was a member of a rightist opposition group within the party which adopted a critical attitude towards the policies of the Soviet Communist Party. He later became a member of the German Socialist Party and after Hitler's takeover of power emigrated to the US.

[5]William S. Schlamm (1904–1978) was an Austrian journalist and writer. He wrote for communist journals until the thirties, after which he went into exile in the US and became a conservative. He is remembered for having coined the saying: "The trouble with socialism is socialism. The trouble with capitalism is capitalists.".

[6]Note that this application dates several months prior to the arrest of Shubnikov and the other UFTI scientists.

[7]The highest legislative body in the Soviet Union from 1922 to 1938.

**Fig. 10.2**  Konrad Weisselberg

It is interesting to read in this document that Shubnikov and other scientists at UFTI had already been singled out for arrest, quite a few months before such arrests were actually made, and that indeed the main accusation, stated under 2 in the above document, against all of them was going to be '*the disruption of defence-related tasks set for UFTI by party and government*'. The rest is all just meant for dressing up this 'principal crime'.

Weisselberg's file contained a copy of an interrogation protocol for Weissberg from 1 June 1937, three months after his arrest, in which he made a confession. His book (Chap. 8 of Ref. [2]) also contains a confession which Weissberg made after being subjected to the 'conveyor' (whereby an accused was kept under interrogation day and night until he broke down). In his book Weissberg writes about this: "*... physical violence towards an accused had first to be sanctioned from above. When the mass arrests began at the end of August 1937 physical maltreatment became general. Up to that period the Conveyor was the utmost physical pressure they were allowed to apply.*[8] *It was enough. It was as painful as any torture. But later on it had one grave disadvantage—it took up a lot of time. Until I experienced it myself I did not believe that a strong-willed man could be made to capitulate by mere interrogation alone. Some prisoners had even held out under torture, but I knew only one man who managed to resist the Conveyor.*

*The Conveyor worked automatically and silently. After a few days a prisoner's limbs began to swell. The muscles in the neighbourhood of the groin became extremely painful and it was an agony to move. The examiner need do nothing. (...) All the examiners had to do was to wait patiently. Time was their ally. For the prisoner suffering the tortures of the Conveyor there was no break. If there were some point at which the torture must cease then a prisoner might be able to summon up all his moral strength and will power to hold out until then. But there was no such point.*

---

[8]This agrees with Conquest (Ref. [5], p. 279) who claims that the change to the so-called simplified interrogation procedures, which involved beatings and other maltreatment, can be dated precisely to 17–18 August 1937, when they suddenly came into force in Moscow, Kharkov and elsewhere.

*I can hold out another night, and another night, and another night, he might think. But what then? What's the good of it? They have all the time in the world. At some point or other I must physically collapse.*" (Ref. [2], pp. 235–236.) Weissberg was subjected to this treatment for almost seven days before he gave in. The date of his confession is not known from the book, but in view of the quotation above it must have been before August 1937.

The protocol of Weissberg's interrogation does not agree with the chapter in his book, which only seems natural as Weissberg wrote his book more than ten years later, but he has also left out a lot that might be damaging for the heroic picture he paints of himself. The book version of the confession only concerns some for the NKVD only mildly interesting stories about his time in Austria and Germany, who had recruited whom, when and where and suchlike. That was not what they had arrested him for. They wanted to hear stories of counter-revolutionary villains at UFTI, but in the book version Weissberg explicitly says that Reznikov (his inter-rogator) "*wanted the names of people who could be arrested by the G.P.U. and no doubt he had my friends in the Institute in mind: Lejpunsky, Shubnikov, Landau, Ruhemann and so on. On this point I was determined not to budge whatever happened: I would not denounce them.*" (Ref. [2], p. 254.) As will be clear from part of the protocol printed below, the truth is sadly rather different. It will be clear that Weissberg's statements in the protocol of his interrogation are extremely damaging for Lejpunsky, Shubnikov, Komarov and Landau, something which Weissberg has chosen not to tell us about in his book.

After some introductory questions about his activities in Austria and Germany before his move to the Soviet Union, Weissberg's interrogation continues as follows (Ref. [1], pp. 212–216):

**Question**: With whom were you organisationally connected as a rightist in the USSR?

**Answer**: I did not have any open organisational connections with rightists. As a rightist I was close to Lejpunsky, Aleksander Ilich, director of UFTI, member of the Communist Party, double dealer, who had taken up anti-party positions.

In a conversation I declared to Lejpunsky my rightist views on various questions with which Lejpunsky agreed. I can separately give a detailed statement on Lejpunsky.

**Question**: You are not speaking the truth. In the USSR you had connections with rightists. I demand truthful statements from you.

**Answer**: I maintain that in the USSR I was not organisationally connected with rightists.

**Question**: How are the relations between you and Lejpunsky, the director of UFTI?

**Answer**: The relations between me and Lejpunsky are very good. In many questions that were extremely difficult for me Lejpunsky gave me great help and support. This support was appre-ciable in the period of my impending dismissal from UFTI by former director Davidovich and Gej, his deputy, when Lejpunsky took the matter up with Armand,[9] the head of the Scientific-Research Sector of Narkomtjazhprom, defended me in front of him and ensured that I

---

[9]Aleksandr Aleksandrovich Armand (1894–1967) was head of the Scientific-Research Sector of Narkomtjazhprom from 1933 to 1937. After Ordzhonikidze's death Armand lost his job, was expelled from the party, but no further harm was done to him.

remained deputy head of OSGO instead of being dismissed. In 1935 I was going to be expelled from the USSR. Upon my request Lejpunsky defended me with the organs of the NKVD, as a result of which the prospect of me staying longer in the USSR again improved. Lejpunsky showed considerable confidence in me and told me top secret NKVD information known to him and concerning me. From Lejpunsky I knew that the organs of the NKVD had proposed to expel me from the USSR. From him I learned and obtained a warning on the necessity to leave the USSR, since I would else be arrested. I obtained this warning from Lejpunsky at the end of 1936.

As a result of an ultimatum issued by Martin Ruhemann to the effect that, if I were expelled, he would also leave UFTI, Lejpunsky and Komarov went to the Scientific-Research Sector of Narkomtjazhprom, where they again defended me in front of Armand and for the second time secured my stay in the USSR to work at OSGO. Under pressure because of my prior summons to the NKVD before my arrest, about which I informed Lejpunsky in detail, telling him that the NKVD had charged me with espionage and subversion, Lejpunsky was forced to dismiss me, after first having again talked to Comrade Mazo, the head of the district administration.[10] But, apparently, his defence of me did not give any positive results. At my dismissal and before my departure, for which I started to prepare, Lejpunsky gave me a positive reference. In August 1936 Lejpunsky gave me a bonus of 1000 roubles. To complete the characterization of my relation with Lejpunsky I want to say that we visited each other at home, where we exchanged opinions on various questions. We spoke, albeit seldom, on political topics, in which respect Lejpunsky knew about my anti-Soviet and counter-revolutionary views.

**Question**: You declare that "Lejpunsky is a double dealer, who had taken up anti-party positions". What is the basis for your statements? Answer!

**Answer**: That Lejpunsky is a double dealer and has taken up anti-party positions I know from conversations with him on a number of political questions. For instance, in 1933 at Lejpunsky's apartment we discussed the situation in the rural areas and in Ukraine. Exchanging opinions Lejpunsky told me that the death-rate in the country had reached 11%. To my astonished question, "how do you know this?", Lejpunsky answered that he knew this from Bukharin.

In 1936–1937 I discussed the question of the Constitution with Lejpunsky. Upon my ironic comments and mistrust of the freedom of printing in the USSR Lejpunsky reduced this issue to the possibility of criticising the director of an institute or factory, while I had clearly ironically referred to the freedom of printing. In February 1937, after my summons to the NKVD, I told Lejpunsky a clearly counter-revolutionary slanderous anecdote in respect of the leadership of the USSR, to which Lejpunsky reacted with a smile without giving me a rebuff. In conversations with Lejpunsky one always felt that he did not agree with the formulation of the question on class vigilance and was extremely dissatisfied with the organs of the NKVD.

Upon his return from Moscow Lejpunsky told me that the atmosphere was so distrustful there that when two friends met they expressed their astonishment to each other that they had not yet been arrested. About the NKVD Lejpunsky said that they would prosecute him if he were to ask Balitsky[11] to investigate what is happening at UFTI. That was at the time when the counter-revolutionary group at UFTI became active.

---

[10]Solomon Samojlovich Mazo (1900–1937), head of the Kharkov District of the NKVD, committed suicide in 1937, leaving a note saying "Comrades, come to your senses! Where does such policy of arrests and extraction of information from the accused lead to?".

[11]Vsevolod Apollonovich Balitsky (1892–1937), People's Commissar of Internal Affairs of the Ukraine, executed in 1937. Like Shubnikov, the son of a bookkeeper and a gymnasium graduate. Arrested in June 1937 and shot in November, similar to Shubnikov. A difference is that he has not been rehabilitated. A petition to that effect was rejected in 1998. For one year the stadium of Dynamo Kiev was named after him; this stadium was then for a very short period named after Nikolaj Yezhov, and is now named after the football manager Valery Lobanovsky.

In February-March 1936 I expressed to Lejpunsky my counter-revolutionary fabrications that since the time of the murder of Comrade Kirov there had only now been established some democratic forms (creation of a complaints office and suchlike) in the USSR, to which Lejpunsky answered: "Yes, the comrades of the NKVD complain that the procurator checks them at every step." It should be noted that Lejpunsky avoided having openly counter-revolutionary conversations with me. To my numerous anti-Soviet remarks Lejpunsky did not at all react, nor did he give me any rebuff.

**Question**: What were Lejpunsky's relations with Bukharin?

**Answer**: I do not know very much about the relations of Lejpunsky with Bukharin.

**Question**: Tell us what you know.

**Answer**: The relations between them were good. In 1931–1932 I was present at one of Bukharin's meetings with Lejpunsky. From the nature of their conversation, the content of which I do not know since at the time I did not understand Russian, I gathered that the relations between them were good. Lejpunsky met Bukharin repeatedly as head of the Scientific-Research Section of Narkomtjazhprom. I do not know the nature of their conversations. Lejpunsky was especially close with Bukharin's deputy Ziskind, who considered him a close friend.

(…)

**Question**: Mention all members of the counter-revolutionary group founded by you and the persons recruited by you for counter-revolutionary activity.

**Answer**: As a confirmed rightist and not agreeing with the line of the Bolshevik party and the measures of the Soviet government on the above-mentioned issues, I founded a group of like-minded persons from the foreign specialists at UFTI. To this group I attracted: Fritz Houtermans, member of the German Communist Party, German subject, scientific worker at UFTI; Martin Ruhemann, English citizen, left liberal, scientific head at UFTI; Barbara Ruhemann, English citizen, scientific worker at UFTI; Charlotte Houtermans, member of the German Communist Party, German subject, works as secretary of the editorial staff and as corrector of the USSR physics journal, published at UFTI. Apart from the above-mentioned persons, Konrad Weisselberg and Tisza from the foreign specialists were also close to this group, but they did not enter the organisation and were not structurally connected with this rightist group I set up. In order to attract him to the rightist counter-revolutionary group I started to work on the member of the Bolshevik party, Pëtr Frolovich Komarov,[12] with whom I started to have political conversations in the middle of 1936. At first Komarov did not at all agree with me, later in the autumn he started to have political talks with me on his own initiative, trying to establish contact with me.

**Question**: What did Komarov say to you?

**Answer**: When in Moscow together with me on OSGO business he told me that Comrade Stalin was a vain man and did not react to numerous greetings at his address; later, at the time of the trial, Komarov told me that from all the accused he most liked Kamenev, who is an old party worker. Komarov also told me that when in the countryside in the period 1932–1933 he had an order to arrest everybody who was counting the dead, the number of whom in his opinion could have been significantly less through appropriate measures. On my question to Komarov whether it is permissible for a member of the Bolshevik party to speak like that, he answered in the affirmative, declaring that in party circles this was accepted.

**Question**: Why did you not establish contact with Komarov?

---

[12]His arrest was authorised in the same telegram as the arrest of Shubnikov and Rozenkevich.

**Answer**: In spite of these counter-revolutionary conversations with Komarov, I did not trust him and decided to wait in order to check him out more thoroughly.

**Question**: When did you set up the counter-revolutionary rightist group at UFTI?

**Answer**: The rightist group at UFTI I mentioned was set up by me in 1935 in May-June. Before that I had taken Ruhemann under my political influence and in many conversations with him implanted my rightist views in him. In the spring of 1935, when Houtermans and his wife arrived in the USSR, I used the situation at the institute in connection with the counter-revolutionary activity stirred up by Landau's group and also attracted Houtermans. Beforehand I had informed Houtermans in detail about the situation at the institute, after which I attracted him to my group.

**Question**: On what basis did you set up the counter-revolutionary group of rightists at UFTI?

**Answer**: The basis for our association were the political conversations with Ruhemann and Houtermans and their wives on the position of foreigners in the USSR, on the lack of trust in them, the absence of a proper democracy in the USSR and suchlike. My expulsion from UFTI by the former director Davidovich also dates back to this time, which was used at the institute for starting counter-revolutionary activity and the disruption of scientific and industrial tasks given to the institute.

**Question**: Speak concretely, which tasks were given to UFTI in 1935?

**Answer**: In 1935 a number of defence-related tasks was given to UFTI by party and government. I cannot say concretely which topics the institute was precisely charged with, since I was not made familiar with secret work. Professor Slutskin, and the scientific workers Rjabinin, Sinelnikov and Shubnikov[13] were engaged in this work on defence-related topics at the institute.

**Question**: How did the counter-revolutionary sabotage activity of the rightist group at UFTI manifest itself concretely as regards the disruption of defence-related tasks?

**Answer**: The counter-revolutionary sabotage activity of the rightist counter-revolutionary group at UFTI manifested itself in:

1  Support of Landau's group, which actively participated in the struggle for the disruption of defence work at the institute.
2  In the struggle against the management of the institute, in particular against the director of UFTI Davidovich and his deputy Gej. This struggle resulted in the disorganisation of the work of the institute and diverted the attention of all scientific workers from the fulfilment of their everyday tasks.

**Question**: Who was in Landau's group? Mention the people who led the struggle against defence work at the institute.

**Answer**: I know that among the scientific collaborators of UFTI there was a group of people who were opposed to defence topics, which group in practice led the struggle for the disruption of defence tasks UFTI was charged with by the government. This group included Landau, Lev Davidovich, scientific head at UFTI, Korets, Moisej Abramovich, aspirant of Landau, Shubnikov, Lev Vasilevich, scientific head of the cryogenic laboratory at UFTI, and Rozenkevich, Lev Viktorovich, scientific collaborator at UFTI. This group led by Landau had influence on the scientific collaborators of the institute and, chiefly, on the Soviet young cadres – the theory group which was headed by Landau.

**Question**: What is the political direction of Landau's group?

---

[13]Shubnikov's name seems to be out of place here.

**Answer**: Judging from Landau's political convictions, the group led by him is a counter-revolutionary Trotskyist group.

**Question**: What do you know about Landau's Trotskyist convictions?

**Answer**: I know Landau to be a convinced Trotskyist since 1932. In numerous personal conversations with me in the course of 1932–1936 Landau openly expressed his counter-revolutionary Trotskyist views. He expressed a lack of confidence in the possibility of constructing socialism in the USSR because of the contradictions that existed in his opinion between the working class and the peasants. Landau explained the difficulties in the collectivization of the land by the opposition put up by the peasants against the Soviet state when carrying out this measure. Landau thought that the politics of the collectivization of the rural economy was a failure and would shortly be stopped. According to Landau there existed forced labour at the *kolkhozes* and this explained their low level and low harvest. Landau also said that the forced labour at the *kolkhozes* was also the reason for the high death-rate in the Ukraine in 1932–1933. Landau considered the Bolshevik party a bureaucratic party, a system which excludes the possibility of keeping revolutionaries in it. Democracy is totally absent in the USSR, one cannot freely criticise, express one's opinion, one must always wait for directions from above – that is how Landau always spoke. In the period of the past trials of the Zinoviev-Kamenev Trotskyist group and the anti-Soviet parallel centre Landau expressed his lack of faith in the reports published in the Soviet press and followed a clear line of incomprehension of these events. Landau always spoke of Comrade Stalin with irony, neither did he call him the "beloved *vozhd*", "father of the people" or suchlike.

**Question**: In which question did you show solidarity with the Trotskyist Landau?

**Answer**: I showed solidarity with Landau on questions of the Bolshevik party, on the absence of democracy in the USSR and on the question of the trial of the Trotskyist-Zinovievite group and the parallel anti-Soviet centre; I disagreed with Landau on the question of the peasantry or collectivization. Nor did I agree with his ironic attitude towards Comrade Stalin.

**Question**: You do not speak the truth. You yourself told Lejpunsky a clearly counter-revolutionary anecdote on the *vozhd* Stalin in an even sharper spirit than Landau's irony in respect of Stalin.

**Answer**: Yes, I confirm that. However, with me that was an isolated case, and Landau's ironic attitude towards Comrade Stalin was systematic.

**Question**: What do you know about the counter-revolutionary Trotskyist attitude of the other members of Landau's counter-revolutionary Trotskyist group?

**Answer**: Korets, Moisej Abramovich, who was under Landau's ideological influence showed complete solidarity with the latter's Trotskyist views. I cannot characterize the political convictions of Shubnikov and Rozenkevich. I know that Shubnikov is a reactionary and an anti-Soviet inclined person, alien to the measures of Soviet power and the party. Both Shubnikov and Rozenkevich are under Landau's influence, who is a great authority to them.

**Question**: What was the common political platform of the rightist counter-revolutionary group founded by you and Landau's Trotskyist group at UFTI?

**Answer**: There was no political platform uniting these two directions. In practice both groups acted against the management of the institute, demanding from the highest Soviet party organisations that the management be removed. In the course of the struggle Landau posed the question that the institute should only engage in questions of pure science, and not in applied technical defence-related work. In these questions the two groups at UFTI acted together, fully agreeing with the demands of Landau's group.

**Question**: Which defence-related work was sabotaged at UFTI in 1935 and what did the joint counter-revolutionary Landau-Weissberg sabotage group achieve in practice?

**Answer**: I do not know which defence-related work was sabotaged at UFTI in 1935; as I already have said, I had no access to secret defence-related work. In practice our joint activity led to the creation of an atmosphere at the institute in which it was impossible to work. As a result the total plan of scientific work at UFTI in 1935 was wrecked. The director of the institute Davidovich and the management were removed. There was the persecution of Professor Slutskin who engaged in defence-related topics, and the expulsion of the specialist worker Rjabinin from the institute.[14]

**Question**: Which role did Lejpunsky play in the joint struggle of your group and Landau's counter-revolutionary group against defence-related work and what was his attitude to this struggle at UFTI?

**Answer**: In the course of 1934–1935 in the period of this struggle Lejpunsky was on a foreign *komandirovka* in England; he was kept informed by me and Landau, mainly, on the question of Davidovich's directorship. In his replies, Lejpunsky advised me and Landau to turn to Comrade Armand at the Scientific-Research Section of Narkomtjazhprom. Upon his return Lejpunsky decided to keep Landau among the cadres at UFTI. Lejpunsky did not understand the political nature of our struggle and thought that there was no counter-revolutionary group acting against defence tasks at the institute.[15] It is important to note that one of the demands of our joint group was the return of Lejpunsky from abroad and his appointment as director of UFTI.

Weissberg confesses to be a 'confirmed rightist' and not to agree with the Bolshevik party and the Soviet government on a number of questions and to have founded a counter-revolutionary group of like-minded persons from the foreign specialists at UFTI: Fritz Houtermans, Martin Ruhemann, his wife Barbara, and Charlotte Houtermans, while he mentions Konrad Weisselberg and Tisza from the foreigners as being close to this group without being actual members.

The designation 'counter-revolutionary organisation' or 'counter-revolutionary group' appears very frequently in NKVD documents. One may wonder what is actually meant by these terms. In his book, *Conspiracy of Silence* (Ref. [2]), Weissberg gives an idea on page 149 where he lets his interrogator Reznikov say: "*In our country counter-revolutionary organisations with an elected chairman and proper membership cards and so on are impossible.*" He then goes on to explain: "*... if three of you are in a room talking about something or the other* (of a political nature) *and a fourth man comes in and then you change the subject, you three belong to the organisation and he doesn't.*" The activity such an organisation engages in is in any case counter-revolutionary (or can be considered counter-revolutionary) if it concerns any activity mentioned in the notorious Article 58 of the RSFSR penal code (Article 54 of the UkSSR penal code) which covered all forms of remotely political crime, but was then draconically interpreted

---

[14]This would suggest that Rjabinin had already left UFTI before Weissberg's arrest while in his book *Conspiracy of Silence* he says to have seen him at UFTI a few weeks before.

[15]This agrees with the information in Chap. 7, namely that in letters to Lejpunsky Landau and Shubnikov had written that they seriously considered moving to Moscow, and that Lejpunsky did not really appreciate the seriousness of the situation at UFTI.

following a Supreme Court ruling of 2 January 1928: counter-revolutionary offences were committed *"when the person who committed them, although not directly pursuing a counter-revolutionary aim, wittingly entertained the possibility of this arising or should have foreseen the socially dangerous character of the consequences of his actions"* (Ref. [5], p. 283).

Weissberg suspects that the interrogator presented the idea of an illegal organisation as a comparatively harmless matter in order to make him confess that he was a member. For, who isn't or has never been in a situation as described by Reznikov? Nevertheless from all the interrogations that will be quoted in the following pages it appears that for the NKVD such a fairly harmless gathering is indeed a counter-revolutionary organisation, at least when the discussions of the group are of an anti-Soviet nature.

Weissberg further says in his statement that the counter-revolutionary sabotage activity of his group manifested itself in support for *"Landau's group, which actively participated in the struggle for the disruption of defence work at the institute and in the struggle against the management of the institute, in particular against the director of UFTI Davidovich and his deputy Gej. This struggle resulted in the disorganisation of the work of the institute and diverted the attention of all scientific workers from the fulfilment of their everyday tasks."* He also mentions Korets, Shubnikov and Rozenkevich as members of Landau's group. He has known *"Landau as a convinced Trotskyist since 1932. In numerous personal conversations with me in the course of 1932–1936 Landau openly expressed his counter-revolutionary Trotskyist views. He expressed a lack of trust in the possibility of constructing socialism in the* USSR *because of the contradictions that exist in his opinion between the working class and the peasants."* And *"Shubnikov is a reactionary and anti-Soviet inclined person, alien to the measures of Soviet power and the party."*

In general the events recalled by Weissberg in the interrogation protocol agree with the history of UFTI in the thirties as described in Chap. 7 and it also reveals what the NKVD meant by 'counter-revolutionary', Trotskyist or sabotage activities. It is not as absurd as it seems, nor something they have invented, but is indeed related to real events. It all relates to the conflict that was simmering under the surface at UFTI from early on and came into the open around 1935 when the institute was given defence-related tasks to perform, which some of the scientists did not like, and indeed actively opposed (interpreted as sabotage). Perhaps they did this with good reason from their (or our current-day liberal) point of view, but it cannot come as a surprise either that those in power at the time in the Soviet Union did not like such an independent attitude and thought it necessary to stamp it out. It is also interesting to see that the arrests of Shubnikov and the other UFTI scientists had already been decided upon and that from the very beginning the line of the NKVD was the exposure of opposition to defence-related work at UFTI. It is a theme that returns in most of the interrogation protocols.

Weissberg spent the next three years in prison in Kharkov, Kiev and Moscow. On 31 December 1939 a special session of the NKVD took the decision to evict him as an undesirable alien from the USSR, after which he was handed over to the

Gestapo as part of the prisoner exchange in the Nazi-Soviet pact. After some time in prison the Gestapo released him into the ghetto of Krakow in Poland, from where he escaped into the Polish underground. He survived the war using the identity papers of a certain count Cybulski who stayed in the US and whose papers he had obtained from Cybulski's wife, who later became his partner (that is why he started to use the name Weissberg-Cybulski). It is likely that the Soviets treated him reasonably well compared to other prisoners since prominent physicists such as Albert Einstein, Jean Perrin and Irène and Frédéric Joliot-Curie wrote letters on his behalf to the Soviet authorities.

In 1958, in connection with a revision of Weisselberg's case, a summary of Weissberg's file was prepared by a KGB official (Ref. [1], pp. 216–217). He states that Weissberg confessed to being guilty of engaging in provocative action and of collaboration with the German intelligence services, to whom he betrayed Austrian and German communists. *"In 1931 on the instruction of the agent Houtermans of the German secret service he was transferred to the Soviet Union for espionage work against the USSR. After his arrival in the USSR he started to work at UFTI. He worked at UFTI while collecting espionage information for German intelligence"* and built a counter-revolutionary, and subsequently a terrorist group. The document also states that on 1 June 1937 Weissberg testified that at UFTI he had set up a counter-revolutionary group consisting of the persons mentioned above, but also that on 28 May 1939, so two years later, he retracted in full the earlier testimony on his counter-revolutionary activity and the counter-revolutionary activity of the persons he had mentioned. *"My testimony of 1 June 1937 neither corresponds with the truth, nor do I confirm it. The statements were given by me under the effect of prison conditions... I did not recruit anyone from the foreigners into a counter-revolutionary group and I had no disagreements with measurements of the Soviet state."* One wonders what had happened between June 1937 and May 1939.

Konrad Weisselberg, the other foreigner arrested early in March 1937, lived in the same apartment as Weissberg and was one of his best friends.[16] They knew each other from Vienna. He had come to the USSR and to Kharkov in May 1933 on a tourist visa, at which point Weissberg had promised to secure work for him in the USSR. In July 1934 he returned to Kharkov on the invitation of the director of the Research Institute for Carbochemistry in Kharkov,[17] where he worked as a scientist. In July 1936 he was dismissed under the pretext of staff reduction, although a short time before this he had earned himself the distinction of 'Stakhanov worker', i.e. shock worker or model worker. In one of the interrogation protocols he says that he was dismissed at his own request. This agrees with the fact that there does not seem to have been any party political reason behind his dismissal, as the Communist

---

[16]The person called Marcel in Weissberg's book is in fact Weisselberg. The statements made about 'Marcel' by Weissberg and the entry about him in [4] are about all we know about Weisselberg who has all but been blotted out from the history of mankind.

[17]It had been founded in 1930 and still exists as the Ukrainian State Research Institute for Carbochemistry.

Party leadership had a high opinion of him and pleaded for him to stay in the Soviet Union. After some months of unemployment (he had been denounced to the NKVD and nobody wanted to hire him) he found employment as a kind of free-lance worker at UFTI. In 1937 he became a Soviet citizen, as he feared expulsion and did not want to leave his Ukrainian wife and young son behind. He was accused of having entered the country from Austria as a political immigrant in 1934, of having established criminal contacts with the German intelligence agent Weissberg and of having become a member of the counter-revolutionary group led by Weissberg. It is stated that *"it has been established that Weisselberg arrived in the Soviet Union with the aim of carrying out subversive work against the Soviet Union. In his criminal activity he was connected with a number of other persons"* (Ref. [1], p. 217), without elaborating what this subversive work or criminal activity actually was. Weisselberg's interrogation protocols show that the NKVD wanted to establish a connection between Weissberg/Weisselberg and the Anti-Soviet Trotskyist Centre of Pjatakov and others, who had been executed early in 1937 after the second Moscow show trial in January 1937. Some of the defendants in that trial, in particular Stanislav Antonovich Ratajchak (1894–1937), who from 1932 was Deputy People's Commissar of Heavy Industry, had had dealings with Weissberg. The visits of George Placzek at UFTI were also a frequent topic. During these visits Placzek, who was called a Trotskyist by the NKVD, had sharply criticised the show trials and other events in the Soviet Union, criticism which had been reported to the NKVD. The interrogations resulted in Weisselberg being accused of the crimes outlined in Articles 54-4, 54-6, and 54-11 of the Penal Code of the UkSSR, although during the interrogations, as reflected in the interrogation protocols in Ref. [1], he does not seem to have been accused of anything in particular. In the concluding indictment he is then suddenly accused of being a member of Weissberg's counter-revolutionary group:

(…) He arrived in the USSR in 1934, took Soviet citizenship in 1937. Upon his arrival from Austria to the USSR in 1934 Weisselberg contacted the Austrian citizen Weissberg (arrested) whom he knew from Austria and has been exposed as an agent of German intelligence. According to statements of the Trotskyist Anders[18] arrested in Voronezh, Weisselberg was a participant of the counter-revolutionary group of foreign specialists at UFTI. The group was led by the spy Weissberg. In his statements Weissberg has confirmed the existence of the counter-revolutionary group of foreign specialists, stating that Weisselberg was close to this group. Weisselberg has personally stated that he has participated in counter-revolutionary conversations with the Trotskyist physicist Placzek who had come to the USSR and told him about his connections with Bukharin. According to a declaration of the German secret service spy Fomin (arrested), Weisselberg was part of an organisation spying for Germany and carried out sabotage work at the Carbochemistry Institute when he worked there. Weisselberg only admitted to participation in anti-Soviet conversations with the Trotskyist physicist Placzek; as a participant of a counter-revolutionary group he is exposed by the statements of Anders and partly by the statements of Weissberg, and as a German spy by the statements of Fomin.

---

[18]Nothing is known about this person, nor about any statement he made.

On the basis of the above Weisselberg, Konrad Bernardovich, born in 1905, Bârlad, Romania, Jew, son of a large timber merchant in Austria, non-party, citizen of the USSR, from 1922 to 1925 a member of a bourgeois youth organisation, from 1928 a member of the association of socialist students, and from 1928 to 1934 a member of the social-democratic party of Austria, up to his arrest working as a scientific collaborator at UFTI, is charged with being a participant of a Trotskyist organisation and with engaging in espionage activity, i.e. of the crimes laid out in Articles 54-4, 54-6, and 54-11 of the Penal Code of the UkSSR

Article 54 of the Penal Code of the UkSSR corresponds to the infamous Article 58 of the USSR Penal Code. It shows up in the indictments of all the accused.[19]

Article 54-1 defined counter-revolutionary activity: A counter-revolutionary action is any action aimed at overthrowing, undermining or weakening the power of workers' and peasants' Soviets [...] and governments of the USSR and of Soviet and autonomous republics, or at undermining or weakening the external security of the USSR and the primary economic, political and national achievements of the proletarian revolution.

Article 54-11 reads: Any kind of organisational or supporting actions related to the preparation or execution of the above crimes is equated with the corresponding offenses and prosecuted by the corresponding articles.

On 28 October 1937 Weisselberg was, without further ado, condemned to death (the death warrant was signed by the General Commissar of State Security Yezhov and USSR Procurator Vyshinsky[20]). He was executed on 16 December 1937, just 32 years old and the only foreigner (although he had become a citizen of the

---

[19]Article 54 had 14 basic components: Article 54-1: high treason; Article 54-2: bourgeois separatism and nationalism; Article 54-3: being an accomplice to the enemy; Article 54-4: being an agent of the world bourgeoisie; Article 54-5: inciting a foreign state to declare war on the USSR; Article 54-6: espionage; Article 54-7: conducting subversive activities; Article 54-8: terrorism; Article 54-9: committing acts of sabotage of the transport, communications or water supply system; Article 54-10: conducting anti-Soviet propaganda and agitation; Article 54-11: being a member of an anti-Soviet organization; Article 54-12: not informing the Soviet authorities about forthcoming or already perpetrated counter-revolutionary crimes; Article 54-13: committing crimes against the working class or revolution movement; Article 54-14: committing sabotage and not fulfilling duties in order to weaken the Soviet power. (http://www.volhynia.com/his-nkvd.html)

[20]It is somewhat remarkable that all sentences in the UFTI cases have been signed by Yezhov and Vyshinsky, the highest officials in the land. On 30 July 1937 Troikas were set up on Stalin's instructions with the power to impose the death penalty. They soon became Dvoikas ('out of revolutionary urgency') consisting merely of two members, one from the NKVD and one from the Prosecutor's Office. All provinces and republics had such Troikas or Dvoikas established which, in principle, dealt with all local cases. Yezhov and Vyshinsky fulfilled this role of Dvoika at the centre. For UFTI it would have been more logical if the local Ukrainian Dvoika had passed the sentence in these cases. Apparently, UFTI was considered so important that Moscow remained closely involved. See Ref. [5], p. 286.

USSR[21]) to suffer this fate in the UFTI affair, but why he had to suffer this harsh fate remains unclear, as he had not played any role at all in the conflict at UFTI and could not, objectively, be blamed of anything even remotely anti-Soviet or of the slightest disloyalty to the Soviet regime. On the contrary, out of idealistic motives he had exchanged his comfortable life in an affluent family in Austria for a meagre existence in the Soviet Union. Nor is it in any way clear why he was so swiftly and harshly dealt with, while Weissberg was kept in prison for more than three years before being forcibly removed from the country.

More than twenty years later, in 1959, Weisselberg's sentence was declared void by the Tribunal of the Kiev Military District. The relevant part of the decision (Ref. [1], p. 224) reads as follows

> (…) Weisselberg was arrested by the NKVD for the Kharkov District on 4 March 1937, and on 28 October of the same year sentenced by the NKVD and the USSR Procurator to the highest measure of punishment. This decision was death by execution. According to the concluding indictment Weisselberg was charged with having arrived in the USSR in 1934 as a political emigrant from Austria, with having established a criminal connection with the agent of German intelligence Weissberg, and with having entered a counter-revolutionary group led by Weissberg. A check carried out on the case has established that in 1937 Weisselberg was groundlessly convicted for the following reasons: as can be seen from the material of the case, the accused Weisselberg did not plead guilty and the entire accusation was based on non-concrete and doubtful statements of Fomin and Anders, arrested in other cases. Fomin and Anders-Kon declared that Weisselberg entered a counter-revolutionary group which was led by Weissberg. From material of the archived inquiry file on the accusation of Weissberg it can be seen that he first pleaded guilty, after having declared that Houtermans had recruited him for espionage activity for Germany. But in an interrogation in 1939 he fully retracted the statements given by him. Houtermans, disregarding the testimony of Fomin and Anders, neither pleaded guilty. The KGB for the Kharkov District also has no compromising materials in relation to Weisselberg. According to a communication in the secret Central State Archive, Konrad Weisselberg, doctor of chemistry, is mentioned in a secret list, compiled by the apparatus of the central administration of the Imperial State Security of Germany as a person who presents a serious hazard to fascist Germany and in the view of German state security should be arrested. The information that Weisselberg belonged to an agency of the German secret service was not checked. From the material of the case it is clear that in carrying out the inquiry the NKVD of the Kharkov District violated socialist legality, no accusation was brought against Weisselberg and the accused was not familiar with the material of the completed inquiry. On the basis of the above I would propose:
>
> To submit an application to the judicial organs for quashing the decision of the NKVD and the Procurator of the USSR of 28 October 1937 in respect of Weisselberg K.B. and to close the case on him.

Hence, there was no ground for his conviction as the entire accusation was based on doubtful statements of defendants in other cases and the NKVD had no compromising material on Weisselberg. Moreover he was listed by German state

---

[21]Becoming a citizen of the USSR did not work in someone's favour as regards treatment by the NKVD it seems. There are many cases of foreigners who became citizens of the USSR and were subsequently murdered, while those who didn't, such as Weissberg, were in the end expelled from the country.

security as a person who presents a serious hazard to fascist Germany and from the viewpoint of German state security should be arrested. The information that Weisselberg belonged to an agency of the German secret service was not checked, the NKVD had violated socialist legality, no accusation had been brought against Weisselberg and the accused was not familiar with the materials of the completed inquiry.

For what it's worth, the result was that Konrad Weisselberg was posthumously rehabilitated. Nothing is known about the fate of his wife and son.

## Testimonies by (Former) UFTI Staff

How careful the NKVD was planning the UFTI case is also apparent from the fact that in April/May 1937, while the Weissberg/Weisselberg case was in progress, the NKVD called in employees of UFTI for questioning. The protocols of the inter- rogations of Aleksandr Akhiezer, Stepan Trofimovich Shavlo, and Vladimir Gej are reproduced in Appendix 2. There must have been many more testimonies, but only the testimonies discussed below were included in the files. Rozenkevich was also interrogated but in a memorandum in his file (see Appendix 3) it is stated that this had already happened in April 1936: "*[o]n 11 April 1936 Rozenkevich was secretly detained and interrogated by us*", so even before the trouble with Landau's dis- missal from Kharkov University started. This date is probably not correct, since if it is, Rozenkevich was interrogated in the aftermath of the Korets affair, who had been released just half a year before, when the NKVD were still licking their wounds from their defeat in that affair. In the following I will suppose that it is a misprint and should be April 1937, although the date is repeated in a memorandum on Shubnikov.

Akhiezer's testimony, dated 25 April 1937, is exceptional as it does not really incriminate anybody and is in essence a rather meek and innocent statement. He does not volunteer any information, states only well-known facts and is very summary in his answers. There is no talk in his testimony of any Trotskyist or counter-revolutionary conduct or organisations. He only relates his own partici- pation and that of Shubnikov, Gorsky, Lifshits, Pomeranchuk and Brilliantov[22] in the collective protest against the threatened dismissal of Landau from his profes- sorship at Kharkov University. He admits that in the terms of the NKVD this protest is seen as an outright anti-Soviet demonstration. He also recalls that they were called to Kiev to talk with the Ukrainian Commissar of Education, Zatonsky. All these are facts which cannot have come as a surprise to the interrogator who must have been fully familiar with them already. It also seems that his interrogator did

---

[22]Brilliantov is mentioned in several statements by various people, but he was never arrested; he was probably not targeted as he was not the leader of a group, laboratory or department. In 1937 Obreimov sent him to Kapitsa's institute in Moscow for safety reasons. That action possibly saved him.

not press him very much or threaten him with unpleasant consequences if he did not volunteer any 'useful' information. He was interviewed by the security officer Vesëlyj, a rather odd name for a Chekist as it means cheerful in Russian, but he may actually have been a very pleasant man. Akhiezer, who was only 26 at the time, was further left in peace and went on to become one of the most important Soviet theoretical physicists of his generation.

As mentioned in Chap. 7, in 1934 at an institute meeting discussing the outcome of the investigation into the UFTI party organisation, Shavlo had declared that he could not see the difference between the Hitler and Stalin dictatorships, an insight which at that time was not given to many, or if it was it was certainly not made public. Now, on 26 April 1937, the NKVD undoubtedly reminded him of this and urged him to testify 'truthfully', so that his earlier careless utterings could be forgiven. He did not disappoint them and his testimony was probably the centre-piece in the imminent case against Gorsky. He declared that "*at the institute Gorsky was always seen as an anti-Soviet, anti-social individual*" and that that opinion was well-founded, claiming also that in Leningrad Gorsky was part of the so-called 'jazz-band' of Landau, Ivanenko and Gamow, a perfectly harmless group which, according to Shavlo, had to be disbanded by the organs of the GPU. Gorsky, according to Shavlo, was always applying 'strike' methods to obtain what he wanted from the management, such as protesting against the insufficient living space of his co-workers. Shavlo has Gorsky say all kinds of anti-Soviet statements, such as:"... *Down with Soviet order in science.*" At the Kharkov Machine-Building Institute in 1934–1935 Gorsky was forced to stop teaching because of the poor quality of his lectures and he just declared:"... *for boors I will not read lectures*". At the end of 1935, together with Landau, Shubnikov, Brilliantov and others, Gorsky acted against the introduction of socialist competition in the work of the institute, declaring that "... *socialist competition in science is unacceptable, go to the factories and compete there...*" In 1935 when Korets was dismissed from UFTI and arrested by the NKVD, Landau allegedly stopped work at UFTI in protest, and Gorsky held an "Italian" strike (*work-to-rule*) for several days, coming to work, but doing nothing, and wholly agreed with Landau. Gorsky joined the collective protest made to Narkomtjazhprom against Korets's dismissal. Gorsky also took part in organising help for the family of the arrested Korets by collecting funds from 'common friends'; such help to the family of a disgraced person, and 'enemy of the people' was disapproved of and often meant trouble with the security forces. Gorsky declared to Shavlo in the presence of his wife: "*Trotsky gave the correct slogan that you need to sandpaper communists and now they ask us, non-party people, to give the sand and we will.*" The protest against Landau's dismissal is also mentioned by Shavlo. Shavlo claims that Akhiezer declared to him that Gorsky gave him directions to play a double deal, that is, to admit that the collective protest was an error, but in essence not to back down from his convictions. Akhiezer does not mention this in his own testimony. Shavlo goes on to say that Gorsky, Landau and Shubnikov systematically harassed members of the Ukrainian Communist Party and the Komsomol. Thus, Landau got rid of Pjatigorsky because the latter acted in the court case against Korets as a witness of the prosecution. Gorsky forced

Braude[23] and others to leave UFTI although they were capable people. Shavlo also alleges to have personally been treated badly by Gorsky when in 1936 Gorsky did not allow him to use the X-ray equipment for a year, charged him to do metal work, and spread rumours about Shavlo being unsuitable for scientific work and so on, impeding the completion of his diploma work in all kinds of ways. For two years Gorsky harassed Strelnikov, the member of the Ukrainian Communist Party, who was the first in the Soviet Union to construct a powerful X-ray tube. Gorsky spread rumours that Strelnikov was a dimwit, an untalented person, and in 1936, using Strelnikov's blueprints, wrote and published a paper, issuing it as his own, on the work of Strelnikov's X-ray tube. This plagiarism was unmasked and published in the journal "*Uspekhi Fizicheskikh Nauk*" in 1936.[24] As a result Strelnikov was forced to leave UFTI.

At the time of Gej's interrogation he had already left Kharkov for Leningrad, after the disaster with Davidovich's directorship, to whom he acted as deputy until Lejpunsky's return. He was now working at Scientific Research Institute no. 9.[25] The interrogation (about a month later than the two above) was conducted by Reznikov, a second lieutenant of the State Security Administration who was also involved in Weissberg's case and according to Ranjuk [6] the interrogation took place in Gej's apartment in Leningrad. As is clear from the statement, Gej was very eager to testify and implicate as many people as possible. He was given the opportunity to vent all his frustration and anger on the people who had made his life miserable a few years before and had forced him to leave Kharkov for a much lower position in Leningrad. The investigator and witness worked very fruitfully together and Gej's testimony must have been a gold mine for the NKVD. It is possible, and actually very likely, that the result of the interrogation was a joint effort of the interrogator Reznikov and Gej to make sure that the testimony would agree with the 'theory' the NKVD had in mind or wanted to build up about UFTI. So Gej believed that at UFTI two identical counter-revolutionary groups existed, the very same that Reznikov had already created in his interrogations of Weissberg. They consisted of the usual suspects: Landau, Korets, Obreimov, Shubnikov, Trapeznikova, Rozenkevich in one group led

---

[23]Semën Jakovlevich Braude (1911–2003), pursued his higher education at Kharkov University, receiving his undergraduate degree from the Physics and Mathematics Department in 1932. He then joined the staff of the Laboratory of Electromagnetic Oscillations (LEMO) at UFTI, and also began graduate work at Kharkov University. His mentor was Abram Slutskin. In an interview with him in 1988 he does not mention Gorsky at all, nor that he at some time had to leave UFTI for one reason or another. Upon a question from the interviewer A.A. Kostenko as to how he can explain that Slutskin's laboratory was the only one that was spared any arrests in 1937–1938, he gives a rather evasive answer (rian.kharkov.ua/library/exposition/file/braude.doc).

There are a few other scientists-engineers called Braude (notably the brothers Girsh (1906–1992) and Boris Vulfovich (1910–1999)) who worked on radio and television technology, but were never at UFTI.

[24]I have been unable to find anything about this in UFN.

[25]It had originated in 1935 as part of the Leningrad Electrophysical Institute (directed by the electrotechnician Aleksej Alekseevich Chernishev (1882–1940)), which in its turn had split off from LFTI in 1931.

by Landau, and in the other group the foreign specialists Weissberg, Houtermans, Martin and Barbara Ruhemann, Charlotte Schlesinger,[26] and Tisza. This second group was led by Weissberg. The two groups were assigned proper labels, one being counter-revolutionary rightist and the other one counter-revolutionary Trotskyist. They were closely connected and jointly carried out counter-revolutionary sabotage work. Their aims included: *"the disruption of the government tasks obtained by UFTI on defence-related topics"* and *"directing the work at UFTI in the future towards purely theoretical research, and no practical activity on applied and defence-related topics"*. He goes into considerable detail, naming the tasks Rjabinin of Shubnikov's group and Slutskin's laboratory were charged with, some of which were frustrated by the sabotage action of the two counter-revolutionary groups, and he concludes that *"[t]here is no doubt that sabotage at UFTI was significant"*. Everything the members of these two groups did was interpreted negatively by Gej and Reznikov, even the use of legal means such as appealing to state and party bodies was interpreted as sabotage. They are accused of harassing pro-Soviet scientific workers and people who carried out defence work, in particular also Slutskin and his collaborators. Gej gives negative personal assessments of the various members of the two groups, also mentioning Shubnikov's sailing trip in 1921 and his subsequent stay in Germany. For further details see Gej's testimony, which has been reprinted in Appendix 2. But again there are no outright lies in his statement; there was certainly no love lost between Slutskin's group and those of Weissberg and Landau. There was a conflict between these two sides, as well as with individual workers at UFTI. Examples are Strelnikov and Rjabinin, who indeed were forced to leave. At one point the situation between Landau and Rjabinin became so bad that they even came to blows. It is very unfortunate that no testimony is available from Rjabinin, who was involved in much of the splendid work, such as the discovery of type-II superconductivity, that was performed in Shubnikov's laboratory in those years. This is actually very surprising as one would have expected him to be one of the first to be called in for questioning. His testimony might have been very useful to the NKVD and could have shed some light on the conflict he had with Shubnikov and Landau, but perhaps he refused to cooperate. He indeed left UFTI after this conflict, but must have done so only in 1937, as Weissberg claims in his book to have spoken to Rjabinin just before his arrest (Ref. [2], p. 43). Weissberg also says that he is not sure that Rjabinin is not an informer of the NKVD. There is however no evidence for this.[27]

---

[26]A playful touch is that Gej mentions Charlotte Schlesinger as a member of the second group. She was not a foreign specialist, but a distinguished pianist, who played in the Kharkov philharmonic, and had come to Kharkov with Houtermans and his wife Charlotte Riefenstahl (Ref. [7], p. 37). She was not working at UFTI, and although living on the premises in Houtermans' apartment can hardly have been in a position to wreck the work at UFTI.

[27]However Dubovitsky (Ref. [8], p. 216) states that in 1936 Rjabinin first went to the Institute of Physical and Chemical Research in Leningrad, and only after the liquidation of that institute in 1940 to Semënov's Institute of Chemical Physics, where he worked in the department of July Khariton. The institute moved to Moscow in 1943.

# References

1. Ju.V. Pavlenko, Ju.N. Ranjuk and Ju.A. Khramov, *"Delo" UFTI 1935–1938* (The "UFTI" Case 1935-1938) (Feniks, Kiev, 1998).
2. A. Weissberg, *Conspiracy of Silence* (Hamish Hamilton, London, 1952).
3. V.V. Kosarev, Fiztekh, Gulag i obratno (belye pjatna iz istorii leningradskogo Fiztekha) (*Phystech, Gulag and back (white spots from the history of the Leningrad Physico-Technical Institute*) in: *Chtenija pamjati A.F. Ioffe* (Readings in memory of A.F. Ioffe) (Nauka, St. Petersburg, 1993), 105–177.
4. Barry McLoughlin and Josef Vogl, … *Ein Paragraf wird sich finden. Gedenkbuch der österreichischen Stalin-Opfer (bis 1945)* (Dokumentationsarchiv des österreichischen Widerstandes, Vienna, 2013).
5. Robert Conquest, *The Great Terror: A Reassessment* (Pimlico, London, 2008).
6. Ju.N. Ranjuk "Delo UFTI" Istoricheskie Kommentarii k knige Aleksandra Wajsberga "Obvinjaemyj" (*"The UFTI Affair" Historical Comments to Alexander Weissberg's book "The Accused"*), Journal 22, no. 117 (Tel-Aviv, 2013).
7. Edoardo Amaldi, *The Adventurous Life of Friedrich Georg Houtermans, Physicist (1903–1966)* (Springer Verlag, Heidelberg, 2012).
8. F.I. Dubovitskij, *Institut khimicheskoj fiziki: ocherki istorii* (Institute of Chemical Physics: Historical Essays) (Chernogolovka, 1992).

# Chapter 11
# The UFTI Affair: The Case of Shubnikov, Rozenkevich and Gorsky

On 6 August 1937 Lev Shubnikov and Lev Rozenkevich were arrested. Vadim Gorsky was arrested more than a month later, on 21 September. According to Pavlenko et al. (Ref. [1], p. 226), the case of Shubnikov, Rozenkevich and Gorsky is without doubt the most impressive of all the cases preserved in the Kharkov archives, both because of the tragic nature of its outcome and because of the abundance of documentary material available from their files. The documents from the KGB files of Shubnikov, Rozenkevich and Gorsky have been collected in Appendix 3.[1]

The green light for the arrest of Shubnikov and Rozenkevich was given in a telegram dated 24 July 1937 and sent by V.I. Mezhlauk,[2] who after Ordzhonikidze's suicide had for a few months become the head of Narkomtjazhprom, and deputy prosecutor G.M. Leplevsky.[3] Both Mezhlauk and Leplevsky would survive Shubnikov by less than a year. Their telegram (Fig. 11.1) was definite proof that the wind blowing from Moscow had changed direction. The other person mentioned in the telegram is Pëtr Komarov, who was a party member and Weissberg's deputy and successor as head of the construction of OSGO. In Weissberg's confession discussed in the preceding chapter he was also mentioned as being involved in his counter-revolutionary group. Not much is known about him, only that he died in prison (Ref. [3], p. 136).

The basis for the telegram is not clear. Did the Kharkov NKVD need permission from Moscow for the arrest of these three UFTI employees, and from their employer, the head of Narkomtjazhprom? That seems unlikely but, if so, some

---

[1]All official documents from the files of the UFTI affair reprinted in Appendix 3 here have been translated by the author from [1]. They are also available on the Internet at http://www.ihst.ru/projects/sohist/document/ufti/ufti.htm.
[2]Valery Ivanovich Mezhlauk (1893–1938), Chairman of the State Planning Committee (Gosplan) from 1934 to 1937, before becoming Ordzhonikidze's successor for a few months. He was a native of Kharkov. Arrested on December 1, 1937; accused of treason, industrial sabotage, contacts with the German government (his mother was German), and heading a Latvian counter-revolutionary terrorist organization (his father was a Latvian nobleman).
[3]Grigory Moiseevich Leplevsky (1889–1938), deputy prosecutor of the USSR; his younger brother Israil (1893–1938) was also a high official in the NKVD.

© Springer International Publishing AG 2018
L. J. Reinders, *The Life, Science and Times of Lev Vasilevich Shubnikov*,
Springer Biographies, https://doi.org/10.1007/978-3-319-72098-2_11

**ТЕЛЕГРАММА**

Из Киева записано по пров.
Харьков, НКВД, Шумскому.

По делу «Катод-Кредо» – Шубникова Льва, Комарова Петра и Розенкевича Льва
арестуйте. Аресты согласованы Москве НКТП, Межлаук, прокурором Леплевским. НР
45804 Иванов.

Передано 24.07.37, 7 час 26 мин.

**Fig. 11.1** Telegram dated 24 July 1937 sent from Kiev to the NKVD official Shumsky in
Kharkov ordering the arrest of Shubnikov, Komarov and Rozenkevich and indicating the
agreement of Mezhlauk and Leplevsky

information from the Kharkov NKVD as to why an arrest was necessary must have
been communicated to Moscow. One would have thought that if permission for the
arrests were needed, it had to come from Yezhov's office. If the Kharkov NKVD
had needed permission for any arrest that was not run-of-the-mill their work would
have been virtually impossible.

Shubnikov, Rozenkevich and Gorsky were all three laboratory heads. Shubnikov
of the low-temperature laboratory, Rozenkevich, who after his spell as head of the
theory division had turned to experimental physics, was the head of the laboratory
for radioactive measurements, and Gorsky of the X-ray laboratory. They were all
accused of participating in a counter-revolutionary Trotskyist organisation of
wreckers. The documents reprinted in Appendix 3 follow a pre-determined pattern.
In all three cases it starts with a decision by a security officer about the desirability
of the person's arrest.

## The Shubnikov File

Trapeznikova recalls that in the spring of 1937 there were some signs of something
being amiss, probably because of Weissberg's arrest on 1 March. But that something was wrong should already have been clear from the turmoil surrounding
Landau's dismissal from Kharkov University half a year earlier. The fact that she
mentions spring 1937 also means that, if Rozenkevich was indeed 'secretly seized
and interrogated' in April 1936, so about a year before, as stated in some NKVD
documents, he must have kept silent about his interrogation by the NKVD.
Otherwise Shubnikov and Landau would have known already much earlier that
something was brewing. In any case, in January 1937, Landau himself had opted
for a safe way out and moved to Moscow, undoubtedly as he understood that things
were starting to go badly wrong at UFTI. He and Shubnikov must have talked about
it, but perhaps they were lulled into a false sense of security by the praise that was

**Fig. 11.2**  Last photograph of
Shubnikov (*from* B.I. Verkin
et al. (1990))

heaped on the low-temperature laboratory and its research at the session of the
Academy of Sciences in January. Trapeznikova writes: *"In the spring of 1937 Lev
Vasilevich and I submitted documents for obtaining our scientific degrees; his
degree was doctor of science and mine candidate. At the time the award procedure
was not the same as now. An authorisation of VAK[4] was obtained, and all papers
for a whole group of people, including us, went to the scientific council of Kharkov
University. When the award took place, it turned out that neither Lev Vasilevich,
nor I was in the list. It was explained that for some reason they had forgotten to
send our papers to the university."* Since all around them people were arrested, she
became worried and proposed to Shubnikov to leave for a distant part of the
country. But he refused to leave his cryogenics behind and went on holiday with
Landau to the Crimea. While he was absent, strangers started to call her on the
telephone; they all asked where Lev Vasilevich was. The telegram mentioned above
had probably arrived, it was dated 24 July, and the NKVD were now looking for
Shubnikov (Fig. 11.2). Trapeznikova answered one of the callers' questions in too
much detail. He then advised her not to talk in detail over the telephone with
strangers. *"In the morning of 6 August Lev Vasilevich returned from the Crimea.[5]
Suddenly at dinner time the phone rang. He was urgently summoned to the institute.*

---

[4]*Vysshaja attestatsionnaja komissija* (Higher Attestation Commission). Degrees were awarded
without the defence of a thesis.

[5]In the version of the story as told by Kosarev (Ref. [3], p. 134) it is suggested that Landau was
also in Kharkov at that time, having returned with Shubnikov from the Crimea, and even stayed at
his own apartment. After Shubnikov's arrest Landau was allegedly warned by Trapeznikova, so he
moved out and stayed in hiding at the apartment of his future wife Kora Drobantseva for a few
days after which he managed to escape to Moscow. If true, the NKVD missed a chance here and
were not as competent as they were claimed to be. But it is doubtful that this version can be
trusted. It would have been very careless of Landau to appear in Kharkov, the town he had left in a
hurry just half a year before to escape from the NKVD; moreover he had already started work at
Kapitsa's institute and probably no longer had an apartment in Kharkov. It is also unlikely that he
could stay in hiding for four days at his future wife's apartment without the NKVD pursuing him.
In her memoirs Kora Landau-Drobantseva (Ref. [4], Chap. 12) speaks of just one night after which
Landau went by night train to Moscow. Moreover Kosarev seems to muddle up the dates of
Landau's departure to Kapitsa in Moscow and Shubnikov's arrest.

*I said "Eat first". But he answered: "No, the matter is urgent". He jumped up and went to the institute. He never returned home"* (Ref. [5], p. 289).

On the day of the arrest his apartment was searched in the presence of his wife. A number of items were confiscated, including correspondence with Wiersma and with Trapeznikova from 1926–27 when he was alone in Leiden. These documents, which could contain interesting information about the connection between UFTI and Leiden in the early thirties and on Shubnikov's first year in Leiden, have never been returned and must still be in the KGB archives.

The first important document in Shubnikov's extensive NKVD/KGB file is a memorandum dated 5 August 1937, i.e. the day before the arrest, and signed by three NKVD officers. On the basis of this document it was decided to arrest Shubnikov. It states that Shubnikov should be kept in custody as he is *"a participant in a counter-revolutionary group of specialists at UFTI led by Professor Landau"* and that *"the task of this counter-revolutionary group (...) is to disrupt defence-related research at the institute and to disorganise all the work"*. Independent of how it is viewed, the latter accusation, although stretching the truth somewhat, is certainly partly true as has been shown in Chap. 7; Landau and Shubnikov were actively opposing such work. Subsequently he is accused of active participation in a counter-revolutionary group of foreign German specialists (engaging in subversion and espionage). Then the sailing trip to Finland in 1921 is mentioned, his stay in Germany and his return to Leningrad where he was closely involved with the most 'reactionary' part of LFTI, that is with Landau, Ivanenko and others. This is actually not true. Shubnikov hardly knew Landau in Petrograd, and only got to know him better when Landau came to Leiden in 1928 (see Chap. 4).

The NKVD seemed to know everything, even about events going back to the chaotic times of the civil war, either from their own files, since there can be no doubt that upon Shubnikov's return from Germany to Leningrad a file on him has been opened, or from Vladimir Gej's testimony. We have already seen from the statements of Gej and Shavlo that the NKVD had been preparing this case already for some time. Here they also state that *"in April of 1936* (should probably be 1937) *we secretly detained and interrogated the member Rozenkevich of the counter-revolutionary group, who confirmed the existence at UFTI of a counter-revolutionary group led by Professor Landau"*. Unfortunately no record of this interrogation of Rozenkevich is available, nor do we know why Rozenkevich was picked out for this particular interrogation. Was he perhaps considered the weakest link? In any case he seems to have said things which he later regretted as the memorandum says that later he retracted parts of his statements, but adds that *"[t]he repudiation by Rozenkevich of part of his statements on UFTI is of no decisive significance for the case, i.e. similar information except for the information on LFTI we also have from intelligence. It is quite clear that Rozenkevich retracted his statement on the instruction of Shubnikov and Landau, with whom he undoubtedly shared the outcome of his visit to the NKVD."* That he talked to Shubnikov and Landau is, however, not certain in my view, for in April 1937 Landau had already left for Moscow and, if he had talked about it to Shubnikov, the latter would have known already around April 1937 that the NKVD were building a case on him, and could perhaps have taken appropriate measures. In that case Trapeznikova's urging to leave, as recalled above, would also have been more pressing, but in any case Trapeznikova does not mention that she or her husband had heard anything from Rozenkevich. It remains a mystery why Rozenkevich gave in so easily to the NKVD. One would assume that he could have behaved as Akhiezer did, but he of course also had a wife and a young child and it may well be that he tried to protect them.

The memorandum also states that Rozenkevich mentioned himself and Shubnikov as members of this counter-revolutionary group, and that the group goes back to 1930 at LFTI where it had been organised by Ivanenko, Frenkel and Gamow. So here the connection with the LFTI case is established and again Frenkel is mentioned as one of the leaders. A further accusation is that this group caused a row with the director Davidovich and tried to get rid of those workers who were prepared to carry out defence-related work. The memorandum even mentions specific projects that were sabotaged by them, such as devices for high-altitude flight, and non-flammable gas for balloons.

A final accusation is that Landau and Shubnikov, when teaching at Kharkov University, allowed a number of anti-Soviet attacks against dialectical materialism. They refused to admit these errors and in protest resigned from the university together with a number of other UFTI employees who were teaching at the university. As set out in Chap. 7, this is not far from the truth and this conduct was qualified as an anti-Soviet strike.

The memorandum concludes that all of these events warrant Shubnikov's arrest. A similar memorandum also exists for Rozenkevich giving details of his interrogation in April 1937 and naming Landau, Frenkel and Ivanenko as members of the group that already existed at LFTI. In addition to the accusation of participating in a counter-revolutionary Trotskyist organisation of wreckers, Shubnikov was also accused of spying for Germany, for which the NKVD used his involuntary stay in Germany in 1922–1923, recalled in Chap. 3.

On 7 August Shubnikov was interrogated for the first time and denied all charges: "*I deny being guilty of participation in a counter-revolutionary-espionage-subversive organisation. I do not wish to give any testimony, since I do not consider myself to be guilty of this*", but on the very same day Shubnikov changed his decision and agreed to give testimony, admitting to be 'guilty':

> With this statement I admit my guilt before Soviet power and, if I am given the opportunity, I want in future through honest and conscientious work to make up for my guilt. Moved by this desire, I declare that I am a member of a Trotskyist sabotage group working within the walls of UFTI. I promise to testify honestly and exhaustively about my activities as well as about the activities of others. I hope that my voluntary confession will reduce the heavy punishment a little and will allow me to return to my work, which in our country is a matter of valour, honour and heroism.[6]

As Pavlenko et al. write, it is impossible to ignore this statement written in his own hand. So the question is what kind of pressure forced this honest and

---

[6]This confession by Shubnikov is actually very similar in tone and spirit to Zinoviev's plea in a letter to the Central Committee in 1933. See Ref. [2], p. 116.

courageous man to write what he wrote? It is unlikely that it was physical pressure, since Shubnikov 'confessed' already on the second day of his imprisonment. Shubnikov was a well-known scientist, he was not free from ambition. Is it possible that the 'engineers of the human soul' promised him scientific oblivion in case he did not confess? (Ref. [1], p. 233.) At that time his wife Olga Trapeznikova was in the last month of pregnancy of their first child, a son, who was born on 31 August and whom Shubnikov never saw.[7] It may be that threats were made at her address if he were not giving in; that she would also be arrested, and that the child would be taken away and put in a children's home under a different name, as was common practice with young children of 'enemies of the people', and would never know who his parents were. The NKVD is known to have used such tactics on other prisoners. Whatever the tactics were, for some reason he must have come to the realization that it was futile to resist, but why so quickly? Trapeznikova visited him once in prison, fairly soon after his arrest, and brought him some books. There was almost nothing they could talk about, she writes, it was impossible. Then they led him away. He went holding his arm behind his back. They could not say goodbye; could not even shake hands, nor come close to each other. She could only look at him, and he at her. It was the last time she saw him; she did not hear anything from or about him afterwards (Ref. [5], p. 290).

The confession was followed up the next day by a more extensive and detailed statement, in which he also names Obreimov, Lejpunsky, Korets, Landau, Rozenkevich, Gorsky, Brilliantov and Usov[8] as members of the group, naming himself and Landau as the prime movers of the group. Parallel to this group there existed, according to the statement, a second group of mainly foreigners led by Weissberg and also including Martin and Barbara Ruhemann, Houtermans and Tisza. The existence of these two separate groups with the composition as mentioned here is the same as in Vladimir Gej's testimony. The aims of the two groups mentioned by Shubnikov include: "*the development at the institute of theoretical work at the expense of work of a technical and defence-related nature*"; "*unwillingness to carry out work of a technical and defence-related nature*", "*grandiose construction of OSGO* (Weissberg's deep-cooling research station), *on such a scale that the premises and equipment cannot be fully used in the coming years*", and some other related activities. Again these are not fantastic, imagined undertakings, such as murdering party leaders, complicated espionage plots, blowing up the country or trying to overthrow the government, but rather detailed acts of steering the work at the institute into a direction that did not fit in with the perceived needs of the country at the time of the industrialization and the first five-year plan. They agree more or less with what we learned in Chap. 7 about the history of UFTI in the

---

[7]Pavlenko et al. (Ref. [1], p. 234) state that they have spoken to a witness, the wife of the glass-blower Petushkov, who claims that Trapeznikova told her that Shubnikov visited the maternity ward, having been driven there in a police van. Trapeznikova does not mention this in her reminiscences in Ref. [5]. She only says (p. 290) that she spoke with Shubnikov twice for two minutes on the telephone after the birth.

[8]This name only appears here. Nothing is known about this person.

early thirties, albeit of course that they are now criminalized. Then, on 11 August, an additional statement follows in which Shubnikov admits to espionage activities starting from the time of his yachting adventure in the early twenties and for which he says he was recruited in Berlin by Ilja Dessler. See Chap. 3 for details of this yachting trip and the subsequent events in Berlin. It is clear though that if he were a spy, he was not worth much in that capacity, for Dessler did not hear from him for three years, at least there is nothing to suggest that they were in contact between 1923 and 1926. It is rather unlikely that, if Dessler were really running him as a spy, he would have waited patiently for so long without carrying out his threat of reporting to the Soviets what Shubnikov had told him about the work at the Optical Institute. Only in the autumn of 1926, when he was on his way to Leiden, can he have met with Dessler in Berlin to provide him with information "*about a method for obtaining large crystals of salt from molten states to produce fine-grained emulsions*".[9] In the second statement he adds to this, probably because it was realised that the spy story as it stood was indeed not very credible, that Dessler had promised to communicate with him in the USSR through a trusted person, but that nobody had showed up. He further states that his espionage activity continued until he left for Leiden, so after this first meeting with Dessler it was all over.

All this is worked out further in a few very long statements, the last one dated 13–14 August. The first statement starts with factual information about all the people involved in counter-revolutionary activity; where they worked before coming to UFTI, who hired them, and so on, without levelling any charges against them. Then follows for each of these individuals the details about their sabotage activity in the form of the "*disruption of technical physics problems and related defence work*" under the slogan of the "*incompatibility of scientific work with applied work*"; the harassment and bullying of people who were prepared to carry out defence-related work. The statement reads as if it has not been drawn up by Shubnikov, as it is mainly written in the third person. Only later in the statement it becomes clear that Shubnikov is (or supposed to be) the author where he declares: "*As a result of my counter-revolutionary wrecking activity, the low-temperature laboratory, which contributed a number of discoveries of promising scientific value,[10] did not contribute anything to the economy and the defence of the country,*" and "*I informed Weissberg about the progress of work carried out by the laboratory. Politically I shared the counter-revolutionary views of the group on resolutions of the party, government and higher powers, and also on the interrelations between Marxist philosophy and physics. I was an active participant of the organised protest strike against L.D. Landau's dismissal from Kharkov University at the end of 1936.*" Then follows a list of the actual wrecking activity of the counter-revolutionary group:

---

[9]I also give the Russian here, as it is not clear to me what this relates to (probably still top secret): *o sposobe poluchenija bol'shikh kristallov solej iz rasplavlennykh sostojanij dlja poluchenija tonkikh èmul'sij.*

[10]It is hardly conceivable that this formulation was suggested by the NKVD interrogator. Why would such a person credit the accused with positive achievements, namely "discoveries of promising scientific value"?

(1) *The creation of an atmosphere of quarrelling and mutual mistrust among the scientific workers of UFTI. (...) This has also to do with the harassment of people engaged in special subjects. At the institute the opinion was put forward that Slutskin and his collaborators are inferior and ignorant physicists. Finally this relates to the expulsion of specialists who worked on special subjects. As an example I can point to my quarrel with Comrade Rjabinin (...). As a result of the quarrel Rjabinin was transferred to another laboratory while undoubtedly I had at my disposal all the means for resolving the conflict in a peaceful way.*

(2) *The closure of a laboratory that had an applied purpose. In 1936 the laboratory of ionic transformations, which was of great interest for modern electronics, was closed.*

(3) *The transmission of important discoveries of a defence nature abroad through publication in the scientific press.*

(4) *Deliberate delay in the utilization of achievements at the institute. Scientific achievements were not delayed, communication about them went quickly to the printer. But technical achievements were usually not brought to an end or they were delayed. Furthermore Strelnikov's X-ray tube, which was of great interest for factories, industrial laboratories and scientific research institutes, did not go into production either. It is true though that in this case the tube was not constructed to full perfection.*

(5) *Sabotage in the training of cadres, in particular for OSGO.*

(6) *Grandiose scale of deception of party and government, expressed in the well-known telegram announcing the splitting of the lithium nucleus.*

(7) *Strikes, disruption of work and other anti-Soviet actions.*

(8) *Furthermore, one of the main tasks of the institute is to be a school for training qualified physicists for factory laboratories and other institutes of the Union (...) The institute has the possibility to train at least 10 qualified physicists per year, but in fact does not train a single one. The institute has done enough as regards scientific research and practically nothing as regards technical research.*

(9) *Recruiting into the counter-revolutionary group of UFTI.*

As discussed in Chap. 7, the first point has much truth in it. There may, however, have been sound reasons for closing the laboratory mentioned in point 2. Choices must be made, each choice has its pros and cons, but it is obvious that not everything can be done at UFTI. Point 3 likewise seems to be a red herring; all publications had to be approved by the scientific council. Of course mistakes can be made, but there is no reason to see this immediately as wrecking (although the latter was indeed common practice in the Soviet Union in the late twenties and thirties); moreover, if such papers were written, then it proves that defence-related work was actually not shunned by the members of the counter-revolutionary group. It also applies to the NKVD that it cannot have its cake and eat it, as it were. Although, it seems that it can when reading the quote from Stalin given by Conquest (Ref. [2], p. 275): "*No wrecker will go on wrecking all the time, if he does not wish to be exposed very rapidly. On the contrary, the real wrecker will show success in his work from time to time, for this is the only means of staying on the job, of worming himself into confidence, and continuing his wrecking activity.*" So whatever you do, you are done for.

Regarding point 4, there may have been perfectly valid reasons for the fact, if there was one, that practical achievements could not be put into practice quickly. While a paper can always be written, it is clear that there must have been something amiss with Strelnikov and his X-ray tube as he was indeed forced to leave UFTI for a few years (which he spent in Dnepropetrovsk). The details of what exactly happened are not known. In this respect point 4 seems to be related to point 1. The training of cadres (points 5 and 8) was indeed a problem. At the January 1937 session of the Academy of Sciences at UFTI the unsatisfactory training of cadres was mentioned. Everybody wanted to come to UFTI, but no one wanted to leave in order to work somewhere else. At some point the institute must have been full. Obreimov had proposed a certain regime for this whereby after one or two years of training people would have to leave and only the most talented would remain at UFTI. This met with strong opposition, however, not from those accused here of wrecking and sabotage, but from the ones that were afraid of losing their UFTI privileges.

Point 6 seems completely out of place here as it was Landau who protested sharply against the sending of the telegram to the leadership of the country, and Shubnikov undoubtedly agreed with him. It concerned the following. In October 1932 the April 1932 Cockcroft-Walton experiment on the splitting of lithium nuclei was repeated at UFTI by Lejpunsky, Valter, Sinelnikov and Latyshev.[11] The achievement, certainly non-trivial as it came within six months of the original experiment, but at the same time not earthshattering either as it was after all a repetition, gained wide coverage in the Soviet press (with a front-page article by Obreimov in *Pravda* of 22 October 1932[12]). A telegram was sent to Stalin announcing this 'shock work' as if they were the first in the world to achieve fission. The work was duly awarded a Stalin prize and elicited scathing remarks from Landau, who saw it as self-aggrandisement. The telegram to Stalin read: "*The Ukrainian Physico-Technical Institute in Kharkov as the result of shock work on the occasion of the XVth anniversary of October has achieved the first successes in the shattering of the atomic nucleus. On 10 October the high-voltage brigade shattered the nucleus of lithium; work is continuing,*" showing that the UFTI physicists had perfectly mastered the bombastic Soviet writing style. So Landau proposed to send another telegram to Stalin including the text: "*The sine was differentiated, the cosine was obtained; work is continuing.*" So, for sure, if there was 'grandiose deception of party and government' it was not through any action of Shubnikov and his allies. It is very odd that this point emerges here. It is very unlikely to originate from Shubnikov, but it seems neither appropriate for the NKVD to bring it up.

---

[11]Georgy Dmitrievich Latyshev (1907–1973). Was at UFTI from 1930 to 1941.

[12]See Ref. [1], p. 151 for a reproduction of this front-page.

The strikes and other anti-Soviet actions of point 7 refer, of course, to the action taken after Landau's dismissal. I suppose it is sufficiently clear that in the end all those involved regretted their action when they became aware of what the consequences might be. It shows that they did not realise the uncompromising and brutal attitude of their own government. In that respect it can be said that they were a group of rather naïve people who had little notion of the real world, living as they did in their own privileged scientific world, thinking that the rest of society could not but admire and support them. Point 9 does not add anything new to the list of accusations; once one of the former points has some truth in it the recruitment into the counter-revolutionary organisation is after all a fact in the NKVD's frame of mind. The worrying aspect is that Shubnikov admitted to being guilty of every charge and in every aspect, even that he had recruited Landau into the group, while Landau was supposed to be (and had already been) the leader in Leningrad.

He then continues with describing his flight to Finland, conditioned by his "*negative attitude towards the Soviet system*". The October revolution was met with hostility, he says, by his age group and by the teachers at the gymnasium he attended, as well as by students and professors at the university. This attracted him to go abroad and when he was offered the chance to do so he seized the opportunity with both hands. He then relates the story of his arrival in Finland and subsequent travel to Germany, which agrees with the story already told in Chap. 2.

Shubnikov also states that in Petrograd the political situation at Ioffe's institute, where he went to work after his return from Berlin, was on average significantly more pro-Soviet than at the physics institute of the university, but on the whole still very anti-Soviet. In particular, Obreimov and his co-worker Strelkov[13] were anti-Soviet, so his surroundings did not give him the possibility to reconsider his political views, which remained very much against Soviet power. It seems that Shubnikov voluntarily mentions these names here. The mentioning of Obreimov is perhaps to be expected as he was a colleague at UFTI. But why bring up Strelkov here? Did he want to show how sincere he was in his promise to cooperate with the NKVD? For Strelkov it could have had serious consequences.

He then relates his *komandirovka* to Leiden and his meeting in Berlin with Dessler, who had recruited him for espionage in 1923. In an earlier statement he had specifically mentioned which information he provided to Dessler. Now he says that he told Dessler about more work that could have had practical value, upon which Dessler allegedly said that "*such activity is not what he requires and that upon my return to the Soviet Union he would force me through his people to work differently.*" Here, I suppose, the interrogator prepares the ground, with Shubnikov's cooperation, to be able to later make the connection with Weissberg.

So slowly but surely Shubnikov is dragged down deeper into trouble. Each time something is added or displayed in a slightly different light in order to heighten his anti-Soviet attitude. He drags in ever more people, like Charlotte Houtermans and

---

[13]Pëtr Georgievich Strelkov (1899–1968), who became one of the first members (from 1936) of Kapitsa's institute in Moscow.

Valentin Fomin, seemingly without any reason. Alternatively, these names must have been suggested to him by the interrogator. He is even enticed to confess things that never occurred, such as his desire, perhaps existing in his head but not communicated to anybody apart from his wife perhaps, that while being idle at UFTI for almost a year he entertained the idea of trying *"to return to Holland where before my departure to the USSR I was offered very good working conditions (...) intending to stay abroad for permanent work in case my position at UFTI would not change."*

Later in August 1937 a personal confrontation with Weissberg was arranged. The latter states in his book that on that occasion Shubnikov declared in a toneless voice (Ref. [6], p. 296): *"In 1932 Weissberg came to our institute. He came from Germany. The Gestapo had recruited him. His task was to organise sabotage and spying work at our institute. He wanted to recruit me for this. Since I had already been a German agent since 1924[14] I declined Weissberg's proposal. From that moment we have been working in parallel, but without any contact.*

*Shubnikov avoided my eyes as he muttered this pitiful rubbish. What had they done with this once vigorous and strong-willed man? He looked exhausted, but there were no signs of any direct physical maltreatment."* Weissberg did not understand what had come over Shubnikov. He knew him as an unyielding person, full of willpower and energy.

Although it is not of great importance, it is still worthwhile to note that this version of the story by Weissberg does not agree with Shubnikov's own statement to the NKVD interrogator, which reads: *"My acquaintance with Weissberg dates from 1932, when he arrived from abroad. Approximately by 1934 we knew each other rather well and my anti-Soviet attitude was clear to Weissberg. Once he told me that my sinful connection with German intelligence was known to him, as well as my anti-Soviet attitude and that I therefore should get involved with him in general counter-revolutionary work. He instructed me that within the walls of the institute I should fight against carrying out technical and defence work and should spread the idea of the incompatibility of scientific and applied work."* He further stated to have agreed to carry out these tasks.

In the second typed statement by Shubnikov, dated 13–14 August, everything is repeated once more, but not fully consistent with the first statement. Weissberg's role is much bigger now; he is clearly the head of the counter-revolutionary organisation. Shubnikov was recruited by him in 1932, and not in 1934 as before. Nor did Weissberg in this second version know about his connections with German intelligence, but was informed about this by Shubnikov himself. The two organisations, one around Landau and one around Weissberg, that have more or less consistently existed in all former statements, are now much less clear. A separate group around Landau no longer seems to exist. Shubnikov says that Weissberg told him at the end of 1934 of the involvement of Lejpunsky and Obreimov. Weissberg also told him that the foreign specialists Houtermans, Martin Ruhemann, Barbara

---

[14]Should at least have been 1923 in order to be possibly true.

Ruhemann and Tisza were recruited by him into the organisation. On Weissberg's instruction he recruited Landau, Gorsky and Brilliantov, while Landau in his turn recruited Rozenkevich and Korets. So the whole setup of the counter-revolutionary organisation has changed, a change that probably has to do with a new interrogator, who had not carefully studied the statements the defendant had made earlier.

For good measure he throws in a second attempt at recruitment for espionage for Germany, now from a shareholder of the Berliner Credit- und Handelsbank for whom he worked. This shareholder was also Russian and had emigrated after the October Revolution. This probably fitted in well with the picture of the NKVD that every Russian who had left the country after the revolution must by definition be a spy against Russia.

An interesting point is that Shubnikov says that during the time of his *komandirovka* in Leiden, from 1926 to 1930, he was only in Germany on two occasions, once by request of Obreimov in connection with the purchase of equipment for the new Ukrainian Physico-Technical Institute, and once on holiday. So he does not mention, as Trapeznikova does, that he went to Berlin every six months to renew his visa. I cannot see any reason why he would hold back that information, as it was completely innocuous and would have been known to the Soviets already since he probably also visited the Soviet embassy for these extensions of his visa or, more probably, his *komandirovka*.

In the final part of his second long statement he relates in detail how the sabotage activities took place in the various fields of research. In the field of nuclear research he belittles the successful splitting of the lithium nucleus recalled above with the words: "*Based on the mere repetition of experiments on splitting the atomic nucleus of lithium, which had already been achieved at that time in Cambridge, Lejpunsky as head of the atomic-nucleus laboratory and director of the institute in that period, and Obreimov sent to the leaders of party and government a telegram with false content on an outstanding achievement of UFTI, thereby achieving large allocations for the large-scale construction of high-voltage installations, which did not contribute anything useful to the institute.*" Ever bigger installations were built and demolished again for no purpose whatsoever. He also again mentions Strelnikov's case: "*In 1934 the UFTI engineer Strelnikov developed an X-ray tube with a power ten times larger than achieved before. Strelnikov's invention should have been widely used in industry for eliminating defects in metals and for the X-ray investigation of structures. Implementing a clear sabotage line in physics to disrupt work that has any technical application, Obreimov and Lejpunsky, in order not to give Strelnikov's invention to industry, dismissed him from the institute under the pretext that it was not possible to continue applied work at the institute.*" But this was sabotage activity by others, especially Lejpunsky and Obreimov. Shubnikov himself committed direct sabotage in the field of low-temperature physics, jointly with Martin and Barbara Ruhemann. "*We committed sabotage in this most important field with significance for defence under the direction of one of the most active participants of our organisation, connected with German intelligence, the foreign specialist Weissberg, with Lejpunsky being fully aware of this. The UFTI low-temperature laboratory, one of the largest in the world and comparable in size to the Leiden laboratory, should, apart from making discoveries in physics, have researched practical applications of low temperatures for separating coke gas with the aim of separating the nitrogen-hydrogen mixture and helium (...). In fact nothing was done on this, including that a number of large technical problems, which could have had an effect on the national economy, were not solved. At the same time Rjabinin who carried out successful work on the application of liquid hydrogen as fuel for aviation engines was first chased out of the laboratory, and subsequently from the institute via harassment and artificial quarrels by me, Weissberg, Landau and Rozenkevich. After Rjabinin's departure from the institute the work he had partly completed was not further developed. Driven to despair Rjabinin had beaten up Landau.*"

Similarly in respect of OSGO, the deep-cooling research station, which according to Shubnikov in his statement "*was constructed for exclusively technical and defence-related purposes, mainly questions of the liquefaction of gases and the separation of gas mixtures by the method of deep cooling,*" the sabotage consisted of a situation where everything was in the hands of foreign specialists who were agents of German intelligence and successfully delayed the construction, such that "*it cannot be used in the coming years in the direction needed for the country.*"

In the end he summed it all up with the words: "*During the entire period of its existence the institute did not yield a single invention and proposal useful for industry and the defence of the country, in spite of the fact that unlimited resources and opportunities for carrying out scientific research and technical work were granted by the country. Moreover, in individual cases when on the initiative of a number of co-workers technical and defence-related work was proposed, such initiative was strangled by participants of the counter-revolutionary organisation, and the co-workers were chased from the institute by creating an atmosphere of quarrels and harassment. Laboratories carrying out useful technical work were closed. The fight against technical and defence work was carried out under the slogan of the incompatibility of scientific and applied technical work. That this slogan is only a cover for wrecking work is clear from the fact that as regards scientific work the institute has performed insignificantly little compared to the means that were spent on it. Colossal resources allocated by the country were spent wastefully (sabotage-like) for the set-up of installations nobody needed and which subsequently did not work and, when not giving any results, were demolished. This wrecking activity was carried out on a large scale at the atomic-nucleus laboratory, which failed to yield any results after having swallowed up 10 million roubles. The low-temperature laboratory led by me had all possibilities for research work that is extremely important for industry and for the defence of the country in the field of the liquefaction of gases and separation of gas mixtures by the method of deep cooling. The laboratory did nothing in this direction, having as a blind person switched over exclusively to scientific work.*"

All this was declared by Shubnikov while in actual fact in this period UFTI was the leading research institute in the Soviet Union, yielding the most important results on physics (purely scientific, but also applied results) of all the institutes in the Soviet Union.

What had come over Shubnikov, one wonders, that he engaged in this kind of absurd self-accusations? Where was his pride in all that he had achieved in just a few years in his laboratory? Did he agree with his interrogator like Rubashov in Koestler's novel *Darkness at Noon* agreed with his? But unlike Koestler's protagonist, Shubnikov was not a repentant Bolshevik who had to do a last service to the party or who was convinced that the party was always right. But still, like everybody else living in the Soviet Union at that time, he could not avoid being influenced by the Bolshevik ideology. He admits time and again that he nurtured an anti-Soviet attitude, but that is completely untrue. If he had, he would never have returned to the Soviet Union after his *komandirovka* in Leiden, but would have taken the opportunity offered to him and have stayed there or moved on to another place, for instance the United States. There is no evidence whatsoever that he ever even entertained that idea. On the contrary he was very eager and enthusiastic to go to Kharkov and start building up a Leiden-type low-temperature laboratory there.

A parallel can be drawn here perhaps with Bukharin, who repented after his interrogators had threatened to kill his wife and new-born son. Perhaps the same was also enough for Shubnikov in order to agree with every absurd statement the interrogators wanted him to make, but he was not physically tortured like Bukharin was. When physical violence is used an explanation can be rather simple for at some point violence becomes unbearable; every human being has his limits in that respect.

Shubnikov's first admission of guilt may have been obtained by the threat of harming his wife and child or, as Pavlenko et al. suggest, by the threat of scientific oblivion (which may have been for Shubnikov what expulsion from the party was for Bukharin). After all, that is what he talks about in his first signed confession: to be allowed to return to his work. As Solzhenitsyn (Ref. [7], Vol. I, p. 118) says, he may perhaps have felt quite pleased that he had not sold anyone out, that although he had assured his own ten years in the camps, his ribs were in any case still whole, his wife and child were safe, and the affair had been worked out quite sensibly. But after this statement the interrogators set about the task of destroying Shubnikov's objective sense of his actions as a scientist at UFTI. In a different hypothetical case Solzhenitsyn explains very vividly how interrogators would set such process in motion, slowly ensnaring the accused by twisting his statements, telling him that evidence has been given against him by others, that they also have been arrested, and so on. He gauged his actions as a scientist against world 'bourgeois' science, but was that the proper thing to do? His interrogators may have convinced him otherwise; that his duty lay first and foremost with his country, and that he should measure his actions against their usefulness for his country. In that case although they may have had some pitiful scientific value, they must actually be seen as a waste of talent and resources, in essence as criminal actions of sabotage that only reinforce foreign anti-Bolshevik forces and hence serve foreign intelligence. So, although his actions in the field of science were subjectively honest, objectively in a Soviet context they amounted to sabotage or treason. In the Moscow show trials the accusations went much further still. There the accused actually confessed to crimes they had not at all committed, that had never happened, and that were completely fictitious, while in Shubnikov's case the task was much easier. He only had to be convinced of the fact that his actions, which in the context of world science were perfectly laudable and represented a positive contribution, were actually criminal and outright negative when viewed in a national Bolshevist context. Once the first step in that direction has been set, it was a fairly easy run to the end.

Merleau-Ponty has analysed these questions of self-accusation within the framework of the Moscow show trials of the thirties and looked for an answer in *"the relationship between the subjective intentions of man and the objective sense of man's actions; between the external, objective view of man and the internal, inward, subjective view of man; in general, the relationship between the objective and the subjective"* (Ref. [8], p. 229). The shortcoming of his analysis is that it is done within the context of Marxist philosophy. For Shubnikov's case that is not a priori very useful as he was not a Marxist, and had not acknowledged the historical mission of the proletariat, its leadership and the logic of history, at least not in the way the communists tried in the Moscow trials or Koestler's Rubashov had done. Still, a similar mechanism can also be discerned here, with science playing the role of the 'objective truth'. The Marxists in the Moscow trials had to accept their guilt or stop being Marxists which, for most, was impossible. In some sense here the opposite is true, as this 'objective truth' of science is subsequently unravelled and the determination of an objective scientific fact established by painstaking experiments at a laboratory in Kharkov becomes an anti-Soviet action if while doing so the actors lose sight of the local context.

On 27 August, after two weeks in which he had been given the opportunity, it seems, to think everything over once more, Shubnikov is asked for the last time whether he admits his guilt and he does so unreservedly: *"I fully admit to being guilty of the charges presented to me"*.

## The Rozenkevich File

The file on Rozenkevich (Fig. 11.3) contains a similar set of documents as Shubnikov's. A memorandum dated 5 August 1937 states that there is no doubt about his participation in the counter-revolutionary activity at UFTI, partly since he admitted to this when interrogated in April 1937, and that his arrest is necessary. The memorandum also states that a few months later, so in the summer of 1937, Rozenkevich retracted part of his statements, allegedly under the influence of Landau and others. As in Weissberg's case, the NKVD may have called him in for questioning several times over a period of a few months before proceeding to the actual arrest. But the NKVD strictly forbade people it had interrogated and sent home to talk with others in any way about what they had experienced and, as discussed above in connection with Shubnikov's arrest, it is doubtful that he shared his experience with either Landau, who already was in Moscow, or Shubnikov.

Next are a resolution on the desirability of the arrest, an order for the arrest and a search of his apartment, very similar to the ones produced in Appendix 3 for Shubnikov. The questionnaire shows that he was born in Leningrad on 11 February

**Fig. 11.3** Lev Viktorovich
Rozenkevich in the early
thirties (*from* Ref. [9], p. 82)

1905, was married to Vera Vladimirovna and had a two-year old daughter Julija. The protocol of interrogation dated 12–13 August 1937, an extensive document which must be the result of several days of questioning, has been reproduced in Appendix 3. It starts with Rozenkevich admitting his participation in a counter-revolutionary organisation at UFTI, not headed by Landau this time, but by Weissberg. As in Shubnikov's case, the existence of two separate organisations no longer seems to be part of the script. Rozenkevich also admits that his counter-revolutionary activity goes back to his student years at LFTI and he mentions Dmitry Ivanenko, now in exile in Siberia, as his fellow counter-revolutionary. The leaders of this organisation at LFTI were Landau and Frenkel.

Rozenkevich did not hold back and admitted his guilt on all counts, including his time in Leningrad and from the very beginning at UFTI. He implicated everybody, but especially Weissberg who in his testimony, like in Shubnikov's, was the key figure of the organisation. *"From Weissberg's arrival at UFTI in 1931 the activity of the counter-revolutionary group was revived and was constantly organisationally strengthened. It also took shape politically as a Trotskyist organisation."* The gist of his statement is the same as Shubnikov's. Again it is stated that large amounts of money were spent but no useful results obtained; applied work of significance for industry and/or defence was not carried out as it was incompatible with 'purely scientific work'; the construction of the high-voltage building was not needed; the construction of OSGO was needlessly delayed for several years; and a number of scientific workers were harassed and forced to leave the institute. New in his statement is a section on theoretical physics, his former speciality, accusing Landau of banning every possibility of providing help in technical questions. He concludes with the claim that "[a] *similar situation exists throughout the entire network of physico-technical institutes, which only by a misunderstanding carry the name physico-technical. Physics in the entire Soviet Union should be thoroughly reviewed and actually put in the service of the country, for which it can and should provide brilliant work."* This advice does not seem to have been followed up.

He too is left to stew for a few weeks, or perhaps in this period he was maltreated, in order to be interrogated again on 26 August on which occasion he simply admitted to all charges: *"Yes, I fully admit to being guilty of the charges presented to me."*

What else do we know about Rozenkevich? Not very much. In 1990 Viktor Frenkel wrote a paper [9] on Lev Rozenkevich in which he gives some details.[15] Rozenkevich was born into the family of an official of the State Duma, who had been elevated to the nobility. In the autumn of 1917 his father moved with his family from Petrograd to Irkutsk, probably to avoid being arrested by the Bolsheviks. In 1920, at age 15, Lev started to study at the Irkutsk Practical Polytechnic Institute, from which he moved in 1922 to the technical faculty of the Moscow Institute of National Economy (currently the Plekhanov Russian University of Economics). After finishing the latter institute in 1924 he worked for a few years at a refinery, combining his work as an engineer with a teaching job. Here he developed an interest in the new physics of quantum mechanics and relativity and requested the State Academic Council of Narkompros to be allowed to do post-graduate work at a university or research institute. The request was granted and in 1927 he was sent to LFTI, where he enrolled in the theory department and became a graduate student with Jakov Frenkel. At first he was not paid a salary and had to earn an income, which also had to support his mother and sister, by giving private lessons. In spite of this he did quite well and soon got a paid position. He wrote a few papers on quantum chemistry in collaboration with Simon Zalmanovich Roginsky (1900–1970), who from 1930 worked at Semënov's Institute of Chemical Physics.

In 1930 he came to Kharkov where, after Ivanenko returned to Leningrad, he became head of the theory department until Landau's arrival in 1932. Upon arrival in Kharkov, he also started to teach at Kharkov University and at the Physics and Mechanics Faculty of the Kharkov Machine-Building Institute, as many other UFTI employees did. He was also heavily involved in the set-up and publication of the journal *Physikalische Zeitschrift der Sowjetunion*, of which he was one of the editors. As far as his research is concerned, he once developed a theory of super-conductivity which however did not become a success. A second unsuccessful venture of Rozenkevich was the formulation of a theory for the scattering of light on light, a problem which had been suggested to him by Landau, and was eventually solved by Landau's smart young UFTI graduates Pomeranchuk and Akhiezer. All this did not make Rozenkevich very happy and he decided to switch to experimental work. According to Viktor Frenkel, a contributing factor to this decision may have been that Rozenkevich wrote a sharply critical paper [10] with Landau on Ioffe's breakdown phenomena of dielectric media, which he later regretted. In this very short paper, of just a few hundred words without a single formula, they gave short shrift of Ioffe's fanciful ideas by bluntly stating that they violate the laws of quantum mechanics and moreover that, supposing that these laws were indeed to fail for the phenomena he is investigating, such failure cannot be determined by the experiments he is proposing, so both theoretically and

---

[15]In this paper Viktor Frenkel says that Ranjuk informed him that Rozenkevich was interrogated by the NKVD in April 1937, which also indicates that the date in the memorandum on Rozenkevich in Appendix 3 is not correct.

experimentally his ideas were faulty. It must have been very hard for Ioffe to be brushed aside in such a summary and almost disdainful fashion. Landau had been venting this criticism of Ioffe's ideas for some time and Rozenkevich's name on the paper may only be due to the fact that he actually wrote down Landau's criticism as the latter sometimes suffered from writer's block.

Already from the end of 1934, Rozenkevich sought contact with the nuclear research laboratory, headed by Lejpunsky. At the end of 1935 he decided to definitively switch to nuclear physics and by the beginning of 1937 he became the head of the UFTI laboratory of radioactive measurements. In this capacity he was a co-author on a number of papers with Lejpunsky, Kurchatov, Sinelnikov and some other experimentalists on the physics of slow neutrons, photoneutrons (i.e. neutrons produced in photon-neutron reactions) and nuclear physics.

## The Gorsky File

Gorsky's case is rather different from the other two. His file contains similar documents, but all of them are fairly short and there are no extensive interrogation protocols. There was no need for them as Gorsky (Fig. 11.4) did not admit any guilt. From the very beginning he stood firm, claiming he did not "*know anything about the existence of a counter-revolutionary organisation at UFTI. Nobody has recruited me into this organisation, I never engaged in sabotage activity, nor do I admit to being guilty of this.*" On 23 September 1937, just two days after his arrest, he was already confronted with Shubnikov who testified, in the presence of Gorsky, that he had recruited him into the counter-revolutionary Trotskyist sabotage organisation at UFTI. Gorsky denied, however, that Shubnikov had ever approached him with such proposal. Two days later he was also confronted with Rozenkevich who testified that "[i]*n 1936 Weissberg and Landau told me about the participation of Gorsky in the counter-revolutionary organisation. Gorsky took part in a number of campaigns, arranged by the organisation, e.g. in 1936 in December in the collective walk-out from the university together with Shubnikov and Landau,*

**Fig. 11.4**  Vadim Sergeevich Gorsky (*Picture taken at Obreimov's 40th anniversary in 1934 at UFTI*)

*in the harassment of the senior engineer Strelnikov, who was taken off the technical problem of X-ray cinematography, which is of great significance for the metal-lurgical industry. On the question of the walk-out from the university a meeting took place between some of the members of the counter-revolutionary organisation, at which meeting Gorsky also took part."* In this event too Gorsky kept a cool head and answered: *"The statements of Rozenkevich are partially true, namely on 26.12.1936 Shubnikov indeed called me and Brilliantov into his office and told us about Landau's dismissal by the rector of the university Neforosnyj, depicting this as a provocation on the part of the rector. However, I was not a member of a counter-revolutionary Trotskyist organisation and did not know anything about its existence."* In view of what happened in the cases of Shubnikov and Rozenkevich, Gorsky's conduct here is quite extraordinary. He withstood all the pressure, admitted only what could not be denied. In the final interrogation on 26 September 1937, however, it seems that he let himself be enticed into using the terminology the NKVD wanted to hear and admitted that he *"was drawn into sabotage activity by (...) Shubnikov"*, which *"sabotage activity manifested itself in the following. On Shubnikov's proposal I took part in a demonstrative presentation of a declaration together with Shubnikov and others on a walk-out from work at the university, which in essence was a strike with an anti-Soviet character. Moreover, we undertook this strike in defence of Landau who had been dismissed from the university because of anti-Soviet speeches, about which I learned later. At the time of the purge of the party at the end of 1934 I took part in slander with the aim of getting rid of the party communist and valuable scientific worker Strelnikov. I took part in the harassment of Strelnikov by Shubnikov, Obreimov and Landau. I ruined the introduction or more accurately the development of a tube by Strelnikov which is of great significance for metallurgy and for other fields of industry."* But he still denied knowing anything about a counter-revolutionary organisation.

So, stripped down to its essentials his 'confession' amounts to admitting to having taken part in the walk-out at the university in defence of Landau and in mistreating Strelnikov. However this was enough for the NKVD to stretch it in the concluding indictment to: *"He is exposed by the statements of the accused Shubnikov and Rozenkevich (arrested). Only confessed to counter-revolutionary activity. Partially confessed: participant of a counter-revolutionary Trotskyist sabotage organisation, wrecker."*

We know even less about Gorsky than we do about Rozenkevich. He was born in 1905 in Gatchina, close to Leningrad, was married to Nina Vasilevna Danilevskaja, who also was a scientific worker at UFTI, had a three-year old daughter, and a seven-year old son from a previous marriage. The NKVD questionnaire on Gorsky in Appendix 3 says that his father was a priest, other information[16] has it that he was a railway engineer (*inzhener putej soobshchenija*). His mother Ljubov Grigorevna Nadezhdina (1881–1914) was the daughter of Grigory Grigorevich Nadezhdin (1851–1921), who was a well-known doctor in Gatchina. In

---

[16]http://kraeved-gatchina.de/data/documents/GLAVNYE-MEDIKI-STAROY-GATChINY-15.pdf.

1914, when Vadim was nine years old and living with his mother in the household of her father, his mother had a quarrel with the husband of her daughter Olga and was shot dead before Vadim's eyes.

Another interesting aspect of his biography is that he had already been arrested once, namely in 1923 in Gatchina, together with an acquaintance whom he helped to drive a sled on which weapons were stored. In the relatively liberal NEP period of the twenties this had no adverse consequences for him. In 1922 he enrolled at the physics-mathematics faculty of the Leningrad Polytechnic Institute and in 1927 became assistant to Ivan Obreimov, whom he followed later to Kharkov. At UFTI he soon became the head of the X-ray laboratory.

In June 1943, when it was still not known to Gorsky's family and friends that he had been executed, Ivan Obreimov, who considered him one of his brightest students, wrote a short essay on Gorsky's work in physics which ended up in the archive of Pëtr Kapitsa [11]. The purpose of this essay was probably to enlist Kapitsa's help in getting Gorsky released. Kapitsa did not do anything with the manuscript as he must have soon found out what the real meaning was of the sentence "10 years without the right of correspondence."

Gorsky's contributions to physics consist of what are now known as the Gorsky-Bragg-Williams approximation (1928) and the Gorsky effect (1935). In 1928 Gorsky published a paper [12] on an X-ray investigation of transformations in a copper–gold alloy. In that paper the atomic mean-field analysis of order–disorder transformation in metallic alloys appears for the first time. Lawrence Bragg and Evan James Williams broadened the scope in 1934–35, and the Gorsky-Bragg-Williams approximation was born. Gorsky's other contribution to physics is the "Gorsky effect" (a theory regarding disordered solid solutions) published by Gorsky in 1935 [13], but only given this name and applied to metal-hydrogen systems from the 1970s. The phenomenon is best illustrated using Gorsky's own words: *"Let us consider a substitutional solid solution of two kinds of atoms with different atomic radii. If we bend such a crystal, it is very natural that the large atoms will diffuse into the stretched layers, and the small atoms into the compressed layers until an equilibrium implying a concentration gradient is reached"* [14].

## Conclusion

There is a concluding indictment among the documents for all three accused together and a separate concluding indictment for each of them. They all say the same thing, namely that the accused engaged in espionage and sabotage activity. Shubnikov is a spy and saboteur by his own confessions and according to the statements of Rozenkevich and the witnesses Gej, Shavlo and others. Rozenkevich is a saboteur according to his own confessions, as well as to the statements of Shubnikov and the witnesses mentioned. Gorsky is a saboteur by the statements of Shubnikov and Rozenkevich and the same witnesses (so in his case a confession is missing, but it will not make any difference). In all indictments it is said "to refer to category 1", which means that they are to be shot.[17] Hence it was the NKVD investigation team that decided (quite arbitrarily) what should happen to the person in question.

The conclusion is that all are guilty of crimes as formulated in Article 54 of the Ukrainian Penal Code and that (for confirmation of the sentence) the cases are now referred to the Main Directorate of State Security according to NKVD order no. 00485. This order, dated 7 August 1937, was entitled *"On liquidation of Polish sabotage and espionage groups and units of the Polish Military Organisation"* and laid the basis for the systematic liquidation of the Polish minority in the Soviet Union. Its goal was the liquidation of absolutely all Poles in the Soviet Union, and although Poles were especially affected by it, it also applied to other nationalities. The order created an extrajudicial sentencing body consisting of two NKVD officials (the *'dvoika'*). It also established the so-called 'album'-procedure whereby cases considered by lower NKVD organs were collected and sent for approval to the USSR NKVD.

The end result of the *Delo UFTI* was that on 28 October 1937 the General Commissar of State Security Yezhov and the USSR Procurator Vyshinsky in three separate decisions based on yet another order (no. 00439 this time, which was dated 25 July 1937 and was the basis for the German operation of the NKVD)[18] confirmed that all three accused had to be shot. The sentences were carried out in alphabetical order of the accused on 8, 9 and 10 November 1937. Shubnikov died on 10 November aged 36; Rozenkevich on 9 November and Gorsky on 8 November, both at the age of 32. Three promising careers came to an end for no purpose whatsoever, just to satisfy the paranoid megalomania of a bloodthirsty, deranged and psychopathic dictator and his henchmen. It had been eleven years almost to the day that Shubnikov had arrived in Leiden as an inexperienced physicist-technician. In those eleven years he had grown into one of the best low-temperature physicists in the world with some splendid discoveries to his name and more promising work to come which never came to fruition. The relatives were not informed of the executions; they were told that the sentence was "ten years of corrective labour camps without the right of correspondence", and it would still take decades before the full truth about their fate came into the open.

On 13 November, just a few days after the execution of their husbands, Nina, Gorsky's wife, and Vera, Rozenkevich's wife, were arrested and received prison sentences. Their ultimate fate and that of their children does not seem to be known; they appear in various lists of victims of the Stalinist repressions, but without any details. Olga Trapeznikova was not arrested, perhaps because of her two-month old son or a promise made by the NKVD to Shubnikov, it is unclear. But she was summoned to the directorate of UFTI and, when she refused to dissociate herself

---

[17]Category 2 would be those who had to be sent to the camps.

[18]One gets the impression that these orders were randomly referred to, just to give the whole process a semblance of legality.

from Lev Vasilevich, they proposed that she voluntarily resign from her job at the institute, which she did. She was turned out of her apartment, but forbidden to leave Kharkov and she and her son Misha lived for some time in a shed in Kholodnaja Gora, a suburb of Kharkov where the prison was also situated. With great difficulty she managed to obtain a certificate from the NKVD that Misha is Shubnikov's son, after which she could register their son as Mikhail Lvovich Shubnikov (Ref. [5], p. 290). Some time later she managed to reach her parents in Leningrad. For two years she could not find any work anywhere. During this time she got help from her parents and from a few friends who were brave enough to help the wife of an 'enemy of the people'. Among them were Georgy Anatolevich Miljutin and Simon Solomonovich Shalyt, former collaborators of Shubnikov at UFTI, and her old friend Aleksandra Vasilevna Timoreva.

# References

1. Ju.V. Pavlenko, Ju.N. Ranjuk and Ju.A. Khramov, *"Delo" UFTI 1935–1938* (The "UFTI" Case 1935–1938) (Feniks, Kiev, 1998).
2. Robert Conquest, *The Great Terror: A Reassessment* (Pimlico, London, 2008).
3. V.V. Kosarev, Fiztekh, Gulag i obratno (belye pjatna iz istorii leningradskogo Fiztekha) (*Phystech, Gulag and back (white spots from the history of the Leningrad Physico-Technical Institute)* in: *Chtenija pamjati A.F. Ioffe* (Nauka, St. Petersburg, 1993), 105–177.
4. K. Landau-Drobantseva, *Akademik Landau—Kak my zhili* (Academician Landau—How we lived) (Zacharov, Moscow, 1999).
5. O.N. Trapeznikova in B.I. Verkin et al. (1990), p. 256–291.
6. A. Weissberg, *Conspiracy of Silence* (Hamish Hamilton, London, 1952).
7. A. Solzhenitsyn, *The Gulag Archipelago* (Collins, London, 1975).
8. Pavel Kovaly, Maurice Merleau-Ponty and the Problem of self-accusations, *Studies in Soviet Thought* 17 (1977) 225–241.
9. V.Ja. Frenkel, Lev Viktorovich Rozenkevich, *Chtenija pamjati A.F. Ioffe* (Readings in memory of A.F. Ioffe) 1990, p. 80–99.
10. L. Landau and L. Rosenkewitsch, Über die Theorie des elektrischen Durschlags von A. Joffé, *Z. Physik* 78 (1932) 847–848; *Phys. Z. Sow.* 2 (1932) 200–201.
11. P.E. Rubinin, Rukopis' I.V. Obreimova iz Arkhiva P.L. Kapitsy (Manuscript of I.V. Obreimov from the Archive of P.L. Kapitsa), *Vestnik Rossijskoj Akademii Nauk* 64 (1994) 243–247.
12. W. Gorsky, Röntgenographische Untersuchung von Umwandlungen in der Legierung CuAu, *Z. Phys.* 50 (1928) 64.
13. W.S. Gorsky, Theorie der Ordungsprozesse und der Diffusion in Mischkristallen von CuAu, *Phys. Z. Sow.* 8 (1935) 443.
14. Olivier Hardouin Duparc and Alexander Krajnikov, *Physics Today On-line*, 13 July 2017.

# Chapter 12
# The UFTI Affair: Other Repressed UFTI Physicists

Following the arrests of Shubnikov, Gorsky and Rozenkevich, Lejpunsky (Fig. 12.1), the director of UFTI, was expelled from the party (for lack of vigilance) and in September 1937 dismissed as director. From Weissberg's book [1] it is clear that he considered Lejpunsky a decent and capable person, who did his best to serve the best interests of the institute, but was caught between the demands of his colleagues and the party. He was finally arrested on 14 June 1938. The immediate reason for his arrest at this rather late date (he had already been implicated by Weissberg in his confession of 1 June 1937 and by Shubnikov and Rozenkevich in August 1937, so why wait for almost a year) is unknown, but in the NKVD decision of 24 June 1938 to keep Lejpunsky in custody he is accused of "*being an agent of the German and English secret services and having carried out espionage and subversive work*" (Ref. [2], p. 312). In an undated statement (probable date on or shortly after 24 June 1938 and written in his own hand) he admitted to having caused enormous damage to Soviet physics as the head of UFTI and having helped the enemy by bringing the spy Houtermans into the USSR, ignoring signals of the hostile attitudes of Weissberg, Landau and Shubnikov, hiding from the NKVD that Weissberg engaged in espionage, and suchlike. All this was due to his political carelessness, rotten liberalism and to overvaluing the significance of the connection with Western European science. The connection with the English secret service was subsequently dropped, but the NKVD still considered him an agent of the German secret service recruited by Houtermans, although he did not confess to this. In a decision of 31 June 1938 on extending custody it says (Ref. [2], pp. 312–313):

(...) Lejpunsky A.I. has been an agent of the German secret service since 1934, has been recruited for espionage-subversive work by the representative of the Gestapo Houtermans, Fritz Ottovich. In 1935 Lejpunsky obtained permission from the Soviet government for Houtermans to come to the USSR and arranged for him to work at UFTI as a foreign specialist. By instruction from the secret service obtained through Houtermans, Lejpunsky systematically collected and passed on to Houtermans espionage information on important scientific work in the field of atomic physics. Using his connection at the Leningrad Physico-Technical Institute he obtained material there on the course of the work on solving novel scientific problems in physics, and passed this material on to Houtermans. (...)

© Springer International Publishing AG 2018
L. J. Reinders, *The Life, Science and Times of Lev Vasilevich Shubnikov*,
Springer Biographies, https://doi.org/10.1007/978-3-319-72098-2_12

**Fig. 12.1** Aleksandr Ilich
Lejpunsky with in the
background the UFTI
building (*From* Ju.F. Frolov,
*Aleksandr Il'ich Lejpunskij:
stranitsy zhizni* (Obninsk
2013))

In spite of these very serious accusations, certainly no less serious than those against Shubnikov, Rozenkevich and Gorsky, he was released in August 1938 and his case declared closed, allegedly for lack of evidence, something the NKVD was normally never short of. In the decision to release Lejpunsky it is stated (Ref. [2], p. 313):

> (…) From the material of the inquiry it has been established that the accused Lejpunsky A. I., as director of UFTI, was in close contact with the exposed enemies of the people, the agents of the German secret service Houtermans and Weissberg, who worked at the same institute. However in the process of the inquiry insufficient data has been collected to bring Lejpunsky before the court, while Houtermans and Weissberg have not given any evidence on the participation of Lejpunsky in counter-revolutionary espionage work, and hence guided by Article 190, paragraph 2 of the UkSSR Penal Code,

### HAVE DECIDED:

> To close the inquiry in respect of Lejpunsky A.I. for lack of evidence, and to free Lejpunsky from custody. (…)

A reason for this surprising turn of events may be that the power of the People's Commissar for State Security Nikolaj Yezhov had been on the wane for some time (since April 1938[1]) and a short period of relaxation had started. Lejpunsky returned to UFTI, where he continued to work as the head of the radioactive laboratory, from 1939 as head of the research on uranium fission and from 1940 on the design of a cyclotron. He played a significant role in the development of nuclear power in the Soviet Union and took part in the work of the Nuclear and Uranium Commission of

---

[1]Early in April 1938 he had been appointed People's Commissar for Water Transport, an extra job which left him less time for the NKVD. This appointment is generally seen as the start of his decline, but as Conquest (Ref. [3], pp. 421–422) points out, this was actually the highpoint of his power, with NKVD men running quite a number of commissariats. Beria's appointment as deputy head of the NKVD in July 1938 can actually only be interpreted as the beginning of Yezhov's decline. When party boss in Georgia, Beria had been targeted by Yezhov, but had managed to save himself; nonetheless, it can safely be concluded that they hated each other. It wasn't until October 1938 that the first overt moves against Yezhov were made. Lavrenty Beria became his deputy in July and eventually his successor. Yezhov was not arrested, however, until early April 1939.

**Fig. 12.2** Ivan Vasilevich
Obreimov (*Wikipedia*)

the USSR Academy of Sciences. In the 1940s he was director of various physics institutes. From 1949 he worked at the Institute of Physics and Power Engineering in Obninsk, from 1950 as scientific leader of the program for the creation of fast-neutron reactors. The institute in Obninsk, which is nowadays a subsidiary of Rosatom, has been named after him.

———

Nonetheless, Ivan Obreimov (Fig. 12.2) did not profit from a relaxation period in the suppression, if there was one. He was also arrested in June 1938, suspected of "spying for Germany and England" and aiding enemies of the people. He was apparently very roughly treated. Weissberg describes (Ref. [1], p. 456) how he was confronted with an old man, without any teeth in his mouth, a shadow of his former self, with the movements as of a dying man, a waxen figure that could talk, but did not utter anything comprehensible. Sitting there utterly broken and speaking like an automaton, Obreimov declared, in Weissberg's presence, that he had been a member of the Gestapo, like Weissberg and Houtermans; that they had met several times to discuss how to enter the Soviet Union in order to strengthen the anti-Soviet activities there, and more of such nonsense. He was held in prison for a rather long time, especially at the Butyrka in Moscow, and eventually in November 1940 was sentenced to eight years in a corrective labour camp. The summary of his case compiled in 1956 in connection with the review of the case on Shubnikov, Rozenkevich and Gorsky states that he is:

(…) accused ex Articles 54–1, sub a, and 54–11 UkSSR Penal Code of being an agent of the German and English secret services and a member of a rightist-Trotskyist organisation.

He was mainly incriminated by statements of people arrested in other cases, Houtermans and Weissberg. In the first interrogations Obreimov pleaded guilty of the charges brought against him and declared that in addition to espionage activity he joined from 1933 to 1937 a group of scientific workers who stood for the recognition of 'pure' science and that only 'brilliant' people should engage in scientific work. According to Obreimov's statement, Landau, Lejpunsky, Shubnikov, Rozenkevich, Korets and others belonged to this group. He did not mention Gorsky.

In 1939 in interrogations in a subsequent part of the inquiry by the USSR NKVD Obreimov retracted his earlier confessions and declared that he had not been a spy and had never carried out anti-Soviet work. But instead he declared that at UFTI there had existed a group of physicist employees who had taken an incorrect position in science speaking out for 'pure' science and, in particular, that he had expressed such views.

On 19.11.1940 by decision of the Special Council[2] of the USSR NKVD Obreimov was convicted to eight years of correctional labour camp for 'anti-Soviet statements'.

In connection with complaints of Obreimov in respect of his case, in 1940 a review was carried out by the organs of the NKVD and it was established that he was convicted without grounds. On 21 May 1941 by decision of the Special Council of the USSR NKVD the case in respect of Obreimov was closed and he was released from prison.

In testimonials from 1940 available in the case given by the Academicians Vavilov, Ioffe, Komarov and others Obreimov was exclusively positively characterized as a specialist in the field of physics. (...) (Ref. [2], pp. 310–311).

The picture of Obreimov as described by Weissberg is not confirmed anywhere else. It is certainly not the picture of a person capable of doing scientific work and yet in 1939, so rather soon after the confrontation with Weissberg described above, Obreimov is said to have finished a number of studies on the application of Fresnel diffraction to physical measurements. Some time later he was sent to a prison camp in Kotlas in the Arkhangelsk oblast in European Russia. The prisoners who passed through Kotlas were generally sent further north to Pechora (in the Komi Republic in North-East Russia), but Obreimov caught the eye of the Kotlas camp commander, who cancelled his dispatch up north and, after having found out that he actually was a leading scientist, even a corresponding member of the Academy of Sciences, procured orders to send Obreimov to Moscow, where he was allowed to continue his work in the Lubyanka prison. Parallel to and independent of these events Obreimov had also written in May 1940 from Kotlas to Kapitsa, upon receipt of which Kapitsa took action to lodge an appeal and on 7 July 1940 wrote a letter to Molotov. The letter is remarkable in the sense that Kapitsa actually writes that "[w]ere it not for a number of valuable and specific, irrefutable results, Obreimov would be at least considered a great eccentric if not outright insane" and "if he is to be put away at all, it would be proper to lodge him in a home belonging to the health ministry and not the internal affairs ministry," i.e. in a psychiatric hospital and not in a prison. In no way does Kapitsa show any personal sympathy for Obreimov in his letter. He must have done so on purpose, as he must have thought (and rightly so, I believe) that people like Molotov would not at all be impressed by any expressions of sympathy. Several other Academicians, such as Semënov, Vavilov, Ioffe and Rozhdestvensky had also written letters on behalf of Obreimov, but it was in the end this letter from Kapitsa to Molotov, combined with the efforts of the Kotlas camp commander, that secured Obreimov's early release in May 1941 (Ref. [4], pp. 570–571). Obreimov continued his career as a physicist and in 1946 was awarded the Stalin Prize of the first degree, ironically for the research carried out in prison. In 1958 he was elected a full member of the USSR Academy of Sciences. He died in 1981.

—

[2]The Special Council had the right to apply punishments 'by administrative means', i.e. without trial.

Many more employees of UFTI were called in, imprisoned and even executed by the NKVD, not only scientists, but also ordinary workers. The precise extent of the repression is not fully known as many files have still not been made available, or their existence is simply not known. Some people in these terrible years disappeared without a trace. Examples are Pëtr Komarov, mentioned in the telegram authorizing the arrest of Shubnikov and Rozenkevich, who is known to have died in the NKVD prison, but no file could be found in the archives. Also, Ivan Maksimovich Gusak, who was a post-graduate student at OSGO and a member of the UFTI party organisation, was arrested in 1937 and died in prison.[3] The same fate was reserved for the administrative employee and member of the UFTI party organisation Konstantin Aleksandrovich Nikolaevsky.

There is one especially tragic case for whom a file was found and some documents have been reprinted in Ref. [2], pp. 283–290: the case of Valentin Petrovich Fomin (1909–1937). Fomin lived for eight years in Germany where his father, a high official in the Ukrainian Vesenkha, was on *komandirovka*. He studied at the Technische Hochschule Charlottenburg in Berlin[4] at least from 1927–1930, at the same time that Houtermans also worked there as an assistant to Gustave Hertz. Fomin mentions Houtermans in the protocol of his interrogation as one of his teachers and as a Gestapo agent who gave him instructions to get a job at UFTI after his return to the USSR. It is indeed perfectly possible that Houtermans met Fomin and told him by way of advice to contact Weissberg upon his return to the USSR if he wanted a job or some other assistance. At that time Houtermans did not yet intend to go to Kharkov, and probably hardly knew anything about it, as UFTI was still under construction. But Weissberg, who had also spent some time at the Technische Hochschule Charlottenburg, was already there.

Fomin returned to the USSR in 1932 and got a job as a physicist-engineer in the high-voltage group at UFTI. He was arrested on 7 October 1937, interrogated quite extensively, sentenced to the 'highest measure of punishment' and executed on 2 December 1937. The concluding indictment summarises 'his crimes' as follows:

(...) The inquiry carried out has established the following:

In 1924 Fomin went from Kharkov to Germany with his father who was sent on *komandirovka* to Germany. His father worked at the time in the economic department of UkSSR Vesenkha. In Germany Fomin finished the Realschule, and subsequently the higher technical school.

In 1929 he was approached by the White emigrant officer Desjatov, an agent of the German secret service. Through him Fomin was recruited for intelligence activity by an officer of the German political police.

On the instruction of the secret service Fomin joined the Union of Soviet students, where he carried out recruitment work for the secret service and investigated the political attitude of

---

[3]He is, however, also mentioned among the UFTI collaborators who died in the Great Patriotic War in [5], although the authors had been unable to find any details.

[4]Now the Technical University of Berlin. The administration of the university has confirmed that Fomin studied there from 1927–1930.

Soviet students. For the secret service Fomin recruited five students who were subsequently transferred to the USSR.

In 1932 on a proposal of the secret service Fomin moved to the USSR and got a job at UFTI through Narkomtjazhprom. The officer who had recruited Fomin had told him that upon joining UFTI he should contact the foreign specialist Weissberg (arrested), which Fomin did.

Fomin entered into intelligence contact with Weissberg and later with the foreign specialist Houtermans, and carried out a number of subversive acts at UFTI:

a. Blowing up an evaporator with liquid air in the Dewar laboratory, which delayed work of special significance;
b. Laying up the oxygen machine, which for a time held up the completion of special work.

Moreover, Fomin recruited UFTI engineers for espionage—Majdanov, Kikoin, Ivanchenko, and Korets, through whom he gathered information on special and defence work carried out at UFTI and at the Kharkov Electromechanical Factory. The information collected by Fomin was passed on to German intelligence via Houtermans and Weissberg.

Based on the above, Valentin Petrovich Fomin, born in 1909, native of Kharkov, Russian, citizen of the USSR, unmarried, until his arrest employee of UFTI

## IS ACCUSED:

of being an agent of the German secret service, engaging in espionage, in subversive activity and in recruitment activity, i.e. of having committed the crimes envisaged in Articles 54–6, 54–9 and 54–11 of the UkSSR Penal Code.

The accused has confessed in full, moreover he is exposed by the statements of Rozenkevich[5] (convicted) as a participant of a counter-revolutionary organisation at UFTI. (…) (Ref. [2], p. 290).

During one of his interrogations he wrote down in his own hand the 'set-up of the Gestapo organisation at UFTI', consisting of:

*Weissberg*—connection with the consulate, organizer, connection with Trotskyists, with Pjatakov.

*Davidovich*—organisation of breakdown of the institute, sending scientific employees to factory, espionage.

*Shubnikov*—espionage on special work, impeding it from being carried out.

*Houtermans*—intermediate step (assistant of Weissberg and his deputy, connection with Moscow organisation).

*Fomin*—helper of Houtermans, recruitment among young people, mediation with the Kharkov Electromechanical Factory, petty sabotage at institute, impeding special work.

*Ivanchenko, Sinelnikov*—Kharkov Electromechanical Factory

*Kikoin A., Majdanov*—espionage of special work, cryogenic and magnetic laboratories.

*Korets*—work to help Davidovich, work at the university (demoralizing young people).

*Ruhemann M.*—work at OSGO in respect of espionage of chemical factories.

---

[5]Rozenkevich had stated that he knew from Landau about Fomin's participation in the counter-revolutionary organisation.

*Ruhemann B.*—idem.

*Weisselberg*—work at the Carbochemistry Institute, espionage.

As will be clear from the history of UFTI recalled in Chap. 7, Fomin must have been fairly confused when writing down or telling the investigator that Davidovich was organizing the breakdown of the institute and was engaged in espionage, and even starker nonsense is his statement that Korets was Davidovich's helper in all this. Davidovich was clearly the villain in the piece, but certainly not a German spy. Perhaps the NKVD thought it useful to have such statement at its disposal for future use against Davidovich.

In his book about Houtermans, Amaldi (Ref. [6], p. 44) gives a rather more gruesome picture of the events leading to Fomin's arrest and execution, namely that NKVD men came to UFTI looking for Fomin to tell him that his brother, a ski instructor in the Caucasus, had been arrested and to ask him to follow them to the NKVD office for some questions. Fomin got permission to go upstairs for a few minutes to collect some personal belongings, during which he drank sulphuric acid from a bottle in the laboratory, jumped out of the window and died a few days later. The latter is in any case not true, as Fomin still lived for almost two months. The story is partly confirmed by a statement of Sinelnikov dated 3 July 1956 within the framework of the rehabilitation of Shubnikov and others (see also Chap. 14). He says that *"Fomin made the impression on me of an abnormal or very frightened person; for some reason he came to me in the laboratory and almost hysterically said that UFTI may 'unexpectedly burn down'. No matter how much I asked, he did not say anything further."* Sinelnikov reported Fomin to the NKVD[6] and to the party organisation, but heard that a few days later Fomin had drunk hydrochloric acid, jumped out of a window and was taken to hospital (Ref. [2], p. 278). A third (partial) confirmation of the story can be found in the diary of Charlotte Houtermans (Ref. [7], pp. 159–160), where the whole business also starts with the arrest of Fomin's brother and an invitation by the NKVD to pay a visit to their office. Instead he went to the lab and drank sulphuric acid, which *"did not kill him, but inflicted such pain and suffering that he rushed home (…) and jumped out of the third floor bathroom window."* He was rushed to hospital and in this terrible state must have been subjected by the NKVD to lengthy interrogations during which he without doubt told them what they wanted to hear. Who wouldn't?

Finally, in the revision of Weisselberg's case in 1958 Fomin's case was also reviewed, which resulted in the following statement:

> (…) Fomin V.P. was arrested by the NKVD Administration for the Kharkov District on 7 October 1937 and convicted on 22 November of the same year by the NKVD and the USSR Prosecution Service to the "highest measure of punishment". This decision was arrived at as follows.

---

[6]This report by Sinelnikov may actually have been the immediate reason for Fomin's arrest, although in the interrogation protocols nothing is mentioned about burning down UFTI, nor of Sinelnikov's report.

**Fig. 12.3** Fritz Houtermans
at his desk in Bern in 1966 the
last year of his life (*From* [6])

Fomin was charged with having been recruited in 1929, when he stayed with relatives in
Germany, as an agent of the German secret service by a German officer through the
mediation of the White emigrant Desjatov. In 1932 he arrived in the USSR and established
at UFTI a connection with Weissberg, engaged in sabotage and recruited for espionage
Majdanov, Kikoin, Ivanchenko and Korets. Espionage information was passed on to
Houtermans and Weissberg.

This accusation was only based on statements by the accused Fomin himself. In the process
of the inquiry Fomin made self-written statements, and subsequently was interrogated by
the investigator ….

It is not clear from the case what served as the basis for Fomin's arrest.

This information has been drawn up in connection with the revision of the case on
Weisselberg, Konrad Bernardovich. (…)

showing that in this case too a completely innocent person, just 28 years old,
was put to death. It was not even clear what served as the basis for Fomin's arrest.

—

The wave of terror that swept over UFTI also caught up with the foreign specialists working at UFTI. Weissberg and Weisselberg's fates have already been
discussed in detail in Chap. 10; those of some of the others will now follow.

After the arrests of his fellow workers at UFTI and his own dismissal from the
institute in September 1937, Fritz Houtermans (Fig. 12.3) went to Moscow to apply
for exit visas for himself and his family. He was arrested on 1 December 1937 at the
Moscow customs house where he went every day in the hope of speeding up the
process. At first he was held in a prison in Moscow, but after a few weeks transferred to Kharkov where he was subjected to the 'conveyor' (continuous interrogation) procedure, as well as to severe beatings,[7] while they threatened him with the
arrest of his wife and children. In actual fact, with great difficulty, but helped by
Kapitsa Houtermans' wife had already managed to get away from Moscow to Riga
(Ref. [6], p. 45). After eleven days of this treatment Houtermans signed the
statements required from him, admitting among other things that he had recruited

---

[7]Frenkel writes in Ref. [8], pp. 92–93 that Houtermans lost all his teeth during these interrogations;
Ref. [9], pp. 70–75 (German translation, pp. 53–57) describes Houtermans' treatment by the NKVD.

Obreimov to work for the Gestapo during the latter's visit to Berlin, that he himself was sent to the Soviet Union by the Gestapo and suchlike. He told his interrogators an incredibly elaborate story, which apparently consists in two versions, one of which is contained in the following extensive protocol dated 29 March 1938 (Ref. [2], pp. 291–197):

**Copy of the protocol of the interrogation of Houtermans,**
**Fritz Ottovich, of 29 March 1938 from case no. 148169**
**in respect of Lejpunsky, Aleksandr Ilich, charged in accordance with**
**Article 54-1 UkSSR Penal Code, arch. 0-21975 UkSSR KGB, Kharkov.**

Houtermans, Fritz Ottovich, born in 1903, native of Danzig, German, German citizen. Son of a bourgeois, director of a branch of a private bank in Germany, doctor of physics. From 1926 a member of the German Communist Party. Until his arrest scientific head at UFTI.

**Question**: In which year and from which country did you arrive in the Soviet Union?

**Answer**: I arrived in the USSR from England in February 1935, I lived in Germany until 1933.

**Question**: Why did you leave Germany?

**Answer**: As a member of the German Communist Party I took part in illegal party work until June 1933. Soon after Hitler came to power I was given notice by the administration of the physics institute of the Technische Hochschule in Berlin, where I worked as a lecturer, that because of my non-Aryan origin I had to give up scientific work. For the same reason my search for other work in Germany remained unsuccessful. In connection with this and with the approval of the corresponding party bodies I prepared for my departure to England. In June 1933, when everything was ready for my departure, I was called by telephone from my work office at the institute into the office of the assistant engineer Shadrin of the institute of technical physics. Entering Shadrin's office, I found there dressed in a uniform coat Senior Lieutenant Schimpf. I knew him as an officer of the Reichswehr, who was sent to study physics at the Technische Hochschule, where I was the head of the practicum class. After having officially introduced us, Shadrin left us after a few minutes, leaving the two of us alone in his office. Senior Lieutenant Schimpf immediately declared to me that he wanted to speak with me about a commission as a representative of the Gestapo. He said that my membership of the German Communist Party was well known to the Abwehr (the counter-intelligence body of the German General Staff) and the Gestapo, as well as the illegal party work I had carried out. When I tried to leave, Schimpf as proof of his knowledge about me mentioned to me the party codename of a person with whom I had party meetings on party work, and my own codename and indicated the place of two secret rendezvous where I had meetings on party work. From this it became clear to me that I was unmasked. Then Schimpf declared that the Gestapo knew about my intention to go with my family to England, but that they would only give me the possibility to leave if I agreed to work by instruction of the Gestapo abroad. In the opposite case, Schimpf explained, my wife and I would be arrested and convicted for high treason. After having said this he gave me fifteen minutes to think it over. I understood that I would not be able to escape abroad since it was clear that I was under surveillance and that the threat of arrest made by Schimpf was serious and real. I decided to agree to Schimpf's proposal to collaborate with the Gestapo. Schimpf received my agreement with satisfaction and proposed that I gave a written undertaking to collaborate with the Gestapo. He required that the undertaking be written in my own hand and the text was dictated to me by Senior Lieutenant Schimpf. The content of the undertaking was roughly as follows: "Presently I, Fritz Ottovich Houtermans, undertake to collaborate with the Gestapo and to follow up the instructions which will be given to me by the Gestapo representative and to precisely fulfil the tasks I will be charged with." Schimpf received the undertaking from me and said that it would be my task to carry out intelligence work in the country where I would be living. He did not give me any concrete task for recruitment. Schimpf only indicated that when I would be put to work abroad, the Gestapo representative would contact me and give me exact indications on carrying out intelligence work for Germany. Schimpf warned me not to attempt to evade fulfilling the obligation I had assumed, as the Gestapo would find me everywhere and would always have available the necessary means to force me to work for them. Upon this we parted. At the end of 1933 I left Germany for England without any difficulties, after having obtained from the German authorities the corresponding documents for departure.

**Question**: At your recruitment by Schimpf did you reveal the contacts known to you and the secret rendezvous of the German Communist Party?

**Answer**: I categorically declare that I did not communicate to Schimpf any information on the content of illegal party work carried out by me in Germany or on my party connections. Schimpf did not ask me about that.

**Question**: Set out your connections with the Gestapo in England.

**Answer**: In the autumn of 1934 the representative of the Gestapo established contact with me in England and demanded that I fulfil my obligations in respect of intelligence work. The following circumstances preceded this. In England with the help of a professor at Cambridge University I got a job as a scientific collaborator in the research laboratory of the firm 'Electric and Musical Industries'.[8] Before the start of my work at the laboratory my wife and I made a short trip to France and Italy. While in Paris, I met a member of the KPD known to me from Berlin, Alfred Kurella, who worked as a secretary to Henri Barbusse. At one time Alfred Kurella worked in a responsible position in the secretariat of the Young Communist International. In 1935 I met him in Moscow, where he worked in Dimitrov's secretariat.

We met with Kurella in Paris and talked with him about personal and general political subjects. I told him where I was going to work, gave him my address and we subsequently parted. Approximately in March 1934 I received in England a letter from Paris from the so-called Institute for the Study of Fascism, which was a communist organisation and carried out work against German fascism abroad. In this letter I was proposed to establish contact with this institute and to render assistance in work to be carried out. In April 1934 at the time of the Easter holidays I again went to Paris, where I met Kurella, who gave me the address of the Institute for the Study of Fascism (INFA), the institute in question. Kurella then introduced me to the head of INFA, who had the party codename Peter (I do not know his family name). Peter asked me to carry out anti-fascist work among the intelligentsia in England. I agreed. In the autumn of 1934, while already in England, I again received a letter from INFA, in which they let me know that a representative of INFA would shortly arrive in England for organizing INFA work and asked me to help him. Indeed I was soon visited by a certain Max (underground codename), who presented me with the necessary papers attesting that he was INFA's representative, and started with my help to lay contacts with the left intelligentsia in England and to develop activity for INFA.

At one of the meetings of representatives of the intelligentsia interested in INFA work, convened by Max, I got to know a man, who pretended to be a German political emigrant under the party codename Hans (I do not know his last name). Hans asked me to meet with him and to talk again about broadening INFA activity. A few days after the meeting I met Hans as agreed in a café in London.

---

[8]Better known as EMI, which was a private company from 1931 to 2012, when it was disbanded. See Wikipedia for its history.

At this meeting of the two of us Hans started his conversation with me by giving the regards of Mr. Schimpf, saying that "it is time for me to carry out work as agreed with me by doctor Schimpf". Taking into account that Schimpf had not given me a password for a contact, Hans, in order to dispel my doubts about his personality, repeated to me word by word the text of my written undertaking which I had given to Schimpf, as well as the content of my conversation with him at the recruitment. Thus having been convinced about his membership of the Gestapo, Hans proposed that I obtain and pass on to him secret material, of interest to the Gestapo, of the television laboratory of 'Electric and Musical Industries'. Hans then warned me that if I did not fulfil this demand, he would tell the English authorities and the firm that I was a communist spy. A day later Hans called me on the telephone and fixed a new meeting with me at the same place. I did not go to this second meeting and never heard anything further about Hans.

**Question**: With what aim did you come to the Soviet Union?

**Answer**: Still before the described meeting with Hans I had accepted a proposal from the Soviet Academician Lejpunsky who was on a scientific *komandirovka* in Cambridge to come to the USSR to Kharkov to work at UFTI as the scientific head of the work on experimental nuclear physics. Since I wished to evade any further meetings with Hans, I speeded up my departure from England to the USSR. On 27 December 1934 I went to Paris, then to Vienna and at the end of February 1935 I arrived in Kharkov, where I was accepted for work at UFTI.

**Question**: You try to hide the actual aims of your coming to the USSR. You are not denying that in Kharkov you carried out intelligence work for Germany?

**Answer**: I do not deny that in the USSR I got into contact with the Gestapo resident Tisza and that I carried out intelligence work for Germany here. But I arrived in the USSR without having any special instruction from the Gestapo for this.

**Question**: Who is this Tisza and how did you establish contact with him?

**Answer**: From the first days in Kharkov at UFTI I got to know the young Hungarian physicist Tisza who lived in the same apartment as me. Tisza tried to get close to me in all possible ways and apparently studied me. Approximately in September 1935 during a walk in the garden at UFTI Tisza declared to me that my relations with the Gestapo were known to him and he identified himself in the same way as Hans had done at the time in London, by knowing exactly the text of the written undertaking given by me to the Gestapo and by details of my meetings with Schimpf and Hans. After this communication Tisza declared that he had instruction to demand from me to carry out intelligence work in the USSR, that I should act in this sector in accordance with the instructions he would give me in the interests of the work of the Gestapo in the Soviet Union. Tisza also warned that if I were not to fulfil this demand he would find means to force me to work or betray me as a spy of the German secret service to the Soviet organs. I remember that Tisza said to me approximately the following: "Do not attempt to expose or betray me. I do not only know people in the service of the Gestapo who work at UFTI, but also in the highest Soviet bodies. And in case of an attempt to undertake anything against me I will get out of the water dry, but you will get hurt." At the end of the conversation Tisza proposed to meet the following day in order to make final arrangements about the work. This meeting took place and Tisza outlined in detail the tasks I was charged with as an intelligence officer of the Gestapo in the USSR. Subsequently I regularly met with Tisza as the resident of the Gestapo; he broadened his instructions to me, and I passed on to him the obtained intelligence material, and informed him about the work carried out by me.

**Question**: Which tasks of intelligence work did Tisza charge you with?

**Answer**: Tisza considered the precise observation of the scientific research work on Fritz Lange's construction of the impulse generator existing at UFTI as the most important and basic task. I should continuously inform Tisza of all possibilities that arose for the technical application of the impulse generator, in particular its military application, for passing on to the Gestapo. This significant interest of the Gestapo in this work of doctor Lange is explained as follows: doctor of physics Fritz Lange works at UFTI as head of the laboratory of shock waves and the manufacture of a high-voltage impulse generator since the summer of 1935. Earlier he worked with the well-known physicist Brasch[9] (now in America) in Berlin, built an impulse generator with a voltage of up to two and a half million volts, and carried out experiments with him on the splitting of the atomic nucleus. In 1935 Lange arrived in the Soviet Union in order to continue this work at an even larger scale with the use of a generator at a voltage of 4.6 million volts. Such an installation had so far not been realized anywhere in the world, and Lange is in this field a leading specialist of global significance and the holder of crucial patents. The realization of the construction of an impulse generator of very high voltage will be of colossal technical significance, in particular, it will be of great significance for the military.[10] It suffices to say that with the practical solution of this problem it will in principle be possible to achieve with the impulse generator an energy equal to 30 million kilowatt-hour from a kilogram of matter, for instance water. The German General Staff were, as I learned at the time from Lange, keenly interested in the possibilities for military purposes of rays that can be produced by means of the impulse generator. I know that the head of the General Staff of the German army, general von Rachenau[11] together with high officers visited Drs. Lange and Brasch in Berlin in their laboratory and discussed with them the possibilities of applying rays of the highest energy for military purposes. For reasons unknown to me Lange stopped with this work in Germany. When giving me the task to observe the course of the work in this field, Tisza explained that thanks to Lange's presence at UFTI and the large resources allocated by the state the constructed impulse generator and atomic nucleus department of UFTI acquire a monopolistic status in the world. This explains the enormous interest of the Gestapo in the scientific work carried out at UFTI and in the construction of the generator. Personally, thanks to my own work and the long-standing connections with Lange and his work, I am especially suited for the Gestapo for fulfilling tasks in this field. Tisza repeatedly emphasized that I should obtain information on the design values of voltage, power and output of the generator to be constructed; on the technical and in particular the military possibilities of applying the highest voltages and on the bodies, in particular military, that are interested in them. He demanded specifically that I observe and communicate material on new ideas of technical and military applications of Lange's work.

In the summer of 1935 the scientific head of the UFTI nuclear laboratory, Valter, constructed an impulse generator with voltages of somewhat higher than 1 million volts and should carry out experiments of secret aim and content. Tisza gave me the task to establish the content of Valter's work. Because of the great secrecy I was unable to obtain this information. I avoided Valter in order not to arouse suspicion. When I spoke about this with Tisza, he stated some time later to me no longer to be interested in this, emphasizing that Lange's work was significantly more serious and important and that it was my task to take care of Lange in the interests of the Gestapo and therefore there was no need to run a risk with my work around Valter.

Subsequently, on Tisza's instruction I should use my long-standing connections among Soviet physicists for broadening and strengthening these connections in order to collect

---

[9]Arno Brasch (1910–1963).

[10]In the case of Shubnikov, Rozenkevich and Gorsky the construction of such high-voltage installations is actually portrayed as wrecking as they are of no significance and just a waste of money.

[11]There was no German head of the General Staff with this name; Houtermans makes this up.

material on their political attitudes, on their work, plans and subjects of work, and work plans. On their connections and role in industry. In other words, I should obtain material to keep the Gestapo informed of everything that happens among the physicists of the Soviet Union. In particular, Tisza charged me with meeting with the following physicists mainly, to clarify their work plans, subjects, and attitudes: Academician Lejpunsky, Professors Shubnikov, Obreimov, Sinelnikov, Valter, Ruhemann, Weissberg, Lange, Kompeters, Gorsky as far as Kharkov is concerned.

I was instructed to seek connections with physicists who had a great name in other cities: with Finkelshtejn in Dnepropetrovsk; with Vavilov, Frank, Tamm, Kapitsa, Landsberg in Moscow; with Ioffe, Semënov, Kondratev, Kurchatov, Mysovsky, Skobeltsyn in Leningrad.

Tisza repeatedly said to me that the Gestapo was very interested in all kind of material on the work of Soviet physicists. Therefore I should in this regard absolutely pass on to him all material. Separately Tisza over-emphasized to me the task of collecting material on their character, necessary for recruitment as an agent of the Gestapo in the USSR. In this respect Tisza pointed out that I personally should not carry out any recruitment, as this could threaten my rapid exposure. For the Gestapo it was important to keep me for the basic work in connection with keeping track of scientific work and the construction of the impulse generator. Tisza also showed a special interest in the quantity of radium present in the various Soviet institutes, the amount of work on its manufacture, export and stocks.

On his instruction, using my connections, I should also obtain data on this.

**Question**: Did you fulfil these tasks you were charged with by the Gestapo resident Tisza?

**Answer**: I did everything in my power and possibilities. As regards the fulfilment of my main task in respect of Lange's work on the construction of the impulse generator and on experiments with it I did the following:

(a) At the end of 1935 I gave Tisza material in the form of a short note, more accurately, a short report on the construction planned at UFTI of a small impulse generator and a detailed description of the technical data of this impulse generator. As regards its application for military purposes I indicated in the report the possibility of using the impulse generator as an X-ray gun, i.e. of using the high-energy X-ray radiation as a weapon of war. I knew, that the head of the General Staff of the German secret service General von Rachenau had shown an interest in this in a conversation with Lange. To a question of Tisza on the propagation range of such rays I pointed out that this had not been verified by anybody in practice, but based on the theoretical calculations I had done I indicated a distance of the order of 500–1000 m. This information I obtained from Lange, who related it to me in confidence.

(b) Early in 1936, in connection with a change of plan of Professor Lange's experimental work, he reported this to the scientific council at UFTI. Lange then planned the construction of an impulse generator of approximately 3.5 million volts. On Tisza's proposal I presented to him the results of Lange's report on this new work of his.

(c) At the end of 1936 or early in 1937 I gave doctor Tisza a summary of another report of Lange to the UFTI scientific council on the results of his work during 1936 and the plan for experimental work on the construction of a new impulse generator with voltages of the order of 4–6 million for 1937. At the same time I gave to Tisza the following material:

(d) A summary of the plans for 1936 and a report on the results of 1935 of the work of Shubnikov, Sinelnikov, Ruhemann, Obreimov, Gorsky and myself.

(e) The same also for 1936 and the plans for 1937.

Carrying out Tisza's tasks, I informed him early in 1936 of the quantity of radium present in the X-Ray Institute in Kharkov and available at UFTI for scientific work. In the autumn of 1937 I went on *komandirovka* to Leningrad, and there, at the Radium institute, due to a lack of discipline and trust I managed to obtain the following information on radium: the quantity of radium on emanation machines of the X-ray institute in Moscow and the Radium institute in Leningrad. On the quantity of pure radium present at the Physics Institute of the Academy of Sciences in Moscow and the amount which UFTI should obtain in 1937. In fulfilment of Tisza's instructions on letting him have character data of prominent physicists I reported to him verbally information on 6–8 physicists. Based on this Tisza wrote a note for himself, on the basis of which he subsequently personally made a report for the Gestapo. As I already explained above, Tisza collected this type of material as it was necessary for selecting and recruiting new Gestapo agents in the USSR. I gave such material on Sinelnikov, Valter, Lejpunsky, Shubnikov, Obreimov, Ruhemann as regards Kharkov; and on Kurchatov as regards Leningrad. I do not know how this material has been used by Tisza, whether he recruited these people for work in the Gestapo. My work as an agent of the Gestapo was limited to this.

**Question**: You avoid statements on your work as resident of the Gestapo in Kharkov. You have to speak about that too. Answer me: did you have connections with Rupp?

**Answer**: I see that I cannot hide myself. In the spring of 1937 Tisza began to prepare for his departure from the country. This he explained by his fear of exposure in connection with the arrest of a number of foreigners. He then charged me with establishing contact with his Gestapo agency in Kharkov for fulfilling the function of resident and for leading the intelligence work of this agency. He pointed out that after his departure from the USSR he would hand over all intelligence material, which I needed and should obtain from the agency, to the representative of the Gestapo who will again come to Kharkov. The code for his connection with me will be a conversation which he will start with me on the study of cosmic radiation with a Wilson chamber and he will subsequently make himself known in the same way as Hans and Tisza did at the time. In accordance with Tisza's instructions, still before his departure from Kharkov I established contact with the following agents of the Gestapo:

*Fomin Valentin Petrovich,* engineer, scientific worker at UFTI; Tisza told me that Fomin has instructions to pass on to me intelligence material on the Electromechanical Institute and the Kharkov Turbine Factory. As a code for making connection with Fomin Tisza told me that I should ask Fomin for a drawing of some piece of equipment that did not exist at the institute, for instance, an air-filled counter of neutron rays. After some time I made contact with Fomin, which was easy, since Fomin was my collaborator.[12] Fomin passed espionage material on to me twice: drawings, constructions of some machines, explanatory memos to them and reports obtained by Fomin on the Kharkov Electro-Turbine Factory. I do not know how Fomin, who did not work at this factory, obtained these materials and through whom. On the instruction of Tisza I demanded sketches of drawings of a speed measuring device on an electromagnetic basis which has an application in aviation. Where Fomin obtained these sketches I do not know.

*Beniamin Margo,* paediatrician, German subject, arrived in the USSR under the guise of being a political emigrant after Hitler came to power. I got to know Beniamin independently of Tisza's involvement in the autumn of 1935, when she arrived in Kharkov from Saratov. She came to me from Doctor Asher, whom I knew from Germany as a member of the German Communist Party. (In 1937 I heard in Moscow that Asher had allegedly been arrested.) Beniamin visited me at home to give me Asher's regards and said to me that he had given her my address and recommended me as a person who could help her setting up in Kharkov. Speaking about Beniamin, Tisza said that she is being used as an agent of the

---

[12]So, Houtermans does not mention here that he knew Fomin from the Technische Hochschule in Berlin.

Gestapo for collecting information of interest to the German secret service on the policy of medical institutions and on political attitudes in party circles. Tisza did not give me a special code word for contacting Beniamin, and explained that since Beniamin visits me at my home as my acquaintance she will be given instructions to pass on espionage material obtained by her during these visits. In May-June 1937 Beniamin visited me and handed over espionage material, whereby I had the following short conversation with her: "Here is material of which you know the significance, give it to you know whom", she said handing me the material. I asked: "Do you have further instructions for work?" To this she said that she knew how to proceed in future. The material presented by her had to do with the situation and course of epidemic diseases in Kharkov.

*Rupp,* his first name and patronymic I do not know, German national, electro-technician, worked at the institute of meteorology and standardization in Kharkov. Tisza warned me that in order to establish contact Rupp would appear at my home under the pretence of appealing to a specialist physicist and as a code he would ask me about a chemical light reaction method for use at frequency measurement stations. After some time Rupp indeed came to my house, used the agreed code and gave me secret material on a frequency meter for radio waves of high frequency. This material had been developed at the institute of meteorology. After Tisza's departure approximately in June or early August Rupp came to me a second time. He again wanted to give me intelligence material he had obtained, but at that time I did not have any contact for passing on the material, since the promised replacement of Tisza had not yet shown up, and I proposed to Rupp to keep this information with him for some time.

*Rozenkevich Lev Viktorovich,* scientific worker at UFTI. Speaking about him, Tisza explained that Rozenkevich is useful for Gestapo work, that preliminary work on his recruitment has been carried out, but that he in fact has not yet been recruited. "It is possible," Tisza said, "that he will come to you and will hand over some material. Accept this material and keep contact with him independent of what this material is, perhaps even of a purely business nature. After some time Tisza asked me whether Rozenkevich had handed something to me, I answered that within the framework of my business relations with him Rozenkevich had handed me personal scientific material on a measurement device for measuring neutron speeds. This material was not secret, but Tisza still demanded this work of Rozenkevich from me. I assume that this material when in the hands of Tisza had to be used for recruiting Rozenkevich for work in the Gestapo.

There was still another agent of the Gestapo at UFTI, but I don't know his name.

**Question**: Did you personally attract new people for intelligence work?

**Answer**: In the spring of 1937 when Tisza brought me into contact with the Gestapo agents he said that for work in the Gestapo I had to recruit leading employees of the high-voltage department at UFTI and established that Anton Karlovich Valter, the scientific head of the department, was suited for this. Tisza proposed to me to carry out this recruitment and said at the time that preliminary work with Valter had already been carried out, but the recruitment had not been completed, and he had so far not been attracted for the work. I said to Tisza that presently I could not fulfil this task as I knew Valter only very superficially. Tisza did not insist.

**Question**: Whom do you know from the agents of the Gestapo in the Soviet Union?

**Answer**: From Tisza I know that the foreman of the Ural Machine Factory in Sverdlovsk Rajter, Ritter or Rajter (I do not precisely remember) is also an agent of the Gestapo. In the summer of 1935 I was at the rest house "Sinon" in Sukhumi. There I met and made the acquaintance of this Rajtler or Ritter. Upon my return from the rest house Tisza asked me whom of the Germans I had seen there. When I mentioned Rajtler, he said that Rajtler was known to him as an agent of the Gestapo. On the evening of my departure on *komandirovka* to Leningrad in the spring of 1937 Tisza gave me the task that, if I were to meet a young German physicist in Leningrad, who had worked in England in the field of neutron and nuclear physics, I should establish contact with him, without however revealing myself as an agent of the Gestapo, and help this German physicist to get work at the Leningrad Radium institute. Tisza said that this fellow is an agent of the Gestapo and should carry out corresponding work in the USSR for Germany. But he did not mention his last name, since he was not sure that he indeed was in Leningrad. In addition Tisza also said that I knew him personally. After having arrived in Leningrad, I accidentally met the founder of atomic physics of the Danish academy Oscar Heil somewhat later at a lecture. When I saw him it dawned on me that Tisza's instruction concerned precisely him, Oscar Heil, since I indeed knew him, he indeed worked in England in the field of atomic physics and in Leningrad he was without work. His membership of the Gestapo was moreover clear by the fact that he as a German political emigrant, who allegedly had fled Germany after the fascists had come to power, had subsequently visited Germany several times without let or hindrance and returned from there. I approached Oscar Heil,[13] we spoke about general subjects; I learned that he was looking for work, and for the time being was busy with translations etc. I did not succeed in getting a job for him.

Written down truthfully in my own words, in Russian understandable to me, read through by me and signed.

*F.O. Houtermans*

Interrogated by the head of the third department of the third section of the State Security Administration, Second Lieutenant of State Security

*Pogrebnoj*

Correct: head of the department of the third sector of the UkSSR NKVD, Senior Lieutenant

*Shterenberg*

That a second version of his testimony existed is only known from the protocol of Houtermans' last interrogation (his 14th apparently) dated 3 December 1939. In this interrogation he retracted everything and said that he had invented the entire story, a remarkable feat in view of the treatment meted out to him by the NKVD,

---

[13]There was a person called Oscar Heil (1908–1994), a German electrical engineer and inventor. He studied physics, chemistry, mathematics and music at the Georg-August University in Göttingen and was awarded his PhD in 1933, for work on molecular spectroscopy. He had however, as far as we can tell, nothing to do with the Danish Academy, nor was he unemployed. At Göttingen, Oskar Heil met Agnesa Arsenjewa (Agnesa Nikolaevna Arseneva, 1901–1991; she was also in Leiden for some time), a promising young Russian physicist who also earned her PhD in Göttingen. They married in Leningrad in 1934. Together they moved to the UK to work in the Cavendish Laboratory, University of Cambridge. While on a trip to Italy, they co-wrote a pioneering paper on the generation of microwaves which was published in Germany in the *Zeitschrift für Physik* in 1935. Agnesa subsequently returned to Russia to pursue this work further at the Leningrad Physico-Chemical Institute with her husband. Here Houtermans may indeed have met him. However, Heil then returned to the UK alone; Agnesa, working in what had by then become a highly sensitive subject, was possibly not allowed to leave Russia. Back in Britain, Oskar Heil worked for Standard Telephones and Cables. At the onset of the Second World War he returned to Germany via Switzerland. During the war Heil worked on the microwave generator for the C. Lorenz AG in Berlin-Tempelhof. In 1947 Heil was invited to the USA. After doing scientific work he founded his own company called *Heil Scientific Labs Inc.* in 1963 in Belmont, California. Agnesa remained in the Soviet Union until her death in 1991.

which must have left him in a very poor physical condition.[14] After some preliminary interchanges, he said in answer to a question by the interrogator what he
pleaded guilty to:

**Answer**: I admit that I am guilty of making false statements, in which I slandered both
myself and several other people whom I knew as absolutely honest people. I also admit that
in some private conversations with acquaintances who visited my apartment, I criticized,
from an anti-Soviet position, certain measures taken by Soviet institutions and the government. An example was the law banning abortions. Such criticism is called
"counter-revolutionary" in Soviet circles, so I admit I am guilty here. I later carried on some
anti-Soviet conversations in my prison cell.

**Question**: While under investigation, you gave extensive testimony that you were a
Gestapo agent. Do you really deny that evidence?

**Answer**: Yes, I deny all my previous statements, since I was forced to make them while in
a poor moral and physical condition.

**Question**: What do you mean by poor moral and physical condition?

**Answer**: Before I began to make my statements in Kharkov, I was subjected to continuous
interrogation, with no opportunity to sleep, for about ten days. Moreover, I was placed in
this state in a cell where other arrested people bluntly told me what I should write in my
testimony and mentioned names of people known to me, about whom the investigators had
apparently told them.

In addition, the investigators read to me some statements about my wife. On the basis of
these statements, I was directly threatened that, if I did not make statements, my wife would
also be arrested, and my children would be sent to a children's home. The children would
be registered under new names, so that I would not be able to find them later on.

Subjected to all this, I decided to dream up some statements, the more so since the
investigators assured me that I would not be given any serious punishment by a Soviet court
and that I, as a foreigner, would simply be deported from the Soviet Union.

**Question**: Why do you consider the testimony that you gave at the hearing in Kharkov
false?

**Answer**: At the hearing, I gave testimony in two versions. The first version was that I had
been recruited by Lieutenant Schimpf, who put me in touch with the resident Gestapo agent
Tisza, who at one time worked in Kharkov and went abroad in 1937. I later testified that for
espionage activity Tisza put me into contact with Fomin, my assistant, with Rupp, an
electrical engineer at the Meteorology and Standards Institute in Kharkov, and with
Beniamin Margo, a paediatrician who also worked at a children's institution in the outskirts
of Kharkov.

Later I testified that I had understood from Tisza that the following people were carrying
out espionage for Germany: Rozenkevich, Valter, Shubnikov, Reiter or Rietter, and Oscar
Heil. That is a summary of the first version of my testimony. That testimony was false,
because none of the people I named had, to my knowledge, taken any part in German
intelligence work.

**Question**: Were you acquainted with the people named in your testimony?

---

[14]For some reason this protocol is not included in [2]; the version presented here is a combination
of the English translations in Refs. [10] and [7], pp. 205–209, checked against the original Russian
version in Ref. [9], pp. 93–97 (German translation pp. 66–69).

**Answer**: I knew Schimpf as an officer of the German Reichswehr back in 1931–1932 when he was a physics student at an institute in Berlin, where I was working as an assistant professor. I was not closely acquainted with him; I just remember his name as a student, and I did not know him as intelligence agent Schimpf.

I stated in my testimony that a certain "Hans" attempted to get in touch with me in England on Gestapo orders. I declare to the investigation that that person does not at all exist in reality. I dreamed him up to lay the groundwork for my trip to the USSR on an espionage mission.

Tisza worked as a researcher at UFTI for a year and a half or two years, and then left to take up residence in Hungary. I have no knowledge of him being an intelligence agent and I chose his name so that the primary initiator of the spy ring would be someone who could not be asked about this.

**Question**: Why did you take such a course of provocation?

**Answer**: In the cell, we, the arrested people, constantly discussed forms of giving testimony about espionage. We came to the conclusion that it would be best to name as organisers of the espionage activity people who were already dead, and it was considered very valuable when someone among the arrested could offer the name of some deceased Latvian or someone of another nationality; the other detainees would immediately start to include these names in their testimonies.

**Question**: Does this mean according to you that all detainees gave false testimony?

**Answer**: I should make the reservation that this obviously was not a mass phenomenon, but in the cell I was in it was a fact.

He now also explains why it was necessary to give testimony in two versions:

**Houtermans**: Obreimov's testimony in which he alleged to have recruited me for espionage work in 1929 was read to me. I had however declared to have started from 1933 with espionage. So I had to face the issue how to backdate my espionage experience, since after all I could not have recruited Obreimov for espionage work in 1929 if I myself had not yet been a spy at that time. I had to invent some new recruiter, since Schimpf, whom I had known in 1931–1932, did not fit this scenario. After considering this question, I decided to write a second version of my testimony, in which I would name Professor Westphal[15] as my recruiter. I wrote that Westphal had recruited me for espionage work in 1929 on the grounds that he had learned of my membership in the German Communist Party and by threatening me, one of his assistants, with dismissal he lured me into espionage work.

Houtermans was kept in prison for three years until May 1940 when he was declared an 'undesirable alien' and handed over to the Gestapo, like Weissberg. In Germany he was again put in prison, but freed through the intercession of the German physicist and 1941 Nobel Prize winner Max von Laue. During the war years he worked on nuclear technology in the private laboratory for electron physics of the physicist and inventor Manfred von Ardenne.[16]

---

[15]A German physicist who was a professor at the Technische Hochschule in Charlottenburg.

[16]After the war Von Ardenne (1907–1997) was taken to Russia by the Soviets where he got his own laboratory for industrial isotope separation. He returned to East Germany in 1954; became a professor in Dresden and ran his own research laboratory which after the German unification became Von Ardenne Anlagentechnik GmbH, a manufacturer of glass coating equipment, e.g. for the solar energy industry.

After the German invasion of the Soviet Union Houtermans was asked (or ordered) by the German authorities to take part in a mission to Ukraine to collect scientific equipment and scientists for transportation to Germany. The mission is said to have been organized by the German Air Ministry (the German Luftwaffe was one of the agencies in Germany with an interest in atomic research). He accepted and in October 1941 also visited Kharkov for a few days to study the state of the various institutions in that town, including UFTI. Reports on this visit differ. One version, told by Frish, states that Houtermans appeared at the gate of UFTI *"dressed in SS uniform, the very same person who had spoken so eloquently about his persecution by the fascists"* (Ref. [11], p. 296).[17] Apart from this story there is also a letter written by Pjatigorsky in the 1990s (at any rate before 1993 since Pjatigorsky died in that year) to Jury Ranjuk, one of the authors of [2]:

### Letter from L.M. Pjatigorsky

In the war years there was a scientific research institute on the territory of UFTI, which the employees who remained in Kharkov have christened "Eberti". This name was connected with the fact that the director was the son of Professor Ebert, a German physicist-chemist… Our Nina Mikhajlovna Shestopalova told me that once all employees of "Eberti" were told to appear and form up on the staircase from the first to the second floor [high-voltage building]: a guest was arriving. The employees came, formed up, and to the enormous amazement of those present the person who entered the building and started to climb the staircase was …. Houtermans! He shook hands with each employee and entered the office where before the war the director sat. He asked where Shpetnyi [director of UFTI] was. Houtermans was dressed in a new fascist uniform, and on his head he had a cap with a crown. He said that he had come to help the UFTI employees.

Houtermans arrived in Kharkov a month after its occupation by the German army and stayed almost a month at the institute. His visit served as proof that he was still a spy.

So also Pjatigorsky has it (from hearsay) that Houtermans was a spy, although not a member of the SS as they did not wear a cap with a crown. According to Frenkel (Refs. [8], p. 97; [9], pp. 126–136 (German translation, pp. 86–92)) the story is not true and Frish's story about Houtermans is full of 'annoying inaccuracies', but it has for a long time determined the opinion on Houtermans in the Soviet Union, where he was considered a traitor.[18] Houtermans was actually dressed in a Luftwaffe uniform, which is blue and could have been confused with the (black) SS uniform.[19]

No scientists or equipment were transported to Germany. It is true that most had been evacuated to the east, but there were still some pieces remaining, such as the huge Van de Graaff generator, which was the pride of UFTI and could have been transported to Germany. The episode is also discussed in Amaldi's book (Ref. [6],

---

[17]It is only in this commemorative volume that Frish talks about Houtermans; it does not appear in his memoirs [12].

[18]E.g. by Kapitsa to Born in 1945, see [13].

[19]Houtermans himself has declared that he mainly walked around in civilian clothes in Kharkov and Kiev (Ref. [9], p. 129 (German translation p. 87)).

p. 79)[20] where the daughter of Konstantin Shteppa, who shared a cell with Houtermans when imprisoned by the NKVD and with whom he later in 1951 wrote the remarkable book *Russian Purge and the Extraction of Confession* under the pseudonyms Beck and Godin, tells that she met Houtermans during that time in Kiev where he was looking for her father and came to their apartment. He was not at all a Nazi, but helped them in every possible way he could. Houtermans may indeed not have been a Nazi and there is no evidence that he behaved dishonourably while in Kharkov at that time, but it cannot be doubted that, after his experiences in Stalin's dungeons, he did not come to the Soviet Union as a friend. Perhaps he did not have much say in the matter and was simply forced to go. I doubt that his German employers, knowing his past, would have let him travel unsupervised and/or unaccompanied to various towns. Houtermans has never been very forthcoming with an explanation of this episode, which was also viewed with suspicion in the West (Ref. [14], Chap. 9).

Houtermans has himself also written down the story of his life in 1937–1940. It is presented below (Ref. [6], pp. 49–56):

"On 1 December 1937 I was arrested at the Custom House in Moscow, where I was preparing to have my property inspected for my departure from Russia. I was immediately brought to the Lubyanka prison where I was shown the order of arrest, dated November 27th from Kharkov, on account of Paragraph 28 (political reasons). After a quarter of an hour I was brought to the big Butyrka prison into a cell for 24 men. Gradually this cell was filled until it held 140 men, sleeping on and under wooden boards, about 2–3 men per m$^2$.

While still in Moscow, eleven days after my arrest I was called by an officer of the NKVD to give a full confession of my alleged counterrevolutionary activities on behalf of the German fascist government, but no concrete charge was brought against me; only the names of a number of my Russian and foreign colleagues from the Kharkov Physics-Technical Institute were mentioned as being members of a counterrevolutionary organisation, such as Shubnikov, Landau, Ruhemann, Weissberg, Fomin, etc. I was told that if I gave a full confession I would immediately be sent abroad. Of course I did not make a false confession and denied any activity against the USSR.

On 4th January 1938 I was brought up in a prisoner car by railway to Kharkov and put into the Kholodnaja Gora prison in Kharkov, in a cell which was still more overcrowded than that in Moscow, but without any sleeping accommodation so that we all had to lie on the floor. I remained there till 10th January when I was brought to the central prison of the NKVD, into a cell perfectly clean and not too overcrowded. Here many fellow prisoners tried to persuade me to make a false confession of things of my own invention as they had done themselves, which I would have to do anyhow sooner or later in order to save a lot of trouble. The same day I was again asked to make a confession by an interrogator, named Dresscher, who threatened to beat me and to get anything out of me.

In the evening of 11th January an uninterrupted questioning of 11 days began, with only a short break of five hours the first day and about two hours the second day. No concrete charge was brought against me, as in nearly all cases of people I have seen in Russian

---

[20]Amaldi's book is in general very sympathetic towards Houtermans; the book contains the reminiscences of Edoardo Amaldi (1908–1989), an important Italian physicist and one of Fermi's *Via Panisperna* boys in Rome, about Houtermans whom he met in Bern where Houtermans was professor from 1952.

prisons, and I was told to give all "facts" myself. The only two questions that were asked were: "Who induced you to join the counterrevolutionary organisation" and "whom did you induce yourself?"

Three officials questioned me in turn, for about eight hours each, the first two days I was allowed to sit on a chair, later only on the edge of a chair and from the 4th day on I was forced to stand nearly all day, I was always kept awake, and when I fell from lack of sleep I was brought to by means of [fainted] cold water that was poured on my face. The chief official who led the questioning was named Pogrebnoi.

The night of 22nd January, shortly after midnight, Pogrebnoi showed me an order of arrest for my wife and another order to bring my children into a home for "*besprizorniks*" (homeless children) under a false name so that I would not be able to find them ever again. I was of the opinion that they were all still in Moscow. I have learned since that they had left shortly after my arrest so all I was told was bluffing, but in my state of weakness after nearly ten days without sleep I fell for it.

In this state I fell unconscious nearly every 20–30 min but I was awakened every time and my feet were so swollen that my shoes had to be cut off. I was beaten only occasionally, and not with instruments as many other prisoners I have seen, and I was told by them that the treatment I had to undergo myself was very mild indeed compared with what they had to endure. At the end I declared, I was ready to sign any statement they wanted on condition that my family was to be sent abroad immediately and I would be shown a letter from abroad by my wife telling me her whereabouts, after three months. In case I would not get such a letter I would revoke any statement I had made. I signed a short statement as they asked me, admitting that I was sent to the USSR by the German Gestapo for espionage. Then I was able to eat luxuriously and got tea and was sent to my cell to sleep where I slept for about 36 h. Then I was asked upstairs again and there I wrote a long confession of about 20 pages in German and I was very careful to give only names of people whom I knew to be abroad, or whose evidence against me—of course forced by 3rd degree methods—was shown to me. I had to write about espionage, sabotage and counterrevolutionary agitation and I was absolutely free to invent anything I liked, no corroboration by facts or by evidence being needed. I made nuclear physics the theme of my espionage, though at that time no technical applications of nuclear physics were known, since fission was not yet discovered, but I wrote a lot of phrases that nuclear energy is existent and that it needed only the right way to start a chain reaction as described in popular novels on this matter. Another instrument I wrote I had spied on was an instrument for measuring absolute velocities of airplanes by the number of magnetic lines of force which went through a coil, a device contradicting the law of conservation of energy, and being obviously a perpetual mobile. I intentionally made my confession as stupid as possible in order to be able to testify that it is nonsense in case of a trial, and I put in a short statement in English in ciphered form that I was under third degree torture and that all I wrote was pure invention.

During the last year of my stay in Kharkov many acquaintances I knew perfectly well as being innocent had been arrested already and it was said that they all had given evidence of being guilty. I did not know then about the methods how these statements were forced from people but I had told my wife in case a signature should ever be forced from me, I would leave out the full stop after my signature, and in case my signature were given by my free will I would always put a full stop after my name. I had the opportunity to do so and I left out the full stop in the written confession. My written confession was translated into Russian[21] and I was left alone and was not troubled anymore till August 1938, living till March in a clean prison cell not too overcrowded in the central Kharkov prison.

---

[21]At this point he must have given the (invented) confession printed above involving Tisza and others.

On 17th March I was called again and a letter from my wife dated from Copenhagen was given to me. The same day I was transferred to the Kholodnaja Gora prison in Kharkov, to a small cell, rather dirty and very overcrowded, where I remained till 2nd August. Food was very scarce and we suffered from hunger. The daily rations consisted of 600 g of black bread containing more water than ordinary bread (equivalent to about 500–550 g of ordinary bread), about 15–20 g of sugar, a mug of soup containing little nourishing value and 1–2 spoons of porridge of some kind a day, from fair estimates made by physicians I met and by myself about 500–1000 kcal per day. Food was always given regularly and I don't know of any cases that prisoners were not given their rations. Treatment by prison officials was hard but not sadistic, but there existed cells where conditions were much worse for people who had not given the confession or evidence wanted. I remained there till 2nd August when I was sent to Kiev in a "Stolypin car", a special sort of railway car for prisoners. I remained in Kiev till 31st October 1938 when I was asked to give more evidence especially against a friend of mine, Professor Lejpunsky, a member of the party and an absolutely sincere man. From prisoners in my cell I learned that he was arrested in another cell in Kiev and a man in my cell tried to persuade me to give evidence against him and told me what I should say. No especially hard pressure was used against me then and therefore I did not give any evidence against him or against Professor Obreimov, another member of our institute that I was asked to accuse. Prison conditions in Kiev were much better than in Kharkov, the rooms being very clean and food a little bit better. It was hard though because it was not allowed to sleep during the day time.

On 1st October 1938 I was sent back to Kharkov and put into a clean cell in the central prison. Prisons were not so overcrowded any more at that time, but still here were 1–2 prisoners per $m^2$ of room. I was not questioned again till January 1939 when I was asked to sign an application for Soviet citizenship. For that case they promised me the chair of a big institute for my research, to be built by the NKVD itself, but I did not consider that offer to be sincere, having met foreigners in prison cells who had agreed to such an offer without having been released, and therefore I said I could talk about this matter only after release and after communicating with my family. This was the only time I got some of the things that were sent to me by my wife and by Mrs. Cohn-Vossen, a friend of mine in Moscow.

I got a blanket and a few pieces of underwear. I did not get any letters nor any money that was sent to me from abroad, as I have learned since. This was rather bad because all the time there was the possibility to buy some additional food supply and smoking material for about 20 roubles a fortnight, and this helped a great deal but since I had less than 100 roubles on me when I was arrested I nearly never could make use of this possibility and therefore I had lost about 18 kg in weight and became more and more feeble. I could not consider a revocation of my confession of the year before and when I was asked to give more evidence against persons, such as Obreimov whom I knew to be in the USSR, I declined but I confirmed my former confession, not wanting to have all the trouble over again. On the new evidence they were not pressing very hard.

In February 1939 I was sent again to Kiev where I was put again in the central prison but in an underground cell without any daylight (artificial light was in all cells during the night) which was very humid. I was asked again by a new official to give evidence against Obreimov and Lejpunsky and I was threatened with beatings in case I refused and shown written evidence of both men against me in their own handwriting. I was very weak by then, I could hardly walk about and so I decided to confirm their statements on counter-revolutionary activity about myself and that I knew about theirs. I put in some slight discrepancies concerning dates etc. with their evidence and my evidence was accepted. Again I was told I would be sent abroad.

In May 1939 I was asked by the People's Commissar of the Interior of Ukraine himself to give evidence against Professor Fritz Lange, a good physicist and friend of mine who was working in the Ukrainian Physics-Technical Institute, and also against Professor Landau,

Professor Ioffe and Professor Kapitsa, all of them prominent physicists of the USSR. He told me he knew well that all of these people were active spies and members of a counterrevolutionary organisation and he only wanted me to confirm this. I said I knew nothing about it but did not try to revoke my own statements given earlier. This confirms the fact I had often heard about in prison cells, especially by men who once had been officials of the NKVD themselves, that it is quite usual to collect evidence about counterrevolutionary activity of prominent people in case their arrest should be effectuated later on. Neither Lange, nor Ioffe or Kapitsa have ever been arrested as far as I learned since.

No paper or books were allowed in prison cells and therefore it was nearly impossible to do any work. Yet from the very beginning of my prison time, I decided to work under all conditions and, since it was the only field I could do, I started already at the end of 1937 to think about problems of the theory of numbers. All I knew was Euclid' s proof about the existence of an infinite number of primes and I started thinking on the problem whether there exists an infinite number of the type $6x + 1$ and $4x + 1$ also, while for the $6x - 1$ and $4x - 1$ I could find Euclid's proof to hold with a slight alteration off hand.

I had no writing materials, so I tried to write some numbers with matches on a piece of soap or on places of the wall where it could not be seen, but I had to erase it all every day before leaving the cell for the toilet.

I thought about that problem for more than a year and finally in Kiev in the first days of March I found that any form $x^2 + xy + y^2$ with $x$, $y$ being relative primes cannot contain any other factor than primes of $6x + 1$ type or 3, and the sum of the squares of relative primes contains only primes $4x + 1$ or 2.

After solving this problem I discovered Fermat's theorem (I only learned its name after I left the prison, as with all theorems I found) and quite a number of theorems in the elementary theory of numbers.

When I found on 6th August an elementary proof for Fermat's famous problem for *n = 3,* which as I have learned since is essentially the same as Euler's, by "descent infinite", I got very excited about it, because I did not know Euler's elementary proof existed, and I applied to the People's Commissar of the Ukraine to get paper and a pencil. (I said I wanted to work out an idea of mine on a method in radioactivity which might be of economic importance). When my petition was not granted, I went on hunger strike (only declining food, not water). I was alone in a cell then and succeeded in getting paper and a pencil after 8 days of hunger strike, by which time I was very much weakened since I had been in a bad state when I started. I wrote a number of theorems: I had found the so-called indices of theory of numbers, a theorem of Lucas and a new proof of a theorem of Sylvester which is in course of publication at the Jahresbericht upon the advice of Professor van der Waerden whom I told about my prison studies in the theory of numbers. I could even keep my writing materials when Professor Melamet (a philosophy Professor from Odessa) was put into my cell, so that I could make steady progress in the theory of numbers.

In August[22] all the evidence I had given one and one half years earlier was rewritten and I was summoned up together with Professor Obreimov for a so-called "double questioning" in which he—of course it was all pure invention—stated before my eyes that I had induced him while still in Berlin to do espionage work for the Nazis, though at the time of his visit to Berlin the Nazis were not in power and a quite small party. I affirmed all his statements because I did not want and, in the state of health I was in, could not afford to suffer all the tortures by which I was threatened.

---

[22]This must have been August 1939 and after the confrontation with Obreimov he must have felt the need to invent the second version of his confession mentioned above.

Suddenly on 30th September 1939 I was called out and brought to the station in a closed car and sent to Moscow. I did not know about the war till January 1940. The isolation of prisoners is extreme in Russia, the only source of information being what is told by newly arrested prisoners and I had not seen such people for a considerable time.

In the train I saw that the official, who had questioned me before, travelled with me in the same train and in Moscow I was brought immediately to the central prison of the NKVD on the Lubyanka. While I was still in the shower bath that everybody arriving there has to go through, I was already called for questioning. I was brought into a luxuriously furnished room in which a man in the uniform of a general of the NKVD sat and beside him in civilian clothes a very intelligent looking man who presided and who asked me politely to sit down. Then he asked me what I felt guilty of. I asked: "Do you want to hear what confession I signed or do you want facts?" "Of course facts," he replied. "This is the first time I am asked this question within these walls," I said. "But since you want to know, the only thing I feel guilty of is that I stole a pair of underpants in the Kharkov prison a year ago, removing the prison stamp on them by calcium chloride in the toilet. That's all". "And what about your confession?," he asked. "That's all pure invention!" Then he asked who had forced me to give a confession and by what means I was forced. I gave all the names as far as I knew them and all details. "We are going to get it all cleared up", he said shortly and I was brought back to my cell, a good cell, where I was alone.

I liked it better that way, since I could work. Everything was extremely clean and I got books, very good ones, too, special food in quite sufficient quantity and a package of cigarettes every day.

Though my Kiev manuscript had been taken away from me when I entered the Lubyanka, I got writing materials again without any effort and I went on to occupy myself with what I had heard since to be "Pell's problem" and other things in the theory of numbers. In this cell I remained without being called a single time until the beginning of December 1939, being alone the whole time. After all I had gone through it was a treat.

In the first part of December I was called up again by another official who asked me absolutely correctly about everything and I answered all questions truthfully.[23] When I asked to write or to cable to my family (I supposed them to be in England, from where I had last heard from them in August 1938) he said I would soon be sent out. I then asked especially not to be sent to Germany and he made a note of it.

About a week later I got new clothes and was sent to Butyrka prison, into one of the big cells where I had been 2 years previously, but now it was not overcrowded there. All people in the room were Germans, not all of them foreigners, some had taken Soviet citizenship. Among them I was glad to meet another Professor, Professor Fritz Noether,[24] former Professor at Breslau University for applied mathematics and later a refugee living in Tomsk. He had been arrested as a German—though being a Jew—and was forced to invent an espionage story also. But in contrast to my case a sentence of 25 years of imprisonment had been passed on him. Shortly after my arrival he was removed from the cell and I have never heard about him since. In this cell we all got special food in sufficient quantity and cigarettes and we had the impression that we were kept there because most of us were in a very bad state of health and they did not want to send us abroad like that. Most of the people were German workers, skilled workers, or engineers, specialists and many of them

---

[23]At this stage Houtermans' last interrogation, related above, must have taken place.

[24]Brother of the famous mathematician Emma Noether (1882–1935). Fritz Noether was first convicted to 25 years of imprisonment and served time in different prisons. On September 8, 1941 he was again sentenced, this time to death on the accusation of engaging in anti-Soviet agitation. He was shot in Orel on September 10, 1941 (*Wikipedia*).

former communists. Among them was Hugo Eberlein, friend of Lenin and Liebknecht and former member of the executive control committee of the Comintern, president of the communist fraction of the Prussian Landtag for many years. He had been beaten severely, like nearly all of them. Some were called out and presumably sent abroad, some arrived directly from camps in Siberia and the far North.

In March I was called out alone and asked to sign a paper, agreeing that I would not talk about what I had seen in Russian prisons and that I would agree to do secret work for the USSR abroad. This I signed because I had learned from many people that most of them were asked to sign such a paper, otherwise one would be kept indefinitely.

I again asked as a condition not to be sent to Germany and this was promised to me by the official who made me sign the paper.

On 17th April 1940 some of us were gathered into another cell in the same building and on 30th April we all were called out, a sentence was read to each of us, that we were condemned to be exiled from the USSR by a special court of the NKVD, and we were transported in a prison car to Brest-Litovsk where we all were taken over by the officials of the Gestapo.

We were not set free, but taken to a German prison in Biała Podlaska, a small town near the frontier line and after some days we were all transferred to the citadel of Lublin.

Isolation was not as strong as in Russian prisons, the regime was more military and food and accommodation conditions much worse than the last time in Moscow. Every day we heard the songs and the noise of drunken Gestapo officers below our windows, while we learned that every day about a hundred Poles and Jews were executed in the prison court.

We had passed the frontier on 2nd May 1940 and were transported to Berlin on 25th May. Some of us were brought to a "Nazi-Rückwandererheim" where they were set free after a few days, but some of us, among them also I, were brought to the police prison on Alexanderplatz. By the way the only prison in my experience where there were lice. Here I met people from concentration camps who told me about German camps and a well experienced communist who advised me how to behave in front of the Gestapo.

A week later I was brought to a small prison at the Gestapo headquarters in the Prinz-Albrecht-Strasse where I was asked about my Russian experiences, why I had left Germany and gone to Russia, and about some communist friends of mine in Germany before 1933. I told them I had known those people but I did not know about any illegal activity of theirs, confirming my information nevertheless on such people I knew to be abroad. I was asked to give an account of my Russian experiences, which I did, also mentioning by precaution the paper I had been made to sign, but not the fact that I had asked not to be sent to Germany.

On 16 July finally I was set free. A few days later I met Professor von Laue from whom I learned the whereabouts of my family. As soon as he had heard that I was in Germany in a Gestapo prison he went there himself, brought me some money and did all he could to accelerate my liberation."

19th May 1945 F.G. Houtermans

———

Martin Ruhemann (Fig. 12.4), the head of the second cryogenic laboratory at UFTI, and his wife Barbara were not arrested. Both had British passports and after their dismissal from UFTI managed to get exit visas to get out of the country.

There were still some other foreigners, such as the German nuclear scientist Fritz Lange (1899–1987), mentioned by Houtermans above, who had been a student of Walther Nernst, the Nobel Prize winner for chemistry in 1920. Lange had

**Fig. 12.4** Martin Ruhemann
(*from* Verkin et al. (1990))

participated in experiments in Germany and Switzerland using lightning for
obtaining fast electrons, which could possibly be used to achieve an artificial
nuclear reaction. He was the only one of the foreign physicists who remained in the
Soviet Union until the end of World War II, after which he went to the DDR. He
became a Soviet citizen in 1937, escaped repression during that time as by a
miracle,[25] participated in the Soviet atomic bomb project and returned to Berlin in
1959, where he started to work in biophysics.

Other foreigners were simply dismissed or their contracts were not extended,
after which they left the country.

With all these people gone UFTI as a scientific institution was essentially
beheaded. Apart from Slutskin all department heads had been removed. It had
ceased to exist as a centre for theoretical and experimental physics of European
significance. Anton Valter and Kirill Sinelnikov were essentially the only more
senior of the Leningraders who survived the purge. They had refused to go along
with Landau and Shubnikov's protests and actions and were left untouched,
showing again that the arrests were not random but targeted at those who had
shown independence of thought and displayed 'anti-Soviet' behaviour, at any rate
behaviour that could be perceived as such. This argument is strengthened by the
fact that both Sinelnikov and Valter possessed a number of the 'objective characteristics' that would warrant arrest (Sinelnikov had stayed abroad, and even married
a foreign woman), while Anton Valter, as a member of the nobility, belonged to the

---

[25]A story about this is recorded in Ref. [2], pp. 308–309. "Lange had obtained a summons to the
recruitment office. There he drew the attention of the doctor because of his clumsiness and poor
knowledge of Russian and, of course, his foreign name. At the check for flat feet Lange overturned
a basin with water, which was too much for the patience of the commission. The suspicion on the
conscript was communicated to 'the necessary place', from which 'specialists' slowly started to
arrive and interrogated the scientist in a separate room. All was clear—he is a spy. On the question
whether apart from his passport he had some other documents Lange at first answered in the
negative. But when he was told to get dressed and to proceed to the car, he suddenly found in his
pocket a small booklet and handed it to the Chekists. It was a certificate of identification. One can
only guess what they thought when seeing this document. It was graced with Stalin's signature. He
sometimes did sign such certificates."

'former people' and was moreover an ethnic German. As recalled in Chap. 7 from the letters of his wife to her sister in England, Sinelnikov had a lot of headaches from the trouble at UFTI, but what he did about it remains unclear. He may have been interrogated by the NKVD in the spring of 1937 as many staff members of UFTI were, but no protocol has been made public like the ones for Gej and Shavlo discussed earlier.

Many others left (e.g. Pomeranchuk, Lifshits, Rjabinin) and mostly junior people, like Akhiezer in the theory division, had to continue the research. UFTI essentially never recovered, partly also due to the German invasion in 1941, which definitively terminated the remaining scientific work in Kharkov. But also those who remained were broken in spirit and their initiative paralyzed. In the interactions with colleagues there was suspicion and estrangement. The scientific work was shrouded in a regime of secrecy with its supervisors, denunciations, permits and suchlike, which was unfavourable for the development of science. A postgraduate student of Lejpunsky, Aleksandr Iosifovich Shpetnyj, was appointed director of the institute. He could hardly be called a scientist, but was more a party worker relying on the regional party committee and the NKVD (Ref. [2], p. 317).

# References

1.  A. Weissberg, *Conspiracy of Silence* (Hamish Hamilton, London, 1952).
2.  Ju.V. Pavlenko, Ju.N. Ranjuk and Ju.A. Khramov, *"Delo" UFTI 1935–1938* (The "UFTI" Case 1935–1938) (Feniks, Kiev, 1998).
3.  Robert Conquest, *The Great Terror: A Reassessment* (Pimlico, London, 2008).
4.  P.E. Rubinin, P.L. Kapitza and Kharkov. Chronicle in letters and documents, *Low Temp. Phys.* 20 (1994) 550–578.
5.  V.S. Kogan and V.V. Sofrony, *Sotrudniki UFTI–uchastniki velikoj otechestvennoj vojny* (UFTI employees–participants in the Great Patriotic War) (Kharkov, 2008).
6.  Edoardo Amaldi, *The Adventurous Life of Friedrich Georg Houtermans, Physicist (1903–1966)* (Springer Verlag, Heidelberg, 2012).
7.  M. Shifman, *Physics in a mad World* (World Scientific, 2016).
8.  V.Ja. Frenkel, Novoe o Fridrikhe Houtermans, *Priroda* 8 (1992) 92–99.
9.  V.Ja. Frenkel, *Professor Fridrikh Houtermans–Raboty, Zhizn', Sud'ba* (St. Petersburg, 1997); German translation: *Professor Friedrich Houtermans–Arbeit, Leben, Schicksal. Biographie eines Physikers des zwanzigsten Jahrhunderts* (Max Planck Institute for the History of Science, 2011).
10. V.Ja. Frenkel and Bruce R. Doe, Fritz Houtermans in Bad Times and Good, *Physics Today* 47 (1994) 104–106.
11. S.E. Frish in B.I. Verkin et al. (1990), p. 292–297.
12. S.E. Frish, *Skvoz' prizmu vremeni* (Through the prism of time) (Solo, St Petersburg, 2009).
13. I.B. Khriplovich, The Eventful Life of Fritz Houtermans, *Physics Today* 45 (1992) 29–37.
14. Paul Lawrence Rose, *Heisenberg and the Nazi Atomic Bomb Project: A Study in German Culture* (University of California Press, 1998).

# Chapter 13
# The Landau-Korets-Rumer Case

As mentioned earlier, Landau had escaped to Moscow in January 1937 and possibly again in August at the time of Shubnikov's arrest. He worked for a year at Kapitsa's Institute for Physical Problems and was only arrested on 27 April 1938 (Fig. 13.1). Why was his arrest so late, one wonders? Houtermans was arrested in Moscow on 1 December 1937 at the request of the Kharkov NKVD, so its arm was long enough to reach Moscow. Why did such a thing not happen to Landau, who was considered by the NKVD *the* leader of the anti-Soviet organisation operating at UFTI? It is unlikely that Kapitsa's institute was a no-go area for the secret police, who even had access to the private homes of the high and mighty in the land. Ranjuk also asks this question and says (not very convincingly, although a more convincing reason is not immediately apparent) that the reason must be his world-wide renown and his timely departure from Kharkov to Moscow (Ref. [1], p. 88). In the NKVD documents published in 1991 it is stated that the statements by Shubnikov and Rozenkevich were the immediate reason for Landau's arrest (Ref. [2], p. 152), but these statements had been made more than a year before. So they had not instilled any direct urgency to proceed to action. But why did they need them anyway as Landau had already been singled out as the leader of this anti-Soviet group before Shubnikov and Rozenkevich had been arrested. For some reason all this was not considered enough for arrest and the NKVD thought that more was needed. Something held them back, so perhaps it was indeed Landau's international renown, as Ranjuk suggested, although it is doubtful that the NKVD cared very much about what the rest of the world might think.

A few days after Landau's arrival in Moscow Moisej Korets (Fig. 13.2) also went to the capital. After his surprising release from prison in 1936 Korets had left UFTI and moved to Voronezh were his parents lived. This is clear from a message dated 5 July 1937 sent by the NKVD in Kharkov to their counterpart in Voronezh: "*According to our information, citizen Moisej Abramovich Korets, born in 1908, engineer-physicist, lives and works in Voronezh. Korets was by us under inquiry as a member of a counter-revolutionary Trotskyist sabotage organisation. We arrested Korets in 1935, but his guilt was not fully proven as a consequence of which Korets was not convicted and the case was stopped.*

© Springer International Publishing AG 2018
L. J. Reinders, *The Life, Science and Times of Lev Vasilevich Shubnikov*,
Springer Biographies, https://doi.org/10.1007/978-3-319-72098-2_13

**Fig. 13.1**  Picture of Landau taken by the NKVD while in prison in 1938 (*from* Vinogradov and Mikhajlov (1991))

**Fig. 13.2**  Moisej Korets in 1935 (*from* the website *secrethistory.su*)

*We have now proceeded to liquidate the entire counter-revolutionary sabotage group at UFTI and from material obtained by us in the inquiry it has been established that Korets is one of the active members of the said counter-revolutionary group and a close friend of the leader of this group the Trotskyist Professor Landau.*

*We intend to arrest Korets. Please urgently establish the whereabouts of M.A. Korets and place him under active observation until his arrest and inform us of all materials obtained.*"[1] But they had missed him and were poorly informed, for already in February 1937 Korets had followed Landau to Moscow, where the latter helped him to obtain a teacher's job at the Moscow State Pedagogical Institute. It shows that it was still possible in those days to escape the watchfulness of the NKVD by moving around in the country. Conquest (Ref. [3], p. 259) states that *"moving frequently was a certain protection, since it usually took "at least six months or a year" before the local NKVD paid much attention or accumulated*

---

[1]From the documents of the Korets affair (*Delo Koretsa*) available at http://www.ihst.ru/projects/sohist/document/ufti/korets.htm.

*enough evidence against a figure whom there was no exceptional reason to per-secute. (...) Siberia, in particular, was a good place to go."* So had Shubnikov accepted Trapeznikova's proposal to go away to a distant part of the country (see Chap. 11) he might also have escaped. It is surprising, however, that Korets could get a job at a state institution in the capital without too many problems. He also started to write popular articles on science, a few of which were published in the journal "Technology for Youth" (*Tekhnika molodëzhi*). But the main issue here was that in Moscow Korets continued his deadly political struggle with the communist regime. And about a year after his arrival in Moscow, in April 1938, he had written the pamphlet which led to his arrest and that of Landau and Rumer. (Ref. [4], p. 109.)

The NKVD waited with their arrests until they had heard from informer reports that Korets and Landau were planning to distribute among the public on 1 May 1938 a leaflet with the following text[2]:

"The great cause of the October Revolution has been basely betrayed. The country is flooded with torrents of blood and filth. Millions of innocent people are thrown into prison, and no one can know when his turn will come. The economy is falling apart. Famine is approaching. Don't you see, comrades, that the Stalinist clique has carried out a fascist coup. Socialism only exists on the pages of the patently lying newspapers. In his rabid hatred of genuine socialism Stalin is similar to Hitler and Mussolini. Destroying the country for the sake of his own power, Stalin is turning it into an easy prey for brutal German fascism. The only way out for the working class and all the working people of our country is a resolute struggle against the fascism of Stalin and Hitler, the struggle for socialism.

Comrades, let's organise ourselves! Do not be afraid of the NKVD executioners. They are only capable of beating defenceless prisoners, catching unsuspecting innocent people, plundering the people's property and inventing ridiculous lawsuits about non-existent conspiracies.

Comrades, join the Anti-Fascist Workers Party. Contact its Moscow Committee.

Organise AWP groups in the factories. Use clandestine techniques. Prepare a mass movement for socialism with agitation and propaganda.

Stalinist fascism only rests on our disorganisation. The proletariat of our country that has overthrown the power of the tsar and the capitalists, will also be able to overthrow the fascist dictator and his clique.

Long live the first of May—the day of the struggle for socialism!

The Moscow Anti-Fascist Committee of the Workers Party"

This is obviously very inflammatory material, and it should have been clear to Landau and Korets that distributing such a leaflet would be akin to suicide. So the first question to be asked is: Is it actually genuine? If so, why did it not result in the immediate execution of the suspects? Or is it another NKVD fabrication or provocation? As argued above the NKVD did not really need this leaflet to carry out the arrest, but it seems beyond doubt that it was the immediate reason for the arrest, discovered just a few days before May Day, on which it was to be

---

[2]The Russian text can be found in Ref. [2], p. 146-147.

distributed. Would no arrest have taken place without this leaflet? Was it created by the NKVD to stand stronger against any protests from abroad? Had they perhaps been told to tread with caution as regards Landau and was this leaflet fabricated to prove to their superiors that this Landau should be dealt with immediately? Boris Ioffe (b. 1926), one of the few people still alive who passed Landau's theoretical minimum exam, thinks it is a fabrication. He knew Landau well and one of his arguments is that: "*Only someone set on martyrdom could have elected to prepare and distribute such a leaflet. Landau wasn't that kind of person.*" (Ref. [5], p. 27.) Gorelik [6, 7] thinks it is genuine, and to arrive at that conclusion he follows the development of Landau's ideas on politics over the years, claiming that he now realized that the Stalinist system was no longer the one he had imagined and that he had decided to fight it. That seems a rather naïve approach to take for a rational person like Landau. By fleeing from Kharkov to Moscow, Landau had already realized that he had no chance against the NKVD. He was well aware of the fate of his friends Shubnikov and Bronshtejn, who had disappeared without a trace. And he had kept silent when they were arrested and tried, making himself as scarce as possible. If he had wanted to do something, why not mobilize the physics community against the arrests made in Kharkov, with the help of Kapitsa and others in Moscow? Yet he did nothing. He is obviously not to blame for this and he may indeed have discussed such matters with Kapitsa. But then suddenly, just a year later, he would be prepared to commit a suicidal act? That does not make sense, whatever his political development. Could it be that Gorelik has in some sense become a victim of the NKVD's deception tactics displayed in the documents made available? After all, the entire 'anti-Soviet, counter-revolutionary Trotskyist organisation' was an NKVD fiction. Neither Landau, nor Shubnikov or Weissberg or Bronshtejn, or any of the other scientists were members of an organisation intent on sabotaging Soviet science. They indeed had ideas about the organisation of science and wanted to have some freedom to pursue their own interests in that respect, but were in no way anti-Soviet or counter-revolutionary. Only when you accept this counter-revolutionary organisation as existing, be it only subliminally, this leaflet makes some sense. If there is no such counter-revolutionary organisation, this leaflet just simply appears from nowhere and could well be a fabrication.

But there are more voices claiming that this leaflet is genuine. Another well-known theoretical physicist, Evgeny Lvovich Feinberg (1912–2005), a close friend of Andrej Sakharov, seems to take it for granted in his memoirs that Landau and Korets were involved in writing this leaflet. He describes Korets as the main mover of the idea to distribute a leaflet, who wrote the basic text, as also follows from the protocol of the interrogation of Landau published in *Izvestija TsK KPSS* in 1991 (Ref. [2], pp. 137–146), showed it to Landau and asked for comments. He allegedly also showed it to another member of the 'group', a person Korets had known from childhood, who betrayed them to the NKVD. This would imply that more people were involved. Feinberg claims to know the name of this traitor, but does not reveal it. (Ref. [8], p. 395.)[3] The protocol of Landau's interrogation makes clear that more people were involved, but that Landau had not recruited them and did not know, nor wanted to know, who they were. Feinberg's story reads as if he heard part of the story from Korets himself. Boris Ioffe says about this: "*In the 1970s, already released, Korets got to talking: he said that yes, he had written the leaflet. Possibly, though, it was easier for him to say this than to confess he'd signed the deposition under torture*" (Ref. [5], p. 27).

Vitaly Ginzburg, on the other hand, is not sure that the leaflet was actually written by Landau and Korets, although he also must have known that Korets admitted it after his release. Ginzburg doubts that "*Landau understood the essence of Stalinism much earlier in the same way as we understand it today*" and is inclined to see this leaflet as a provocation by the NKVD, but also accepts Gorelik's investigation into its authenticity (Ref. [10], p. 271). To understand the essence of Stalinism is however not as big a deal as Ginzburg wants to suggest. As mentioned in Chap. 7, Stepan Shavlo, who was later used by the NKVD as a witness for the prosecution in the UFTI affair, had declared at a meeting of the party purge commission at UFTI "*that he could not see any difference between the Hitler dictatorship and the Stalin dictatorship*". So Shavlo understood the essence of Stalinism already in 1934, and he will not have been the only one. Landau may actually have heard his statement at that meeting.

The protocol of Landau's interrogation referred to above is dated 3 August 1938 (supplemented by 'personal testimony of Landau L.D.' dated 8 August 1938 (Ref. [2], pp. 147–149), which essentially repeats the protocol) and is, supposedly, some sort of summary or resume of the interrogations of the preceding months. That there must have been earlier statements by Landau follows also from Jury Rumer's declaration that already on the third day after his own arrest he was shown and asked to confirm a letter Landau had written to Yezhov admitting that he had organised a group of physics professors with the aim to undermine theoretical physics in the country (Ref. [11], p. 212).

In view of the tactics generally used by the NKVD, it seems reasonable to assume that the statements allegedly made by Landau were made under pressure, probably extreme pressure, and have perhaps even been dictated by the investigator of the case. Landau's anti-Soviet attitude is traced back to his Copenhagen days when he was still a loyal Soviet citizen and convinced that the "*Russian revolution had brought individual freedom to many and wanted to believe that it would do so*

---

[3]Feinberg says that "when the war began this man volunteered to the front (those suspecting him thought—to redeem his fault by death) and perished". Korets was in contact with a group of students at the Moscow Institute of History, Philosophy and Literature, to which the poet Pavel Kogan (1918-1942) belonged. He apparently had a copy of the leaflet, given to him by Korets. Kogan was not arrested and a few years later indeed went to war where he died in action in 1942, so it is very tempting to conclude that Kogan went to the NKVD with the leaflet, a conclusion made by the Russian/Israeli historian and journalist Mikhail Heifetz in an Israeli newspaper in 1991 (referred to, but doubted in [9]).

*increasingly*" (Ref. [12], p. 109).[4] Then slowly the image is built up of an anti-Soviet person who becomes involved in a counter-revolutionary organisation, which he joins out of revenge for the arrest and conviction of his father, we are told, and who wants to reorganize the Soviet state on bourgeois-democratic principles. The distribution of such a leaflet would then just be the next logical step in this process. The set-up of the protocol is such that the reader is inclined to at least allow the possibility in his mind that all this could be true. Korets convinced Landau of the necessity to go over to 'mass agitation' (a rather absurd idea in the situation Landau was in, of a rather lonely scientist at an institution that was still very much in the construction stage where there was hardly a soul he could discuss physics questions with, let alone switch to 'mass agitation'). Landau was prepared to cooperate on the strict condition that he would only be involved in discussing the text of the leaflet, but not be given any data on other people involved in the production and distribution of the leaflet.[5] But he explicitly admits to having been informed by Korets of the existence of such kindred spirits.[6]

The protocol contains some rather improbable confessions on Landau's part, such as, regarding the leaflet that it would be stupid to 'directly advocate a capitalistic system' as not much success could be expected from such an approach in the country, nor could it be written as originating from 'rightists' or Trotskyists, but it had to be phrased as coming from an organisation to the left of the Soviet regime. This makes it look as if the leaflet was some sort of gimmick, not reflecting serious convictions of the composers of the text.

In the final statement on the case (*spravka*) compiled by an NKVD official in April 1939, it is stated that, when interrogated anew earlier in April by this official,[7] Landau retracted all his earlier statements. In these statements he had admitted to having joined an anti-Soviet group out of bitterness and as revenge for the arrest of his father in 1930 for sabotage in the oil industry and to having carried out sabotage work at UFTI in Kharkov together with Rozenkevich, Shubnikov, Korets and Weissberg, by disrupting important scientific work of the institute that was of significance for the defence of the country, and to having harassed young talented specialists, such as Rjabinin and Strelnikov. All this is virtually the same as what appears in the statements of Shubnikov and others. Landau also mentioned Kapitsa and Semënov as participants of the anti-Soviet organisation, although he later 'refined' this by saying that he actually counted on their participation, but had not decided to full openness as he did not know them well enough. The leaflet drawn up

---

[4]His loyalty to Russia and its regime in those years is also clear from the interview which he gave to the weekly *Studenten* in Copenhagen in 1931 and has been partly reprinted in Ref. [12] (p. 109 ff).

[5]Since Korets also wrote the leaflet, one wonders why he needed Landau.

[6]As mentioned earlier, Korets was in contact with a group of students at the Moscow Institute of History, Philosophy and Literature who were going to copy and distribute the leaflet. Apart from Korets, Landau and Rumer nobody else seems however to have been arrested in connection with this.

[7]A protocol of this interrogation was not added in [2].

by Korets is mentioned, but does not play a major part in this 'summary of the evidence', in spite of the fact that it is actually the only piece of 'hard' evidence, the rest is just hearsay or originating from his own confessions or those of others. The fact that this leaflet was hardly used may actually be the most important indication that is was indeed fabricated.

Landau never spoke much about his time in prison and what he had gone through, except in a tribute for Kapitsa's 70th birthday in 1964, when he wrote: *"Because of a stupid denunciation I was arrested and accused of being a German spy. Now I can sometimes even find this funny, but then it was no joke. I spent a year in prison and it was clear that I couldn't last another six months—I was simply dying"* (Ref. [13], p. 67). Peculiar in this statement is that Landau speaks about being accused of spying for Germany, which is actually completely untrue. In the documents released in 1991 in *Izvestija TsK KPSS* he is nowhere accused of being or having been a German spy, nor was there any such denunciation. It was Leonid Pjatigorsky (see Chap. 7) who was suspected of having reported Landau to the NKVD as a German spy, accusations which were repeated later in biographies of Landau, but had to be rectified after a court case. It was the intercepted leaflet that had led to his arrest. Landau's statement quoted here is from 1964, two years after the terrible car accident that greatly impaired his faculties and may be the reason for its incorrectness. But in many later documents it is stated as a fact that he was arrested as a German spy.

In this final statement on Landau's case it is also stated that Korets admitted to having been recruited at UFTI into an anti-Soviet group of physicists and having carried out sabotage work. In addition, he also admitted to having been recruited as a spy for Germany by Fomin, which actually agrees with Fomin's statement (Ref. [14], p. 288).

The third person arrested in this case Jury Borisovich Rumer[8] (1901–1985) (in the West known as Georg Rumer) admitted to being guilty of sabotage work and participation in an anti-Soviet group of physicists and to having been recruited in 1929 in Berlin for espionage work by the 'German' Professor Ehrenfest (Ref. [2], pp. 152–154). He was not actually heavily involved in this whole business, but was swept along in the affair. Rumer was educated at Moscow University in mathematics. From 1929–1932 he stayed in Göttingen in Germany, just arriving unannounced and without any grant or other financial support.[9] Max Born, who went to

---

[8]A man with an incredible history which has been described very entertainingly in a pleasantly readable and at places fascinating book by Ryutova-Kemoklidze [11], superbly translated by John Hine from the Russian original *Kvantovyj vozrast* (Nauka, Moscow, 1989). The title of the book is somewhat misleading as it is mostly a biography of Rumer, although there are some gaps, notably as regards his life at Moscow University in the twenties.

[9]Ryutova-Kemoklidze's book [11] contains a remarkable story about how Rumer received some initial support from Baron Warburg.

great lengths to obtain financial support[10] for him and introduced him to all the great physicists of the day, made him his assistant. Rumer worked with Walter Heitler on the newly to be developed field of quantum chemistry and with Hermann Weyl[11] and Edward Teller[12] on independent invariants in vector space. In 1929 he met Landau and Einstein in Berlin, introduced to them by Ehrenfest. He was generally liked and seemed to sail through life rather smoothly. Returning to Moscow University in 1932 he was immediately appointed a lecturer in physics, with the backing of Leonid Mandelshtam and on the recommendation of Erwin Schrödinger (Ref. [11], p. 199),[13] and became a professor in the following year. He was a frequent visitor to Kharkov where he cemented his friendship with Landau and started to collaborate with him, a collaboration that continued after Landau moved to Kapitsa's institute in Moscow. Rumer was a staunch supporter of the Soviet Union, the Soviet regime and communism, who believed in the rightness and justice of all that was done in the country; even the Luzin affair,[14] which broke out against his former professor at the Moscow mathematics faculty in 1936, could not really wake him up, but it did not take long before he was himself swept away in the avalanche of evil a year or two later. He was arrested on his birthday in 1938 together with Landau and spent years working on aircraft design in the *sharazhka*[15] of Tupolev on Radio Street in Moscow and elsewhere until his release in 1948, exactly ten years after his arrest. He was not allowed to go back to Moscow and spent the rest of his life in Siberia (Enisejsk, Novosibirsk and Akademgorodok). As shown by his later correspondence with Born this whole experience did not make him lose confidence in the moral and political superiority of the Soviet system. (Ref. [11], p. 280.)

---

[10]Born had quite a number of protégés like Rumer for whom he managed to get support from various sources, including the Lorentz Fund and the Rockefeller Foundation, but also from (rich) private individuals (Ref. [11], p. 115-116).

[11]One of the greatest mathematicians of the 20th century; successor of David Hilbert in Göttingen and from 1933 one of the first occupants of a faculty position at the Institute for Advanced Study in Princeton.

[12]The future 'father' of the American hydrogen bomb.

[13]The Rumer papers in the Archive of the Russian Academy of Sciences have also been made available on the Internet (*Otkrytyj arkhiv SO RAN, Fond Ju.B. Rumera*) (http://odasib.ru/openarchive/Portrait.cshtml?id=Xc_furs_634919475472910156_1027).

[14]At the mathematics faculty Rumer was one of the students of Nikolaj Nikolaevich Luzin (1883-1950). Difficulties arose for Luzin already from 1930 when his thesis advisor D.F. Egorov, known for significant contributions to differential geometry, was arrested on the basis of his religious beliefs. Luzin resigned in protest, but was further attacked in 1936, among others by his students Aleksandrov and Kolmogorov as an *enemy under the mask of a Soviet citizen* in what is known as the Luzin affair. This affair was characteristic for the atmosphere at Moscow University where many such personal attacks took place in the thirties. In this case the personal attack was related to Luzin's failure to cast a promised vote to elect Aleksandrov to the Academy of Sciences (Ref. [15], p. 205).

[15]The rocket designer Sergej Pavlovich Korolev also worked there, as did Leon Theremin and Jury Krutkov.

And then the punishments of those involved in this case. While Shubnikov, Bronshtejn and others were executed, Landau only spent one year in prison, Korets, although accused of being a German spy, received ten years in a labour camp (later extended by another ten years)[16] and Rumer, also a German spy, who for that matter was cleared of involvement in the leaflet business, obtained a similar punishment, which he spent, as already recalled, for a great part in a *sharazhka* in Moscow. These were very lenient sentences if compared with those of a year before. In this respect Yezhov's fall from power (he had not yet been arrested, but had been replaced by Beria in November 1938 as head of the NKVD) played an important role; the NKVD now showed remarkably less appetite for (immediate) execution than before. Times had clearly changed. During 1937 and 1938 681,692 prisoners (353,074 and 328,618 respectively) received death sentences (nearly 1,000 per day), while in 1939 this number dropped to 'only' 2,552. (Ref. [16], p. 268 and http://www.hrono.info/organ/gulag.html.)

Landau was released on 28 April 1939 following a letter by Kapitsa to Beria. Earlier on 6 April Kapitsa had already written a letter to Molotov in which he stated to be in need of Landau's help in doing theoretical work on "*a number of new phenomena which promise to shed light on one of the most puzzling areas of contemporary physics*". He then continues to defend Landau, not by praising him to high heaven or showing him particular sympathy, as he probably knew that this would not help, but by saying: "*It is true that he has a very sharp tongue, the misuse of which together with his intelligence has won him many enemies who are only too glad to do him a bad turn. But for all his bad character, which I myself have had to cope with, I have never noticed any sign of dishonest behaviour.*" As in the (later) letter for Obreimov, Kapitsa must have worked out how to deal with people like Molotov and Beria; how to portray to them the person he wanted to protect. He ends the letter with requesting that Landau's brains be used for scientific research while in the Butyrka prison. It had some effect, for later in that month Kapitsa was summoned to the NKVD where some high-placed officers tried to show him the fat file on Landau's case. He refused to look at it, however (another clever attitude for what should he have said if confronted with the leaflet printed above: hmm... yes, Hitler, Mussolini, who are they? Never heard of them...) saying that he understood nothing of legal technicalities (as if there were any) and insisting that it would be a disaster for science if Landau were not released. There and then he must have come to an agreement with the NKVD that in return for Landau's release he would personally vouch for his conduct, since the letter he wrote on 26 April to Beria cannot have been spontaneous; he must have had a hint that it would work: "*I hereby request the release from prison under my personal guarantee of the arrested Professor of Physics, Lev Davidovich Landau. I guarantee to the NKVD that Landau will not engage in any kind of counter-revolutionary activities against the Soviet government in my Institute and I will also take all necessary measures within my power to ensure that he should not engage in any such activity outside*

---

[16]He was amnestied in 1952, but remained in exile until 1958.

*the Institute either. In case I should notice any remark of Landau which could be harmful to the Soviet state I shall immediately inform the organs of the NKVD"* (Ref. [13], p. 67 and 350).

Due to his release, following Kapitsa's letter, Landau has never been formally convicted of anything. Once he had been released on bail, his case was archived and only formally closed in July 1990 by a decision of the chief government procurator that the accusation of counter-revolutionary crimes was unfounded. The decision stated that there was indeed a leaflet but that *"as regards its content this document is directed against (...) distortions of the Marxist-Leninist principles of building a socialist society in the USSR, directly associated with the cult of personality of I.V. Stalin, and is not a call to overthrow, undermine or weaken Soviet power or to commit counter-revolutionary crimes"* (Ref. [2], p. 157). Does this mean, one wonders, that it is admitted here by an official body of the former Soviet Union that Stalin's reign was indeed comparable to Hitler's and Mussolini's as stated in the pamphlet? In this formal statement of July 1990 it was not said that the leaflet had indeed been an NKVD fabrication. But as Boris Ioffe says: *"The secrets of Lubyanka remain secrets today"* (Ref. [5], p. 28).

The letters by Kapitsa to Beria and Molotov were not his only ones in defence of Landau. Already on the day after the arrest, Kapitsa had written to Stalin protesting against Landau's arrest. He also mobilized other scientists, notably Niels Bohr, to write to the Soviet authorities. No documents have been published which show that these letters had any effect. Kapitsa's first letter of 28 April 1938 is the first document published in the number of *Izvestija TsK KPSS* [2] referred to above, but no subsequent instruction to the investigators to treat the prisoner in a certain manner. Bohr's letter was only made available to the investigators in November 1938 when the inquiry had essentially been completed, and did not lead to any immediate action. There can be no doubt, though, that Kapitsa's last letter in 1939 to Beria, and his action prior to that letter, saved Landau. Nor can there be any doubt that the action that Kapitsa undertook required a lot of courage as it could easily have gone badly wrong for Kapitsa himself. Landau later paid tribute to Kapitsa and acknowledged that he had saved his life. Until long after Stalin's death Landau was kept under close surveillance by the NKVD/KGB. He was surrounded by informants, among whom must have been physicists whom he trusted and considered his friends,[17] but the KGB also used operational techniques, namely wiretapping and eavesdropping through concealed microphones. In the early nineties a KGB report[18]

---

[17]In this respect there could still be quite a few surprises if the archives were to be opened.

[18]The document had already been declassified in 1992 in connection with the preparation of the legal proceedings against the Communist Party of the Soviet Union. According to Gorelik (Ref. [17], p. 77) the report may have been requested by the Central Committee since Landau had asked permission to go abroad.

regarding Landau was published.[19] It was dated 20 December 1957, so at the beginning of the Khrushchev thaw, addressed to the Central Committee of the Communist Party and contained information on the material collected on Landau until that date. It shows that Landau was still considered an anti-Soviet person, as the report states: "*His political views throughout many years have been those of a man with decisively anti-Soviet sentiments, with a hostile attitude towards all Soviet activity, who finds himself, according to his own declaration, in the position of a 'learned slave'.*[20]" It is clear from the report that he indeed remained very critical of the Soviet regime and considered it outright criminal. He is quoted as having said: "*Our system, the way I know it from 1937 onwards, is a fascist system by definition, it has remained that way and cannot change so easily. Therefore there are two questions. First, to what extent can this fascist system be improved from within... Second, I think that this system will be continuously loosening up. I think that as long as this system exists, one should never nurture any hope that it will eventually transform into something decent, altogether it is even ridiculous. I do not count on it*" and "*Our system is a dictatorship of a class of officials, a class of bureaucrats. I deny that our system is a socialist one because the means of production do not at all belong to the people, but rather to bureaucrats.*" About science and scientists in the Soviet Union he had neither much positive to say: "*Our science has been totally prostituted, and to a higher degree than abroad; there scientists do have some degree of freedom, despite everything. Selling out has become profitable not only for scientists, but also for critics, writers, newspaper correspondents and journalists—these are prostitutes and nonentities. They are being paid, and in exchange they do what is ordered from above*" and "*...In our country science is neither understood, nor appreciated, which by the way is not surprising, since it is run by fitters, carpenters and joiners. There is no room for scientific individualism. Research directions are dictated from above... ...The patriotic approach will harm our science. More and more we are fencing ourselves off from Western scientists and cutting ourselves off from leading scientists and technicians.*"

Especially interesting are Landau's views on special assignments, that is work on atomic weapons and suchlike. Kapitsa's Institute of Physical Problems was also involved in such work, and so was Landau. His task was to carry out numerical calculations, rather than theoretical physics. In this connection he and his team calculated the dynamics of the first Soviet thermonuclear bomb (the *sloyka* or 'layer cake') which was filled with lithium deuteride. Although he was aware that he was helping create terrible weapons for terrible people, he was too afraid to refuse. He confined his work to the tasks he was charged with, without displaying any

---

[19]Originally published in Russian in various publications: *Komsomolskaja pravda* 8.8.1992; *Voprosy istorii* 8 (1992) 112-118; *Istorichesky arkhiv* 3 (1993) 151-161; *Viet* 3 (1993) 126-131 and *Obshchaya Gazeta* of 26 November 1999. It can also be found (including a translation into English) on the website of Vladimir Bukovsky's Soviet Archives.

[20]Reference to slaves as kept by the Romans.

initiative. But as soon as Stalin was dead and a few months later Beria fell from power, Landau abruptly stopped this work. He told Isaak Khalatnikov (who, incidentally, took over Landau's job in the project and was happy to do so): "*That's it. He's gone. I'm no longer afraid of him, and I won't work on this anymore.*" [17, 18] And he quit the bomb project. But already in 1952 the KGB report mentioned above reported about his attitude towards carrying out special assignments: "*A sensible person should attempt to keep as far as possible from practical activity of this kind. One should use one's entire strength in order not to venture into the thicket of the atomic business. At the same time every refusal and distancing oneself from this type of activity should be done with utmost caution. ...Landau believes that the goal of a wise man who wishes, to the extent possible, to live a happy life, is to maximally distance himself from the tasks that the state defines for itself, especially the Soviet state, which is built upon oppression.*"

And in January 1953, so before he had actually stopped such work: "*Had it not been for the 5-th paragraph,*[21] *I would not have done any special assignments, but rather would do only science, which I have now put aside. The special assignment work which I am now conducting, gives me a kind of power... ...But from this there is a long way to having me work "for the benefit of Motherland, etc." which shows through in your letters to me. You can send this kind of letter to the Central Committee, but leave me out of this. You know that it is all the same to me what place is occupied by Soviet physics: the first or the tenth. They have reduced me to the level of a "learned slave" and this determines everything. ...You have been called up to put Soviet physics on the first place in the world. I am not your helper in this matter.*"

Later too he kept these views as was reported to the KGB in April 1955. At the end of March Landau had been summoned, together with Vitaly Ginzburg, to A. P. Zavenyagin (the minister in charge of the atomic project) to discuss special assignments. In the conversation with the source (i.e. the KGB informant) he had said very harsh things about Yakov Zeldovich "*who spreads around all sorts of filth*". He told the source that under no condition would he agree to take up the special assignments again and that he finds it unpleasant to discuss the subject. On the way to the ministry, Landau warned Ginzburg to not even consider mentioning that he needs Landau for the impending work.

Later Landau told the source that the minister received him in a very polite and courteous manner and behaved very well. Landau quickly convinced the participants of the meeting that he should not take part in the special assignments, but, as he expressed it himself, he could not refuse the suggestion to discuss these questions every now and then. "*But in fact, there will be no discussions whatsoever*", Landau said. Landau's views and attitudes on this matter are remarkable since they

---

[21]Euphemism for being a Jew, a reference to the fifth paragraph in the passport stating nationality.

are unique among Soviet scientists, who had been made very docile and meek by the terror of the late thirties and were always very eager to help the Soviet regime get everything it wanted, including weapons of mass destruction. There were only pitifully few, Andrei Sakharov being the most outstanding example among them, who regretted their support after they already had sold their soul to the devil.

# References

1. Ju.N. Ranjuk, L.D. Landau i L.M. Pjatigorsky, *VIET* 4 (1999) 79–91.
2. V. Vinogradov and N. Mikhajlov (eds.), Lev Landau: god v tjur'me (*Lev Landau: a year in prison*), *Izvestija TsK KPSS* 3 (1991) 134–157.
3. Robert Conquest, *The Great Terror: A Reassessment* (Pimlico, London, 2008).
4. B. Gorobets, *Krug Landau* (Letny Sad, Moscow, 2006).
5. B.L. Ioffe, Lev Davidovich Landau in M. Shifman (ed.), *Under the Spell of Landau: When Theoretical Physics was Shaping Destinies* (World Scientific, Singapore, 2013).
6. G.E. Gorelik, ≪ Moja antisovetskaja dejatel'nost'… ≫ Odin god iz zhizni L.D. Landau (*"My anti-Soviet activity" One year in the life of L.D. Landau*), *Priroda* 11 (1991) 93–104.
7. G.E. Gorelik, *"Meine antisowjetische Tätigkeit…" Russische Physiker unter Stalin* (Vieweg, Braunschweig/Wiesbaden, 1995).
8. E.L. Feinberg, *Epoch and Personalities* (World Scientific, Singapore, 2011).
9. M.Ja. Bessarab, *Lev Landau* (Moscow, 2008).
10. V.L. Ginzburg, *About Science, Myself and Others* (Institute of Physics Publishing, Bristol, 2005).
11. M. Ryutova-Kemoklidze, *The Quantum Generation* (Springer Verlag, Heidelberg, 1995).
12. H.B.G. Casimir, *Haphazard Reality: Half a Century of Science* (Harper & Row, New York, 1983).
13. J.W. Boag et al., *Kapitza in Cambridge and Moscow* (North-Holland, 1990).
14. Ju.V. Pavlenko, Ju.N. Ranjuk and Ju.A. Khramov, *"Delo" UFTI 1935–1938* (The "UFTI" Case 1935–1938) (Feniks, Kiev, 1998).
15. Masha Gessen, *Perfect Rigor* (New York, 2009).
16. Emil Draitser, *Stalin's Romeo Spy* (London, 2011).
17. G. Gorelik, The Top-Secret Life of Lev Landau, *Scientific American,* August 1997, 72–77.
18. Isaak M. Khalatnikov, *From the Atomic Bomb to the Landau Institute* (Springer Verlag, Heidelberg, 2012).

# Chapter 14
# Shubnikov's Rehabilitation

There are two aspects to Shubnikov's rehabilitation. First there is the political rehabilitation, the recognition that his conviction and subsequent execution were not based on any facts, and second there is his re-emergence into the world of physics. As a so-called 'enemy of the people' his name could not be mentioned. Soviet policy was to blot out any knowledge of such enemies, as if they had never existed. According to the strict Stalinist rule it was demanded to instantly break off relations even with one's best friend, once said friend had been arrested. (Ref. [1], p. 50.) There was no presumption of innocence until a court had pronounced a sentence, if there indeed was such a court. The work of the victims was either forgotten or assigned to someone else. Abram Kikoin was one of Shubnikov's post-graduate students at UFTI, selected by him personally in the autumn of 1935 in Leningrad at the university. As can be seen from Shubnikov's list of publications, one of Kikoin's first papers, entitled "Optical experiments on liquid Helium II" was published in *Nature* in 1936 with Shubnikov as the first author. In 1938 Kikoin published two further papers in *Nature* on experiments with liquid helium [2], this time with Boris Lazarev as co-author. Shubnikov had published the celebrated work on the proton magnetic moment with Lazarev, who became Shubnikov's successor as head of the cryogenic laboratory in 1938. But Abram Kikoin's son Konstantin is probably right in claiming that not Lazarev, but Shubnikov should have been the co-author on these papers. (Ref. [3], p. 209.) In his reminiscences about the UFTI low-temperature lab (Ref. [4], p. 323) Abram Kikoin confirms that Shubnikov suggested that he investigate the properties of liquid helium II for his PhD thesis. This is not to say that Lazarev has in any way acted dishonourably in this matter. It was simply no longer possible to publish anything with Shubnikov as an author. The decision to change the author on the papers was taken by the "senior comrades" and Kikoin was not consulted about it. It will then not come as a surprise either, I suppose, that in both papers by Kikoin and Lazarev Shubnikov was not mentioned at all, and that there were no references to any of Shubnikov's papers. Even the earlier paper by Shubnikov and Kikoin in *Nature* was not mentioned. Konstantin Kikoin also states in his book that his father took Shubnikov's wife Olga Trapeznikova to the hospital at the end of August when she went into labour for the

© Springer International Publishing AG 2018
L. J. Reinders, *The Life, Science and Times of Lev Vasilevich Shubnikov*,
Springer Biographies, https://doi.org/10.1007/978-3-319-72098-2_14

birth of her son, and that when he came to the institute the following day the other workers "*were shunning him as they would the plague.*"

All this happened in a year or so after Shubnikov's execution, when people were still careful and afraid to do anything that could possibly upset the NKVD organs. Of course, the Soviets could not erase Shubnikov's name from the existing literature, certainly not the Western literature, so the Shubnikov-De Haas effect remained (although it was not yet known under that name), but at least they could make sure that no reference was made to him in Soviet physics literature or other places. This is already apparent from the list of publications in the back of this book, where the six publications of 1938 and 1939 do not bear Shubnikov's name as a co-author for work that had been done under his supervision and guidance before his arrest. Perhaps the two Kikoin papers mentioned above should also be added. It is remarkable that the paper from 1939 actually bears the name of Olga Trapeznikova, Shubnikov's wife, as wives of enemies of the people were normally subjected to the same fate.

But in the long run it was not possible to avoid any mention of Shubnikov or his work in the literature. His contributions to physics were too important for this and people would find ways around it, as can be seen in a paper by Lazarev from 1939 in *Doklady Akademija Nauk SSSR* (Ref. [5], p. 93–99), where the papers by Shubnikov and De Haas were just lumped together and quoted as *W.J. De Haas a. oth.* In this way the explicit mention of Shubnikov's name could still be avoided, but even such practice did not last long, for the same book contains a reprint of a paper from 1944 in *ZhETP* with a full reference to 1936 papers by Shubnikov and Khotkevich. In 1944 it was apparently no longer necessary to avoid quoting his name and in 1950 the Shubnikov-De Haas effect could even be mentioned repeatedly as such in the text of a paper in *ZhETP,* i.e. in the Russian language. (Ref. [5], p. 174–197.) The paper was also full of the De Haas-van Alphen effect, so there was no way to avoid Shubnikov-De Haas. It was an instance where physicists in the Soviet Union continued to show some independence of spirit.

It was Abrikosov, however, who in his 1957 paper drew specific attention to Shubnikov's pioneering work on type-II superconductivity, but that was already after his formal political rehabilitation, which had a rather long history. It was Trapeznikova who already a few years after Shubnikov's arrest, when she still did not know what had happened to him, started to write to the authorities inquiring after the fate of her husband. On 29 April 1939 the Procurator's Office of the Kharkov District told her the lie "*that on 10 November 1937 Lev Vasilevich Shubnikov was convicted and sent to a Correctional Labour Camp. The sentence in this case is final and not subject to appeal.*"[1] Later in the year she again wrote to the Procurator's Office with an appeal to have the case reviewed and received a similar

---

[1]These and other documents were discovered in the KGB archives in Kharkov by the UFTI nuclear physicist Jury N. Ranjuk who published them in collaboration with Ju. A. Frejman in the Journal of Low Temperature Physics [6]. Some of these documents have also been reprinted in [7].

reply: "*your appeal in the case of Lev Vasilevich Shubnikov to the Regional Procurator's Office on 29 October has been considered and in the absence of any grounds for a review of the case and an appeal of the verdict the appeal has not been acted on*".

In an extensive letter of September 1939 to Lavrenty Beria, at the time the People's Commissar of Internal Affairs, in which she vigorously pleaded the innocence of Shubnikov and set out in detail his work and achievements for Soviet science, she again requested that the case be reviewed. She also stated in the letter that: "*I only know that he was convicted on 11 November 1937 and sent to a special regime camp without the right to correspondence*". This was a clause in the sentence of many people convicted for political reasons in the Soviet Union. In a large number of cases it was often announced to relatives that the sentence was "10 years of corrective labour camp without the right of correspondence", while the paperwork contained the real sentence: "the highest degree of punishment: execution by shooting". In 1939 many people, including Trapeznikova, did not understand the official euphemism and incorrectly believed that their relative was still alive in prison. She also wrote that: "*[a]fter the arrest of L.V. Shubnikov the Institute investigated his activities in the laboratory and found that there had been no sabotage. My husband was convicted without a trial and I am perfectly convinced that he is not guilty of anything and that the sentence is a mistake and that L.V. Shubnikov could not do anything that would cause damage to his Motherland; on the contrary, he has done exceptionally important service to Soviet science as the foremost scientist in his field.*" Her petition for a review of the case fell again on barren ground, nor was it a reason for the authorities to come forward with the truth. She just received the blunt and curt reply: "*your appeal has been checked and has been dismissed*". She did not give up, however, and wrote another letter to the Supreme Soviet of the USSR, from which she again received the same reply in June 1940.

It is hard to fathom how the perverse minds of these people work, when confronted with a document marked 'top secret', dated June 1944, concerning Vadim Sergeevich Gorsky, also executed by shooting in November 1937 in the UFTI affair, in which the official administering the case archive requests that his colleague in Sverdlovsk make a note on the case file card to the effect "*that in June 1944 Ivan Vasilevich Obreimov, corresponding member of the Academy of Sciences of the USSR, was informed that V.S. Gorsky, serving a sentence in NKVD camps, died on 25 November 1941 from lobar pneumonia.*" So Obreimov, inquiring after the fate of his colleague Gorsky was told another lie, which also had to be recorded in the files of other NKVD offices and comes as close as anything ever will to the practice of history revision as practised constantly in George Orwell's 1984. Why did they spin these lies, were they ashamed of themselves, and was it a lesser crime on their part when somebody died of pneumonia in a shabby, remote camp exhausted after back-breaking, useless work? After all, catching pneumonia is your affair (we cannot be blamed for the climate, can we?), catching a bullet in the back of your head is not.

Trapeznikova did not give up. After Stalin's death she wrote to the State Procurator of the USSR again in September 1954 asking for a review of Shubnikov's case, even if he were no longer alive. In that letter she also wrote that *"up to 1946 I was unable to learn anything about the fate of my husband. In the autumn of 1946 agencies of the Interior Ministry in Leningrad informed me orally that my husband had died of a heart attack in a remote camp in the autumn of 1945. I did not receive a written notification of his death."* So, she had waited almost ten years since Shubnikov's execution in order to be presented with a further set of lies.

However, now in 1954 her letter to the Procurator provoked the reply that her appeal was being reviewed and that the result would be communicated separately. When this had still not been forthcoming in 1956 she wrote the following extensive letter in March of that year, showing that her fighting spirit had in no way diminished (Ref. [7], p. 274):

To the Chairman of the Party Control Commission
Comrade Razuvaev
from citizen Trapeznikova O.N.
residing in Leningrad
Kirovsky Prospect,
no. 73/75, apartm. 66

My husband Lev Vasilevich Shubnikov, Russian, born in 1901, was arrested by the organs of the NKVD in Kharkov on 6 August 1937. He occupied the position of head of the cryogenic laboratory at the Kharkov Physico-Technical Institute (UFTI) and was professor at Kharkov University. Lev Vasilevich Shubnikov was a first-class experimental physicist, a great specialist in the field of low temperatures. He founded the first cryogenic laboratory in the Union, where for the first time in the Union liquid hydrogen and liquid helium were obtained. In a resolution of the external session of the USSR Academy of Sciences of 24 January 1937 the enormous scientific and technical significance of the creation of the cryogenic laboratory was noted, which was on a par with the best global centres of cryogenic work. But on 6 August 1937 my husband was arrested, in November convicted by a troika according to Article 58 and sent to a distant camp for "ten years without the right of correspondence". Neither at his arrest, nor during the search of both our home and the laboratory was anything found that incriminated him (I kept the protocol of the search), nor was his property confiscated.

Until 1946 I did not succeed in learning anything about the fate of my husband. In the autumn of 1946 the organs of the Leningrad MVD informed me orally about the death of my husband that very autumn in a remote camp due to heart failure. I have not received any notification in writing about his death. In September 1954 I wrote a letter to Comrade Rudenko of the State Prosecutor's Office of the USSR with the request to rehabilitate my husband. On 22 November 1954 I obtained a reply from the Prosecutor's Office signed by the procurator of the section for special cases, first-class jurist Comrade Mikhajlov that my complaint is being verified and that the result of the verification will be communicated to me separately. However so far I have not received any communication from the USSR Prosecutor's Office.

Lev Vasilevich Shubnikov was a native of Leningrad where he finished the Polytechnic Institute in 1926. His entire adult life he lived under Soviet power.[2] All his scientific life he spent under my eyes, i.e. we married in 1925. I am myself a scientific worker and also a

---

[2]A peculiar sentence and only partly true of course as he was already 16 at the time of the revolution, spent almost two years in Germany and Finland (August 1921–Summer 1923) and four years in Leiden (1926–1930).

physicist and I can therefore say with full responsibility that he could not have done anything to harm his motherland, on the contrary he only worked to its benefit. He was a genuine Soviet man, enthusiastic about his occupation, a tireless worker and a splendid organizer.

I request the rehabilitation of my husband, even if he is not among the living. This is not only necessary for me, but also for my son, who was born in 1937, almost a month after the arrest of his father. My son and I had to suffer a lot in life and I do not want that my son now when becoming an adult must pay for a father who is not guilty of anything at all and in order that we do not have to carry the heavy stigma of belonging to the family of an arrested person.

I request that the answer to my request be sent to the address:
Leningrad-122, Kirovsky Prospekt no. 73/75, apartm. 66
Trapeznikova Olga Nikolaevna
26 March 1956
(signature)

And finally, again half a year later, an extensive document dated 15 August 1956 was produced, but not communicated to her (Ref. [6], p. 67–69; Ref. [7], p. 278–280), by some investigator, obviously acting on orders from above, who gave the following summary of the case:

.... Shubnikov, Rozenkevich and Gorsky were arrested in the period August-September 1937 by the Ukrainian NKVD for the Kharkov District on the basis of statements of the witnesses Shavlo and Gej. There only exists an arrest warrant for Gorsky. In the interrogation of 16.03.37 the witness Shavlo declared that he knew Gorsky from joint work at UFTI as an anti-social and anti-Soviet person, who acted against the socialist emulation in science, did work that was aimed at wrecking the qualifications of members of the communist party and the Komsomol, which manifested itself in the fact that he did not allow Shavlo and other students on work placement to have access to the equipment of the laboratory, mocked them and declared "I am not going to teach boors".

At the same time Shavlo declared that others had told him that in 1925 when working in Leningrad Gorsky was a member of an anti-Soviet organization, and during the party purge in 1934 he gave the purge commission false information on communists with the aim of compromising them.

In the interrogation of May 1937 the witness Gej gave further testimony that based on personal observations during his period of work at UFTI he suspected that counter-revolutionary groups of rightists existed there, to which in his opinion Landau, Korets, Obreimov, Shubnikov Lev Vasilevich, Rozenkevich Lev Viktorovich and foreign specialists – husband and wife Ruhemann, Schlesinger, Weissberg and others belonged. These persons, so declared Gej, carried out sabotage at the said institute – frustrated government instructions on working out a number of scientific physics problems which were of significance for the defence of the country and the development of Soviet science by delaying the construction of a number of experimental installations, caused squabbles, harassed scientific workers, chased them from the institute, and held clandestine meetings in the apartments of Obreimov, Shubnikov and Weissberg.

The witness Gej further declared that it was known to him that in 1921 Shubnikov had fled abroad in a yacht under strange circumstances, lived in Finland and Germany, and subsequently upon his return to the USSR had said that he had accidentally found himself abroad, since the yacht had been carried away in a storm. In addition Gej declared that

Rozenkevich and Shubnikov took part in clandestine meetings of the counter-revolutionary sabotage group and in all its hostile activity.

However he did not say from where he had got all this knowledge and because of his death there is no possibility to ask him again.

In April 1937 Akhiezer was also interrogated in the case, declaring that he was an employee at UFTI together with Shubnikov, Gorsky and Landau and that they also taught at Kharkov University. In December 1936 in connection with Landau's dismissal from teaching at the university he together with Gorsky and Shubnikov, in order to show solidarity with Landau, gave out a statement on their resignation from the university, because of which they were called to the UkSSR people's commissariat, where they admitted their error.

In the process of the inquiry in the current case the accused Shubnikov declared in the interrogation of 7 August 1937 upon a question put to him that he had never belonged to a counter-revolutionary espionage-subversive organisation and did not plead guilty. However, on the same day in a statement addressed to the interrogator he admitted to being guilty of counter-revolutionary activity and asked that a deposition be taken from him.

Subsequently in statements written in his own hand and in the interrogation of 13-14 August 1937 Shubnikov declared that in 1921 by sailing on a yacht in the Finnish gulf[3] he fled abroad and lived for some time in Finland and Germany, and in 1923 returned to the USSR to Leningrad and started work at the Physico-Technical Institute.

Before his return to the USSR he was recruited in Germany by a certain Dessler for espionage work for the German secret service and in 1926, when leaving the USSR on a foreign *komandirovka* met with Dessler in Berlin and gave him secret information on the work of the Leningrad Physico-Technical Institute. After having moved to UFTI, so Shubnikov declared, he met Weissberg there in 1932, who had returned from a foreign *komandirovka* and had recruited him for the second time into a counter-revolutionary Trotskyist-espionage organization which existed at UFTI, and on the latter's instruction gave him information on the work at UFTI of significance for defence and industry.

Furthermore, Shubnikov declared that apart from him Landau Lev Davidovich, Gorsky Vadim Sergeevich, Brilliantov Nikolaj, Lejpunsky, the foreign specialists Houtermans, Ruhemann Martin, his wife Barbara, and Obreimov belonged to the said organization and in 1935-1936 he had personally recruited Landau, Gorsky and Brilliantov into the organization.

On the criminal activity of the participants in the counter-revolutionary sabotage organization Shubnikov declared that he together with the aforementioned persons frustrated the solution at UFTI of scientific problems which were of great significance for the defence of the country and the development of industry, by causing squabbles among the employees of the institute and creating impossible conditions for developing their scientific and practical activities with their subsequent expulsion from the institute and the appointment of foreign specialists in their place. The sabotage was carried out in the low-temperature laboratory and at other experimental installations by constantly rebuilding them as a result of which a number of problems set before UFTI by the party and government was not solved.

In addition to this, Shubnikov declared that together with other participants of the anti-Soviet organization he often met in his own and Weissberg's apartment, as well as in Landau's office, where they held counter-revolutionary conversations and moreover that he intended to leave on a foreign *komandirovka* and not return to the USSR.

The accused Rozenkevich Lev Viktorovich declared that working in 1928 at LFTI in Leningrad he mixed with hostilely inclined persons – Ivanenko and others, who raised him

---

[3]This must be a mistake, since the trip was on Lake Ladoga.

in a counter-revolutionary spirit and in 1928 on Ivanenko's proposal he joined the anti-Soviet organization existing at LFTI.

In 1930, after his transfer to work in Kharkov at UFTI he joined the Trotskyist group whose participants were Weissberg, Shubnikov, Lejpunsky, Brilliantov, Korets, Ivanenko, Landau and the foreign specialists – wife and husband Ruhemann and Houtermans. With these persons, so Rozenkevich declared, he engaged in anti-Soviet conversations and in sabotage aimed at wrecking work at UFTI on scientific problems that were of significance for the defence of the country and for industry, in inciting squabbles and creating unbearable conditions for the work at UFTI of young specialists with their subsequent expulsion from the institute. Moreover, Rozenkevich declared that he and Ivanenko discussed the question of leaving on a foreign *komandirovka* and not returning to the USSR.

In interrogations, but also in a confrontation with the accused Shubnikov the accused Gorsky denied his participation in a counter-revolutionary Trotskyist organization, declaring that he did not know anything about the existence of such an organization, and that Shubnikov had not recruited him into it. The materials of the inquiry on the basis of Article 200 of the Penal Code of the UkSSR were not presented to the accused Shubnikov, Gorsky and Rozenkevich and the time of conclusion of the inquiry in the current case is not known.

On 28 October 1937 by a decision of the USSR NKVD and the USSR prosecution service the highest measure of punishment was determined for Shubnikov, Gorsky and Rozenkevich – to be shot. The said decision was carried out in the period 8-10 November 1937.

By instruction of the chief military procurator the grounds of the conviction of the above-mentioned persons have currently been subjected to verification, in the process of which materials of the archived inquiries on Weissberg, Landau, Ivanenko, Obreimov, Houtermans and others, whom Shubnikov and Rozenkevich had mentioned as co-participants in their criminal activity, were studied.

After having been arrested, Weissberg and Landau initially spoke of their participation in the counter-revolutionary organization existing at UFTI, and mentioned as co-participants Shubnikov and Rozenkevich. Subsequently they declared that these statements do not correspond to the truth, since they made them as the result of the application of unlawful methods of interrogation, in connection with which the cases on Landau and Weissberg were closed in 1939. The arrested Ivanenko did not make any statements on the criminal activity. However, taking into account that he was a social hazard as he originated from the family of a nobleman, he was convicted to three years in a correctional labour camp.[4]

The arrested Obreimov and Houtermans at first stated that they were German spies, took part in the counter-revolutionary organization existing at UFTI, and mentioned Shubnikov, Gorsky and Rozenkevich and others as co-participants. In subsequent interrogations they retracted their statements, declaring that their protocols were forged by the investigator, in connection with which the said cases were closed. Houtermans as a foreigner was evicted from the USSR.

---

[4]The 'application of unlawful methods of interrogation' was, however, not the reason for closing the cases of Landau and Weissberg. They were actually not closed. Weissberg was handed over to the Gestapo and Landau was released upon a letter from Kapitsa. Ivanenko had already been arrested and exiled in 1935 shortly after the murder on Kirov. His arrest had nothing to do with the UFTI affair. Here too the investigator wants to give the impression that some due process took place in these cases while that was in no way true.

Thus, the statements of Shubnikov and Rozenkevich about their participation and that of Gorsky and others in a counter-revolutionary Trotskyist sabotage organization allegedly existing at UFTI, and the persons they refer to have not been verified. In addition, the currently examined witnesses Strelnikov, Usikov, Kovalëv, Chernets and others who have known Shubnikov, Rozenkevich and Gorsky for a long time through their work at UFTI have characterized them positively. ...

This summary seems a fair reproduction of what actually happened and the 'facts' remained the same, but the conclusion the investigator drew was completely the opposite, namely that the cases of Shubnikov, Rozenkevich and Gorsky had no basis in fact and he advised that they be dropped for lack of a crime. This conclusion is as incomprehensible and unsatisfactory as the sentence had been twenty years before. Shubnikov had confessed to being a German spy and to committing sabotage at UFTI. Were they not crimes? There may not have been any evidence for these crimes, although he spelled them out in detail as can be read in the documents in Appendix 3, apart from his confession and some dubious testimony, but then the sentence should have been set aside for lack of evidence. Or were the confessions unreliable? Were the accusations against Shubnikov invented from start to finish? Then it might have been of interest to know who had done this and why. Was undue pressure applied perhaps? Was he threatened in any other way? No word about this. Only that 'unlawful methods of interrogation' were used in other cases, for instance in the cases of Landau and Obreimov. But what have these cases, which started only about a year later, to do with Shubnikov's case? He was already long dead before Landau and Obreimov were arrested. It is true that in his statements Shubnikov mentioned them, or more probably it was suggested to him that he mention them, as participants of the counter-revolutionary organisation at UFTI, but that was not the reason for their arrests. To use the fact that these people, after having first confessed, retracted their statements and confessions later as proof for the unreliability of Shubnikov's own statements more than a year earlier and long after his execution, is again a sop, a very unconvincing attempt to give the impression that the inquiry in the case of Shubnikov, Rozenkevich and Gorsky had any resemblance to a judicial process. The investigator wants to suggest that this 'new' evidence, brought forward by the retractions of Landau, Obreimov and others, came up only later. Oh, if we only had known it at the time, he would not have been convicted. *Tant pis.* In fact it is just a cover-up for murder. And apparently nobody was to blame, at any rate nobody is mentioned for having to carry the blame for this travesty of justice.

Prior to the investigator's report several people were heard, including Landau, Shavlo, Sinelnikov and Strelnikov. Their testimony is interesting in itself and has been reproduced in Appendix 4. As can be seen, Landau stated among other things in respect of Shubnikov that "*his patriotism is confirmed by the fact that he voluntarily gave up work in Holland for work at home.*" I believe that this is stretching it a bit, as Shubnikov's main reason for leaving Leiden and his work there was that his *komandirovka* could not be extended. Leiden University wanted him to stay on for at least another year, but he did not get permission. He could have disobeyed and stayed on in Leiden, but that would probably have blocked any future return to Russia.

In these statements it is also for the first time that we hear personally from Strelnikov, the engineer who had invented a novel X-ray tube and was allegedly forced to leave the institute, a short, but powerful statement that he did and does not know anything about hostile anti-Soviet activity of the above-mentioned persons, nor about espionage, subversive and sabotage activity. A longer statement would have been extremely interesting as he was one of the persons allegedly harassed and forced to leave UFTI. Why had he not been heard before, either before of just after Shubnikov's arrest?

Kirill Sinelnikov's statement of July 1956 is especially interesting as he played a dubious role in the conflict at the institute during the thirties, and now in 1956 he declares that he does not know anything about counter-revolutionary activity or about any counter-revolutionary organisation at UFTI or LFTI to which Shubnikov, Rozenkevich and Gorsky may have belonged. However, he affirms that they did engage in strike action, not at the institute, but at the university following Landau's illegal dismissal as recalled in earlier chapters. And he further states that "*we indeed did have indisputable spies – Fritz Houtermans and Weissberg, and also the obscure person Fomin...*" He reported Fomin to the NKVD and to the party organisation, but his anxiety increased when he heard a few days later that Fomin had attempted to commit suicide by drinking hydrochloric acid and jumping from a third floor window. His comments on Landau and the problems with the students at Kharkov University are also interesting and may actually have some truth in them.

Landau, Strelnikov, Brilliantov and Sinelnikov all declare that they don't know anything about a counter-revolutionary organisation active at UFTI in the late thirties, but they neither say what actually did happen in those years. And apparently they all knew what is meant by a counter-revolutionary organisation. Only Abram Chernets's statement is more revealing in that respect. He speaks of the existence of two camps at the institute and of a group of scientific workers, among whom Shubnikov, who acted against the management and the party organisation.

On the basis of the investigator's report a further report, dated 13 October 1956, was compiled by the Deputy Procurator General of the USSR and sent to the Military Collegium of the USSR Supreme Court. The Deputy Procurator General concluded that "*[t]here is in this case no objective proof that the accused have carried out counter-revolutionary activity at UFTI, except for unsubstantiated statements by Shubnikov and Rozenkevich themselves and statements of the witnesses Shavlo and Gej examined in the inquiry.*

*Moreover the witnesses Shavlo and Gej (died in 1955) gave similar statements in respect of Brilliantov, Trapeznikova Olga, Landau, Ruhemann M. and Ruhemann B., and others who were not subjected to arrest or whose cases were dropped.*

*When re-examined, Shavlo declared that in his opinion Gorsky had an anti-Soviet attitude, did not like communists, harassed and persecuted cadres, in particular Strelnikov. However Strelnikov, as can be seen from the materials in the verification, did not testify to any anti-Soviet manifestations on the part of Gorsky.*"

(Ref. [6], p. 48; Ref. [7], p. 281.) So Shavlo stuck to his earlier statements, but was exposed as a liar by Strelnikov's statement. All other depositions by former employees of UFTI, some of which have been printed in Appendix 4, were unreservedly positive in respect of Shubnikov, Rozenkevich and Gorsky. They were all added to the file. The Deputy Procurator General further states that "*[i]n the course of the verification of the case no compromising material whatsoever has been obtained. At the same time it has been established that Rejkhman, who took part in the inquiry in the current case, was convicted for unfounded arrests of Soviet citizens.*" So there we are, the villain has been discovered, it was Rejkhman. How could we have missed that? Comrades Stalin and Yezhov washed their hands of guilt. The Deputy Procurator General concludes by asking the Military Collegium "*to set aside the decision of the USSR NKVD and the USSR Prosecution Service of 28 October 1937 in respect of Shubnikov, Lev Vasilevich, Rozenkevich, Lev Viktorovich, and Gorsky, Vadim Sergeevich, and to close the case on them for lack of crimes.*"

None of these documents were communicated to Trapeznikova. She states (Ref. [8], p. 291) that Lev Vasilevich was posthumously rehabilitated on 11 June 1957, and it is not known which document was precisely issued to her. The documents cited here were all secret and only made available after the collapse of the Soviet Union. They were a further elaborate charade in order to give the impression of a proper judicial inquiry. The archives also contain a certificate dated 1 July 1957 from the Military Collegium of the Supreme Court of the USSR, of which Trapeznikova may actually have received a copy, stating in case no. 44-024554/56 that (Ref. [6], p. 72):

"The case against the accused Lev Vasilevich Shubnikov, up to his arrest on 5 August 1937 scientific head of the Laboratory of Low Temperatures of the Ukrainian Physico-Technical Institute, was reviewed by the Military Collegium of the USSR Supreme Court on 11 June 1957.

The decision of the NKVD and the Procurator's Office of the USSR of 28 October 1937 in respect of Shubnikov L.V. is set aside and the case is dropped for lack of a crime.[5]

Shubnikov L.V. is rehabilitated posthumously.

The Chairman of the Judicial Board of the Military Collegium of the USSR Supreme Court

Colonel Kostromin (*Kostromin*)"

This is accompanied by a death certificate stating that Lev Vasilevich Shubnikov died on 8 November 1945 from heart failure, which was entered into the death register on 22 August 1957 in Kharkov. No place of death or any other information is given.

So, after having staged an elaborate rehabilitation procedure of an innocently convicted and executed man, they immediately continued to add insult to injury by

---

[5]A few standard formulations were used in rehabilitation documents: "for lack of a crime" was the most common, but also "based on previously unavailable information", and in some cases "due to the lack of proof of guilt".

lying about Shubnikov's cause of death, the same as in Gorsky's case. This conformed with the official policy at the time that death sentences should be covered up and natural causes of death communicated to relatives since all hell would break loose, they thought, if it were to become known how many innocent people had been murdered.

It is of course no accident that, at first, Trapeznikova did not receive any answers to her requests since Khrushchev's initially secret speech to the Twentieth Congress of the Communist Party of the Soviet Union on 25 February 1956 was still to come. In that speech *On the cult of personality and its consequences* Khrushchev, who during Stalin's life had been one of his most faithful henchmen and was personally responsible for countless murders and other crimes, exposed Stalin's crimes and especially the personality cult around him. The mass repressions under Stalin's regime had already been investigated and recognised before the speech by a special commission chaired by the old Bolshevik Pëtr Nikolaevich Pospelov (1898–1979). The evidence gathered by the commission was presented by Khrushchev in his speech, which was followed by a period of liberalisation (the Khrushchev thaw), the release of thousands of convicts from the camps and an extensive rehabilitation process for those convicted and/or executed without cause [9].

As is clear from the lie on Shubnikov's cause of death the story was not over yet. On 13 June 1991 Trapeznikova again wrote a letter to the Politburo of the Communist Party, asking for additional information on the fate and death of her husband since she had *"reason to believe that L.V. Shubnikov did not die in 1945, but was shot in Moscow in 1937 by order of Yezhov and Vyshinsky."* The reply was swift this time and read as follows (Ref. [6], p. 49; Ref. [7], p. 281):

4 July 1991

Dear Olga Nikolaevna,

In response to your application to the Politburo of the Central Committee of the Communist Party of 13 June 1991 we inform you that Lev Vasilevich Shubnikov, born 29 September 1901, native of Leningrad, Russian, non-party, head of a laboratory at UFTI, was arrested on 6 August 1937 on the unfounded accusation of participation in a counter-revolutionary Trotskyist group, which carried out subversive work at UFTI.

By decision of the USSR People's Commissar of Internal Affairs and the USSR Prosecution Service of 28 October 1937 Shubnikov L.V. was executed by shooting on 10 November 1937. There is no information in the archive materials kept by the KGB administration on the place of execution and burial. It is not excluded that this was in Moscow.

A certificate on the death of Shubnikov L.V. will be sent to you by the official register office of the Kiev Regional Executive Committee of Kharkov city.

Head of the subdivision of the KGB Administration of the UkSSR for the Kharkov District.

*A.V. Fomin*

It had taken her 54 years to find out the truth, and was in the end helped by the fact that the evil empire was in its death throes.

# References

1. Robert Conquest, *The Great Terror: A Reassessment* (Pimlico, London, 2008).
2. A.K. Kikoin and B.G. Lasarew, Experiments with Liquid Helium II, *Nature* 141 (3577) (1938) 912; A.K. Kikoin and B.G. Lasarew, Further Experiments on Liquid Helium II, *Nature* 142 (3589) (1938) 289–290.
3. K. Kikoin, *Po obe storony svobody – Esse, ocherki, vospominanija* (On both sides of freedom – Essays, sketches, reminiscences) (Philobiblon, Jerusalem, 2011).
4. A. Kikoin in B.I. Verkin et al. (1990), p. 321–329.
5. B.G. Lazarev, *Zhizn' v nauke, izbrannye trudy, vospominanija* (Life in Science, Selected Papers, Reminiscences) (Kharkov, 2003).
6. Ju.N. Ranjuk and Ju. A. Frejman, Rehabilitated posthumously: Documentary history of the rehabilitation of L.V. Shubnikov, L.V. Rozenkevich and V.S. Gorsky, *Fiz. Nizk. Temp.* 18 (1992) 52–73 [*Sov. J. Low. Temp. Phys.* 18 (1992), 34–50].
7. Ju.V. Pavlenko, Ju.N. Ranjuk and Ju.A. Khramov, *"Delo" UFTI 1935–1938* (The "UFTI" Case 1935–1938) (Feniks, Kiev, 1998).
8. O.N. Trapeznikova in B.I. Verkin et al. (1990), p. 256–291.
9. W. Taubman, *Khrushchev: The Man and His Era* (New York, 2003).

# Chapter 15
# Afterword

In the 1990 memorial volume on Shubnikov, Akhiezer gives the following impression of him and of his friendship and cooperation with Landau (Ref. [1], p. 337–338): "*His appearance was absolutely wonderful—shrewd lively eyes, a large bald head and a somewhat reddish complexion which reminded of Mr. Pickwick. Moreover he was extremely mobile: walking with him in the park, I always marvelled at the ease with which he ran up the steep slopes of ravines.*

*Shubnikov and Landau were not only connected by a personal friendship, more accurately stated, their friendship was based on the closeness of their attitude in life: they did not appreciate intrigue, did not appreciate political careerism. In this respect they were without compromise—in the circle of Shubnikov and Landau only value mattered. It was difficult with them, you had to show that you were passionate about science and devoted to it.*

*The scientific symbiosis of an experimentalist and theorist of such class, of L.V. Shubnikov and L.D. Landau* (Fig. 15.1), *is unique; I cannot recall an analogue in the history of physics. Landau was in general glad to listen to experimentalists, who told him of their work. But closest of all was Landau to two great masters of experiment, to Lev Vasilevich Shubnikov and Pëtr Leonidovich Kapitsa. Their experiments inspired him, and the discussion with him helped them. This is also true for the work of Shubnikov on superconductivity, anti-ferromagnetism and for the remarkable work of Kapitsa on superfluid helium. Shubnikov and Landau would often sit until well after midnight in the cryogenic laboratory, discussing the results of experiments, which led to important discoveries. This includes in the first place the discovery of the intermediate state of superconductors, of which Landau formulated a theory.*

*We, young theorists, quite often went to see Lev Vasilevich in the laboratory, exploiting him as an inexhaustible source of knowledge. He always knew everything. When we started a new activity in the field of physics of the condensed state or needed data for writing papers or lectures, Landau said that we should go to Lev Vasilevich and ask everything we needed from him or learn from him what to read, where to get it, etc.*"

© Springer International Publishing AG 2018
L. J. Reinders, *The Life, Science and Times of Lev Vasilevich Shubnikov*,
Springer Biographies, https://doi.org/10.1007/978-3-319-72098-2_15

**Fig. 15.1** Lev Landau
(painting hanging in his
former office at UFTI
(*picture taken by the author*))

In the foregoing the life and work of this extraordinary physicist has been described. His life and short, but brilliant career can be viewed as illustrative for many scientists in the tumultuous and brutal early decades of the Soviet Union. A lot of emphasis was necessarily put on the Ukrainian Physico-Technical Institute (UFTI) in Kharkov and to a lesser extent on Leningrad, which were the two places where Shubnikov worked in the Soviet Union. Other places, and in particular Moscow, have consequently received less attention. The situation in Moscow, although science was practised at first definitely at a much lower level, was much more complex than in Leningrad and became rather tumultuous in the thirties and later. In Leningrad Abram Ioffe dominated the scene, had good relations with the other leading scientists, and there seems to have been little infighting, while in Moscow philosophers, mathematicians and physicists were often at loggerheads with each other. It would be of great interest to shed some more light in the rather confused Moscow situation in those years, which was further complicated by the move of the Academy of Sciences and various affiliated institutes from Leningrad to Moscow in 1934.

## The Nature of the Repression that Devastated the Physics Community in the Thirties

The final question that needs answering is whether the repression that swept through the physics community in the thirties was random or not. Terror as a political strategy always has a random element. But that does not necessarily mean that there are no reasons, or rather events, attitudes or circumstances that were perceived as a threat or potential threat by the regime. The situation with UFTI as described in this book is unique in the sense that a lot is known both about the actual events that triggered the eventual repression and about the NKVD's actions. In many cases where scientists were imprisoned or even executed we do not know what triggered NKVD action, an obvious and outstanding example of this being Matvej Bronshtejn who seems to have been executed for no apparent reason.

In this book it has been shown that from a Soviet point of view there certainly were reasons for a clamp-down on UFTI, LFTI and other scientific centres in the country where people thought that they knew better than the Bolsheviks what should be done, how science should be organised and for what purposes. Arrests were not made mainly or solely on the basis of the scientific views people held, such as an expanding universe or because of physical idealism. Activities by the group of scientists at UFTI can retrospectively be qualified as anti-Soviet or, at any rate in the conflict on splitting UFTI, it could be expected that the losing side would lose a lot. Each side had its supporters. The Leningrad scientists thought that they could rely on the support of high party officials, such as Ordzhonikidze and Bukharin, but when that support fell away they were easy prey for the NKVD.

It is clear that all the arrests, imprisonments and executions discussed here were completely illegal, even according to Soviet law, but it is naïve in my view to suppose that they were purely random, for example for fulfilling quota imposed from above. It is easy to discard the arrests as the actions of a criminal regime headed by a paranoid potentate and leave it at that. I believe, though, that the matter is actually much more complicated. There were 'objective' facts that could be called anti-Soviet, such as the strike at the university in Kharkov and the compilation of an anti-Soviet leaflet, and it is these facts that triggered the arrests.

Actually it all started much earlier. Shubnikov's case was constructed around Landau and Ivanenko, who had come under suspicion after the murder of Kirov in 1934. But once a person had been arrested or accused, the Terror had its own dynamics. A case was constructed and only very seldom did it result in an acquittal. In a bizarre reversal of cause and guilt, the fact of someone's arrest was already 'proof' of guilt. Otherwise you would not have been arrested, would you? The underlying principle of the Stalin universe was that by definition everybody was guilty, or at least potentially guilty or capable of being guilty. The organs had to dig up this guilt in their inquiry. Seemingly innocent statements or criticism, or even genuinely well-meant proposals for setting up a scientific experiment could be construed as wrecking or sabotage, endangering the state. Scientific correspondence with a fellow scientist abroad who was working in the same field was a sure sign of espionage or potential espionage. The logic behind it was impeccable.

The assault on UFTI was a deliberate attempt by the authorities, i.e. by the NKVD, to get rid of a bunch of people who were considered a nuisance, as they were not sufficiently loyal in serving the cause of socialist construction.

That does not mean that the repression was a completely rational affair. There was for sure also a random element. After all, it did not matter if a couple of innocent people would be crushed in the process. The punishments meted out seem in any case to have been totally arbitrary. Why was Shubnikov shot and Korets spared? Much of this had to do with the power change at the top of the NKVD with the removal of Yezhov, but, if anything, Korets was a far more dedicated anti-Soviet person than Shubnikov, who for all intents and purposes was and had always been a loyal Soviet citizen. Korets's 'crimes' were far more severe than Shubnikov's and were indeed directed against Soviet power or at least against Stalin's, while Shubnikov had only shown some disloyalty when he argued against defence-related work and resigned from his teaching position at the university in solidarity with Landau. But he had been abroad and an elaborate case of espionage could be constructed. So in Shubnikov's case his rather small misdemeanour could be blown up by the zealous NKVD investigators into a huge affair. We happen to know a lot about Shubnikov, about his earlier escapade to Finland and Germany, about his work in Leiden, about his activities at the Kharkov institute, and how the NKVD dealt with him, and hence we can fairly precisely see how the whole affair came about. In the eyes of the NKVD, there was a good reason to arrest him. He had obstructed or tried to obstruct defence-related work at the Kharkov institute, there can be no doubt about that.

About Korets we know much less, but he too obstructed such work, but allegedly also drew up a pamphlet (and confessed to having done so) calling for a new revolution and to overthrow the Soviet government, but was still 'only' convicted to a period in the camps. Landau, for that matter, was in prison for only a year and was released, although he had participated in all these 'crimes'. His release can be ascribed to the intervention of Kapitsa, Bohr and others who wrote to Stalin very early in the proceedings.

As regards Shubnikov we have access to quite a number, perhaps even all, documents describing the NKVD investigation, the interrogations and statements by the accused. For some reason not all documents of the Korets-Landau-Rumer case have been made available. But perhaps it is better not to know everything about someone who in the memory of posterity has to remain a saint: *Mezhdu nami zhilo chudo* (There Lived a Miracle Among Us), as is the title of a rather uncritical review of Landau's biography by Majja Bessarab [2][1] in the issue of *Priroda* published in 2008 [3].

The punishment was of course draconian and had nothing to do with justice, but the suppression itself was by no means random. Landau, Shubnikov and others could have seen it coming. They could have known from earlier experience (of Ivanenko, Bursian, Korets around 1935) that they were dealing with a criminal regime and that they were playing with a fire that would eventually consume them. Especially the demand of dividing UFTI into two separate institutes can be viewed as hostile to the state. They played a dangerous game and must have known it was a dangerous game. The stakes were extremely high.

The suppressed physicists can in my view be divided into four categories:

a. Those who were arrested for a 'valid' reason, namely anti-Soviet activity or some behaviour that could be seen as such by the regime. Of course they were not criminal in any sense by today's standards or even by the standards of decent regimes of that time, nor does it mean that these physicists were disloyal to the regime; their only sin was independent thought;

---

[1]Majja Bessarab is a niece of Landau's wife and is apparently making a living from biographical writings about Landau. The 2008 book is a rewriting of her earlier 1971 book.

b. Those who were arrested because they were denounced by others (colleagues), either voluntarily for their own gain or during interrogation extracted by force, as belonging to some counter-revolutionary group;

c. Those who were arrested because they were "have beens", such as descendants of the nobility or bourgeois, or had fought on the side of the Whites in the Civil War. As far as this category is concerned the random element was considerable. Some physicists of noble descent were arrested and packed off to the camps, while others, even close family members, were left alone. Semënov for instance was untouched in spite of his White army past;

d. Those who were 'accidentally' arrested, swept up in some action or other (collateral damage, so to speak) or to fulfil quota, or just randomly.

We have seen quite a number of examples of categories a., b. and c. in earlier chapters. Of category d. very few, perhaps Matvej Bronshtejn was such a case, but too little is known about his particular case to make a definite judgement. That said, I find it very likely that the reason for his arrest was his close connection with the astronomers at Pulkovo and with the physicists at UFTI.

All kinds of reasons have been put forward to explain arrests. For instance, for Bronshtejn the fact that his name coincided with Trotsky's real name. This is insufficient, for if such circumstance were a reason for arrest, why then were his twin brother and other siblings left alone? The NKVD was perfectly able to distinguish between the various Bronshtejns that were walking the Soviet lands.

Nor can the fact that someone had spent some time abroad have been the sole reason for arrest. There are too many who had been abroad that were left in peace. Jakov Frenkel, Abram Ioffe, to name two that spent considerable time abroad, Frenkel even early in the thirties in the United States of America, where he must have been infected with some most unwanted and dangerous capitalist and bourgeois bacilli, and must have been recruited for spying activity. The capitalists would not have missed such a chance, would they?

In Shubnikov's case it is also striking that his stay in Leiden (for four years) did not play any role whatsoever in the NKVD investigation. The whole period is essentially ignored; no spying for the Dutch secret services, if they existed, or spying for Germany while in Leiden. Perhaps the Netherlands did not really exist for the NKVD. After all, it was a tiny country that did not even have diplomatic relations with the Soviet Union.

All these 'objective characteristics' were in most cases not the main reason for arrest. Yet, they were used by the NKVD to spin their plots. The main reason had to be looked for in alleged or potential disloyal behaviour.

If the pamphlet they allegedly wrote is genuine and not an NKVD fabrication, Landau and Korets were obviously anti-Soviet and indeed planning active resistance. The change of power at the head of the NKVD was probably responsible for the fact that they got off fairly lightly compared to Shubnikov, Rozenkevich and Gorsky and as regards Landau there are also Kapitsa's efforts to get him freed. But that was after a year in prison. There is so far no evidence that Kapitsa's earlier letters to Stalin and Molotov had any effect. Shubnikov, Rozenkevich and Gorsky were also guilty of anti-Soviet activity as they participated in what was perceived as an anti-Soviet strike and paid the supreme price for this. Their activity was, however, completely harmless compared to what Korets and Landau had in mind. In the case of Shubnikov et al. the reaction, still under the leadership of Yezhov, was swift and brutal, while Landau and Korets profited from the change of power at the top of the NKVD and stayed alive.

In general I think it can be concluded that a careful study of each case is needed before the actual reason for any arrest becomes clear. The reason has to be looked for in the (potential) behaviour of the individual or group in question, not solely in the documents produced by the NKVD investigators in which they often tell a confusing and not very illuminating story. Shubnikov is a case in point in this respect, since all this discussion about spying and counter-revolutionary behaviour is nothing more than a smoke-screen. It is as if the NKVD does not want future readers of their files to know what the actual truth of the matter is, so they do their best to try to throw them off the trail. It has often been stated that Landau, Shubnikov and others were random victims of the Great Purge, but in this book it has been shown that this is not the case. In the eyes of the NKVD and the party leadership UFTI had to be cleared of 'anti-Soviet' elements, that is of those individuals that did not want to contribute without protest to the construction of the socialist society that those in power had in mind and in the way they wanted this to happen.

Beck and Godin (Ref. [4], p. 89 ff) list a large number of categories of prisoners. Category 5 consists of the Technicians and Specialists. Industrialization demanded, they say, the complete intimidation of the people vital to its implementation. At first the campaign against this group had been directed against technicians of the pre-Soviet era, as those who had grown up under the Soviet system were still very few; a campaign that was stopped very early in the thirties as suddenly as it started since the damage done to society became too large. But now it was thought necessary to drive home the point again and weed out those specialists and technicians who had shown not to be unconditionally loyal to Soviet power, or were suspected of not being unconditionally loyal.

## How Many Physicists Were Suppressed?

In the thirties theoretical physicists were a rare breed and all theoretical physicists in the Soviet Union knew each other. It is alleged that at the conference in 1929 in Kharkov all the theoretical physicists in the Soviet Union were present. There were about 60 participants. A group photo of the conference printed in *Priroda* in 1990[2] shows 35 individuals, a few of whom are foreigners, experimental physicists and wives of physicists. Of the 24 identified Soviet physicists on the photograph twelve were suppressed and of those twelve half were murdered in one way or another. On

---

[2]Printed in *Priroda* 7 (1990), p. 76.

the website of the *Institute of the History of Science and Technology*[3] 43 repressed physicists, both experimental and theoretical physicists, with an LFTI connection are listed, 13 of whom were executed. This also includes the physicists that were working at UFTI.

All in all more than fifty physicists working in fundamental physics have felt at one time or another the heavy hand of the NKVD, while most of the others must have lived in fear of as yet having to experience this fate. If Landau's estimate, given at the March 1936 session of the Academy of Sciences,[4] of a few hundred serious research physicists in all fields around 1936 is reasonable, this means that between 25 and 50% of the physicists working in fundamental physics (theory and experiment) must have had direct unpleasant experience with the secret police. It would be of interest to obtain more clarity regarding the actual numbers (as has been done for geology [6]), but so far nobody seems to have taken the trouble to research this topic in such detail that hard numbers can be given.

## A Final Note on Sources

Soviet biographies of leading physicists tend to read as the lives of saints; unpleasant events or unpleasant characteristics of the person are generally not mentioned; nor is anything said about Stalinist repression. A particularly striking quote from a Landau biography from late Soviet times (Ref. [7], p. 30) about the reason why Landau left Kharkov early in 1937 reads: "*The conflicts in which Landau and some of his friends and pupils became involved led to considerable unpleasantness that was a serious matter. In the end, it was necessary to think of moving to another city.*" This 'unpleasantness' was certainly serious and included the incarceration and subsequent execution of Shubnikov, Rozenkevich and Gorsky, the arrest of many others and the virtual destruction of UFTI. The whole book does not contain a single word about these matters. Although a biography of Landau, it does not even mention that he spent time in prison. The same is to some extent also true for post-Soviet writing on Landau and others. Landau's relatives can perhaps be excused as they want to make a living out of their 'hero', but also serious scientists when writing their reminiscences often lapse into worship of some kind, finding all kinds of excuses for Landau's sometimes intolerable conduct.

There are no or very few serious scientific biographies of Soviet/Russian scientists, exceptions may be the biography by V.Ja. Frenkel of his father Jakov Frenkel and the writings of Vitaly Ginzburg. An example of such a poor biography is the book by Gorelik and Frenkel on Bronshtejn [8]. It seems to be very hard for Russian biographers to tell unpleasant facts about their heroes and even about the country, which makes it all the harder to obtain a clear idea of what really was going on in the dark times of Stalinism.

---

[3]http://www.ihst.ru/projects/sohist/indexmat.htm.
[4]Izv. AN SSSR. Serija fizicheskaja no. 1–2 (1936) 63–162; Ref. [5], p. 295–305.

The story is not fully over yet. The documents and other material seized in the search of Shubnikov's apartment at the time of his arrest have still not been returned. They include valuable historical information, such as letters exchanged between Shubnikov and Trapeznikova when he was alone in Leiden during 1927, which could shed light on his experiences during that time, as well as correspondence with Wiersma after Shubnikov had returned to the Soviet Union. The documents must still be in the archives of the FSB, the successor of the KGB, but are currently not available for inspection.

# References

1. A.I. Akhiezer in B.I. Verkin et al. (1990).
2. M.Ja. Bessarab, *Lev Landau* (Moscow, 2008).
3. A.A. Bjalko, Mezhdu nami zhilo chudo (*A miracle lived among us*), *Priroda* 1 (2008) 90–91.
4. F. Beck and W. Godin, *Russian Purge and the Extraction of Confession* (Hurst & Blackett, London, 1951).
5. P.R. Josephson, *Physics and Politics in Revolutionary Russia* (University of California Press, 1991).
6. V.P. Orlov (ed.), *Repressirovanye Geologi* (Moscow, 1999).
7. A. Livanova, *Landau: A Great Physicist and Teacher* (Pergamon Press, 1980) [Russian 2nd edition: A. Livanova, *Landau* (Znanie, Moscow, 1983)].
8. G.E. Gorelik and V.Ja. Frenkel, *Matvei Petrovich Bronstein and Soviet Theoretical Physics in the Thirties* (Birkhäuser, 1994).

# Appendix 1

**Memorandum** for the People's Commissariat of Heavy Industry in which Shubnikov formulates the tasks of the physical-technical research in cryogenics (Ref. [1], pp. 43–47).

From this memorandum it is clear that Shubnikov was very much aware of the needs of industry and how his research in the low-temperature laboratory could contribute to the development of industrial processes.

## The UFTI Low-Temperature Laboratory

The development of experimental physics in recent years is characterized by the introduction of new powerful research methods. In order to make nature talk the modern experimental physicist sets up experiments at very high stresses, colossal pressures, very low temperatures, and uses devices that have a very high sensitivity or large precision. Thanks to the fact that technology has been forced to work for science and that laboratories can use the best that technology has at present to offer, the development of modern physics proceeds at fabulous speed. Deeper knowledge of the physical processes of nature does in turn have an enormous influence on the development of technology.

In recent years a number of new areas of technology have exclusively been created within the walls of physics laboratories. The area of low temperatures belongs to these extremely powerful possibilities for research in physics. Research at low temperatures is interesting for many reasons. In the first place, every new means of research leads to the discovery of completely new phenomena. One of the most well-known of these phenomena is the superconductivity of metals, implying that at very low temperatures a whole series of metals instantly loses all resistance to the passage of an electric current.

The phenomenon occurs in different metals at different temperatures, while it has been found that the transition from the ordinary state to the superconducting state occurs in a temperature interval of less than $1/100,000°C$, i.e. the transition is as sharp as the melting transition of a metal. In the superconducting state it has so far not been possible to measure the existence of an electrical resistance—it is infinitely small. The following experiment has been carried out: a current was induced in a superconducting coil and its strength was measured for four days. Within the limits

L. J. Reinders, *The Life, Science and Times of Lev Vasilevich Shubnikov*, Springer Biographies, https://doi.org/10.1007/978-3-319-72098-2

of accuracy of the measurements, no decrease of the current was observed, from which it was concluded that the resistance of the superconductor is in any case $10^{20}$ times smaller than of the metal in the ordinary state. A qualitative explanation, let alone a theory of this phenomenon is so far lacking.

Apart from exploring new phenomena, the region of low temperatures also has a specific interest. The fact is that all phenomena that occur in solid matter usually depend strongly on the temperature. Knowledge of temperature dependence is extremely important for analysing physical phenomena in nature and for verifying our theoretical ideas. However, the investigation of the temperature dependence at ordinary temperatures is uninteresting, since thermal motion is very strong and the phenomenon under investigation is complicated by additional phenomena, while investigations at very low temperatures at which thermal motion is small and the phenomenon becomes much simpler are of very great interest. As an example of such temperature dependence, whose investigation at low temperatures gave us much information on the structure of matter and the nature of thermal motion, we mention the dependence of the specific heat on the temperature. In recent years the investigation of the specific heat of gaseous hydrogen at low temperatures has led to an extremely interesting result. Contrary to other gases the specific heat of such a simple substance, like hydrogen, showed a large deviation from the values obtained by theorists. Calculations of a hydrogen molecule on the basis of quantum mechanics showed that ordinary hydrogen consists of a mixture of two different molecules and the anomalies in the specific heat are explained by the fact that we do not have a pure gas, but a gaseous mixture. These two forms of hydrogen—ortho- and para-hydrogen—have now been obtained separately and their properties are being studied.

In spite of the fact that physics only recently penetrated the region of low temperatures, methods for obtaining low temperatures have been transferred rapidly from the physics laboratory to a wide range of technological applications.

In technological applications low temperatures are mainly used for the separation of gases. Of special interest is the separation of air into its constituent parts: extracting oxygen from it, and separating coke-oven gases, mainly extracting hydrogen from it.

The need of the expanding industry for oxygen grows every year. At present oxygen is mainly used for the welding and cutting of metals, for detonation work etc. Air is separated into its constituent parts, subjecting it first to full liquefaction, and then fractionation. In this way it is in principle possible to separate air into its constituent parts: nitrogen, oxygen and noble gases: helium, neon, argon, krypton and xenon. Since the equipment is not yet perfect, air is usually separated into two components: oxygen and nitrogen, while the remaining gases are lost or remain as an impurity. These impurities, as recent research has shown, are often very harmful. For instance, an impurity of argon in oxygen decreases the speed of welding and cutting of metal approximately by a factor of ten. Noble gases find broad application in illumination engineering: half-watt lamps are filled with argon, and neon, krypton and xenon are used for manufacturing glow tubes (beacons, advertising lamps).

Hydrogen is at present only of interest for the chemical industry, since mixed with nitrogen it is used for the synthesis of ammonia—a basic product for the production of mineral fertilizer and the defence of the country. There are two procedures for obtaining hydrogen—the method of water gas[1] and the method of deep cooling of coke gas. The last method was introduced in Germany by the Linde firm[2] in 1923 and has spread ever wider since. In this method coke oven gas is broken down via fractional liquefaction of a higher boiling mixture, while hydrogen itself, which boils at a very low temperature, does not condense.

This area of engineering is very young; both the equipment and the method of separation are far from perfected. Therefore in the near future in this area much work has to be done by engineers and physicists.

In spite of the great interest of present-day physics and engineering in low temperatures, we have only a few laboratories working in this field. The reason for this is that this field of work requires extremely high technical skills, which are unavailable to most laboratories, not only in the USSR, but also abroad. A pioneer in this field is the low-temperature laboratory in Leiden (Netherlands), where Heike Kamerlingh Onnes succeeded for the first time in 1911 in obtaining liquid helium,[3] which boils only at 4 °K, i.e. at −269 °C. In 1926 liquid helium was also obtained at two other laboratories: in Berlin and in Toronto (Canada).

In 1926 Soviet physics could still only dream of its own laboratory of low temperatures. However, three years later the rapidly increasing technical level of the country, the growth of industry and the more serious organisation of scientific work in research institutes gave us the confidence in the necessity and possibility of founding such a laboratory in our country, in the Union. Such a laboratory was organized at UFTI in Kharkov. The size and outfit of the laboratory, as well as the scientific and technical tasks assigned to it put our laboratory on a par with the three laboratories mentioned above. This month it is three years since the birth of the institute, and we will report on three successes which we have achieved in the field of low temperatures.

We obtain low temperatures by means of liquid gases that boil at these temperatures. Our first task was to set up a facility for liquid gases in quantities that could fully fulfil the requirements of the laboratory. In order to be able to work in the entire range of low temperatures we needed an installation for obtaining liquid nitrogen (boiling point −196 °C), liquid hydrogen (−253 °C) and liquid helium (−269 °C). At the end of 1930 we started up a facility for obtaining liquid nitrogen, with an output of 30 l/h. This facility did not present any difficulties. Its operating principle is based on the fact that compressed air, when expanding, cools down. Thus, for instance, when expanding air from 200 to 2 atm. the temperature drops by 40°C. We can use the air cooled in this way for cooling and reducing the initial

---

[1]A synthesis gas, containing carbon monoxide and hydrogen, made by passing steam over a red-hot carbon fuel such as coke. Highly flammable.

[2]Still the world's largest industrial gas company by market share as well as revenue, founded in 1879. The Linde Group has over 600 affiliated companies in more than 100 countries. Revenues were EUR 17 billion in 2004.

[3]Shubnikov is mistaken here. Kamerlingh Onnes liquified helium in 1908. In 1911 he discovered superconductivity.

temperature of compressed air, which by subsequent expansion cools even more. The process is continued until part of the air does no longer convert into the liquid state. The facility consists of two main parts: a compressor, which compresses the air to 200 atm., and a device for liquefaction. By placing the liquid air in a special fractioning device we can extract liquid nitrogen from it.

The installation for obtaining liquid hydrogen works according to the same principle, however for hydrogen when expanding to be cooled, it was necessary to first cool it with liquid nitrogen. Since mid-1931 we have had liquid hydrogen in the laboratory, and the installation now yields 14 l/h. Setting up a facility of such capacity and making it work continuously presents great technical difficulties and requires very experienced, highly-qualified technical personnel that is used to doing experiments. Moreover, work with hydrogen presents a large hazard of explosion and requires special care.

Liquid helium is obtained by the same principle as liquid nitrogen and liquid hydrogen. For helium when expanding to be cooled, it was necessary to first cool it with liquid hydrogen. We obtained liquid helium in 1933.[4] By letting liquid helium boil under decreased pressure, we obtained a temperature of −271.2 °C. In order to obtain liquid hydrogen and liquid helium these gases must be very pure. An impurity of 0.1–0.2% is already sufficient to obstruct the process as snow is deposited in the apparatus. We have therefore constructed special equipment for purifying these gases by a deep-cooling method.

The field of low temperatures is not a particular branch of physics, but a method for doing research in physics, hence the nature of the scientific work of the laboratory is extremely diverse. Furthermore, since the UFTI low-temperature laboratory is currently the only one in the Union, we have also taken on ourselves a large number of physical-technical investigations of which industry is currently in great need.

Scientific work is performed in three directions:

Investigation of the conductivity of metals, in the first place superconductivity. We have only just started this work, since it requires liquid helium temperatures. Currently we are investigating the mechanism of the penetration of a magnetic field into a superconductor. As is well known, the superconducting state can be destroyed by a sufficiently strong magnetic field, while at a certain field strength the resistance of the metal returns with a jump to its ordinary value. The reason for this phenomenon is completely unknown. We hope to penetrate deeper into the essence of this phenomenon by investigating the penetration process and the distribution of the magnetic field in the superconductor. In due course, when developing the technical possibilities of experimentation with liquid helium, we expect to further extend our research in superconductivity.

Investigation of the structure of solids. As is well known, when cooled down below a critical temperature all gases become liquid and when cooled further they solidify. The temperature at which the gas, liquid and solid phases are simultaneously in equilibrium is called the triple point. All substances possess a triple point;

[4]It was actually at the end of 1932.

the only substance for which no triple point has been observed is helium. Apart from this amazing phenomenon helium shows an even more amazing one: namely two modifications of liquid helium. These anomalies in the behaviour of such a simple substance as helium show that we are dealing with a completely unknown phenomenon. We believe that we can get to know the phenomenon better after having finished the currently ongoing measurements of the melting heat of helium at temperatures below 2°K. The second work in this field is also devoted to the investigation of anomalous modifications. From investigations of the specific heat and dielectric constant it is known that at low temperatures a number of crystals undergoes a transition into a peculiar modification. The characteristic of these modifications is that the transition is not connected with a change of the crystal lattice, nor with a discharge of heat. This anomaly is probably connected with the fact that at a certain temperature the molecules in a crystal can start to rotate. To answer this question we are currently investigating the expansion coefficient of these crystals in X-rays and by means of fast electrons we can determine their structure before and after conversion.

The investigation of solids by the optical method is of great interest. In the laboratory it has been found that almost all transparent crystals at low temperature have no continuous absorption band, but a linear absorption spectrum. We know how much data on the structure of molecules has been obtained in the course of time from spectral analysis data. Continuing the work on the investigation of the absorption spectra of crystals, we hope to understand the structure of molecules, the nature of their thermal motion and the electric forces acting between molecules.

A third area of work is research on magnetism. It is known that at sufficiently high temperature all ferromagnetic substances lose their ferromagnetism and undergo a transition to a paramagnetic state. By cooling down a suitable paramagnetic substance to a sufficiently low temperature it is in principle possible to transform it into a ferromagnetic state. Many paramagnetic substances at low temperatures show anomalous behaviour as a function of the magnetic field and temperature. We believe that these anomalies must be attributed to a specific phenomenon of ferromagnetism in these solids. We are currently investigating these phenomena, which are of great interest and will increase our knowledge about the nature of the ferromagnetic state.

We are only just starting research on magnetism as, apart from low temperatures, this work also requires strong magnetic fields. The construction of such magnets is at present being completed at the institute. In this area we are currently investigating the cryomagnetic anomaly shown by a number of substances at low temperatures.

All work at low temperatures requires a precise determination of the temperature at which the experiment is carried out, since 1/1000 °C at low temperatures plays the same role as several degrees at ordinary temperatures. We are currently preparing and calibrating the only thermometers in the Union with which the whole region of low temperatures can be measured with great precision. These thermometers are made from platinum wire and they operate on the basis of the change of the electrical resistance as a function of the temperature.

Carrying out an experiment at low temperatures is technically very challenging and requires a large amount of special equipment. Most of this equipment we construct and prepare ourselves in the laboratory. For instance, the manufacture of metallic vacuum vessels for storing liquid hydrogen, apparatus for purifying gases, all vacuum apparatus and apparatus at high pressure etc. The physical-technical work of the laboratory mainly concentrates on investigating phase diagrams of gases, gas mixtures and liquid gases. Knowledge of these constants is extremely important for industry, since an accurate rating of equipment for gas separation is only possible on their basis. We have currently measured a whole series of these constants and developed an accurate rating method of apparatus for the separation of air. However, in spite of the fact that a significant part of the laboratory is engaged in physical-technical measurements, we are not in a position to satisfy the ever increasing demands of the ever growing industry, since an accurate rating of apparatus requires a huge number of constants, which so far have nowhere been published in the literature. Moreover, apart from physical constants the accurate rating of apparatus requires a number of empirical constants, since the processes in the apparatus are in fact very complicated. To determine these constants we have to work with models which resemble the industrial apparatus.

Since this work goes far beyond the work of a scientific research institute and requires special equipment, it is advisable to organise a special research station for deep cooling. UFTI is charged with setting up such a station. The research station will be constructed in 1934 in Kharkov at the experimental coke ovens of the Coal Chemistry Institute.

This research station will consist of a research facility for the separation of air, a research facility for the separation of coke gas and a physics laboratory.

1933

# Appendix 2
# Interrogations of Members
# of the UFTI Staff (Ref. [2], pp. 260–269)

## Interrogation of Aleksandr Akhiezer, member of Landau's group.
### Protocol of Interrogation
### *25 April 1937*

I, Security Officer Vesëlyj, have interrogated
1.      Name: Akhiezer
2.      First name and patronymic: Aleksandr Ilich
3.      Date of birth: 1911
4.      Place of birth: Cherskoe, BSSR
5.      Address: Kharkov, Tchaikovsky Street 16, apartment 12
6.      Nationality: Jew, citizen of the USSR
7.      Passport no: 209485
8.      Place of work, function: engineer of the theory group of UFTI
9.      Social origin: father—doctor

..................................................................

**Question**: How long have you known Professor Landau?
**Answer**: From 1934 at UFTI, where I worked with him.
**Question**: You also worked with Landau at Kharkov State University?
**Answer**: Yes, I worked with Landau at Kharkov State University in the academic year 1936/37.
**Question**: Did you take part in the collective protest against Landau's dismissal from the State University because of anti-Soviet manifestations early in 1937?
**Answer**: Yes, I did. That was at the end of 1936.
**Question**: Who, apart from you, took part in this collective protest?
**Answer**: Professor Shubnikov, Gorsky, Lifshits, Pomeranchuk and Brilliantov. All these people were at the time workers at UFTI and also worked at the State University.
**Question**: Tell us in detail about this incident.
**Answer**: On 13 December 1936 the rector of the university declared to Professor Landau that for a whole series of reasons he had to resign from his job. Landau refused to do so, upon which the rector warned him that in that case he would personally be forced to dismiss him. Landau arrived at UFTI straight from the rector and told me, Lifshits and Pomeranchuk what had happened. He was very upset by what had occurred.
We, that is I, Pomeranchuk and Lifshits fully shared Landau's indignation and declared to him that out of solidarity with him we would also hand in notice of resignation from the university. The following day Professor Shubnikov declared to us that regardless of our decision he already had prepared a statement of resignation and intended to hand it to the rector that day. This led to the situation that each one of us, that is I, Pomeranchuk, Lifshits,

© Springer International Publishing AG 2018
L. J. Reinders, *The Life, Science and Times of Lev Vasilevich Shubnikov*,
Springer Biographies, https://doi.org/10.1007/978-3-319-72098-2

Brilliantov, Gorsky and Kikoin also wrote separate statements to the rector with the request to release us from work, since we considered that in connection with Professor Landau's dismissal we were robbed of qualified leadership. Lifshits took all these statements to the university.

Apart from me, the rector of the university called everybody to him to give an explanation. Two days later I and the others were summoned to Kiev to the People's Commissar Comrade Zatonsky. In Kiev we were received by Zatonsky in whose presence we fully acknowledged our error and agreed to continue our work at the university.

After returning from Kiev, I acknowledged my mistake at a general meeting of UFTI workers. At the same time I was excluded from the trade union, but was reinstated by the district committee. When I appeared at the university, I learned that my dismissal was ordered because of disruption of work. I protested against this formulation to Comrade Mezhlauk and was reinstated to work at the university, where I work until the present day. None of the others who had handed in statements together with me were reinstated to work at the university.

**Question**: Are the actual reasons for Landau's dismissal from the university known to you?
**Answer**: From Landau I know that the reason for his dismissal was that he was alleged to have a poor relation with the students and was an idealist. In the journal "*Novy Mir*" the aspirant Lvov,[5] who works in Leningrad or Moscow, had placed an article on that question. I did not have any other sources (information) on the reason of Landau's dismissal from the university

**Question**: Who took the initiative to hand in statements of resignation from the university in connection with the incident with Landau?
**Answer**: There was no initiator as such, but as I already said the final push for this action was Shubnikov's statement of his decision to leave the university in protest against Landau's dismissal.

**Question**: Shubnikov supported your decision to follow his example?
**Answer**: Yes, he did. I do not now remember what he said in that connection.

**Question**: How did Landau react to your decision to hand in a solidarity statement?
**Answer**: Landau did not discourage us from our decision, but he did not directly encourage us to do this either.

**Question**: You understand that your participation in the collective protest at the university is in our terms an outright anti-Soviet demonstration?
**Answer**: Yes, I do.

**Question**: And did you understand this when you handed in your statement?
**Answer**: No, at the time I politically underestimated this demonstration.

**Question**: What was your relation with Landau?
**Answer**: I worked all the time under Landau's supervision. I visited him a few times at his house, but there were no close relations between us.

**Question**: Did Landau ever express his political convictions to you?
**Answer**: No, never.

**Question**: It is known to us that you and Landau are so close that you address each other with "ty". Is this indeed true?
**Answer**: Yes, it is.

**Question**: And yet in spite of this you deny being close to Landau?

---

[5]Vladimir Evgenevich Lvov (1904-2000), graduated in the early 1920s from the physics and mathematics faculty of Leningrad University, contemporary of Shubnikov and Landau. He became well known for his popular writings on science. He played a rather unsavoury role in the struggle against idealism and directed his criticism especially at Bronshtejn and Landau, who had called him the "scribbler Lvov" (*pisaka Lvov*).

**Answer**: Yes, since Landau used "ty" in his entire group regardless of the official position of the workers.

**Question**: What was the relation between Landau and Shubnikov?

**Answer**: Between Landau and Shubnikov very close relations existed.

All written down truthfully in my own words and read by me. *Akhiezer*

Interrogation carried out by Security Officer of the third section,

Sergeant of State Security *Vesëlyj*

## Interrogation of Stepan Shavlo dated 26 April 1937.

### Additional Protocol of the Interrogation of the Witness Shavlo S.T. on 26.04.1937

**Question**: What do you know about the UFTI worker Gorsky?

**Answer**: I have known Gorsky since 1934 from joint work at UFTI. At the institute Gorsky was always seen as an anti-Soviet, anti-social individual. When I got to know Gorsky more closely, I convinced myself that this opinion on Gorsky has a well-founded basis.

In January 1937 at a general meeting of UFTI workers Professor Sinelnikov, who still knows Gorsky from Leningrad, spoke and declared that in 1925 when working at LFTI Gorsky was in a counter-revolutionary organisation with the name 'jazz-band', disbanded by the organs of the GPU.

Gorsky's anti-Soviet nature manifested itself in the fact that he always systematically applied 'strike' methods in order to obtain what he wanted from the management of the institute. I know for instance that in 1932 Gorsky organised a 'strike' in protest against insufficient living space of his co-workers. Gorsky said: "… Down with Soviet order in science." After this Gorsky and his co-workers Kan, Rjabinin and a few others did not show up for work in protest.

In 1934–35 Gorsky, who taught at the Kharkov Machine-Building Institute, also stopped his work there in response to an admonition from the management on the poor quality of his lectures, after which he was excluded from teaching. Gorsky then declared: "… for boors I will not read lectures". I know about anti-Soviet actions by Gorsky in 1932 from the former secretary of the UFTI party organisation Shepelev and the member of the Communist Party Strelnikov, formerly working in Dnepropetrovsk.

At the end of 1935, together with Landau, Shubnikov, Brilliantov and others, Gorsky acted in an organised way against the introduction of socialist competition in the work of the institute, whereby Gorsky declared that "… socialist competition in science is unacceptable, go to factories and compete there…" The UFTI party committee gave a judgement about this at the time.

In 1935 when the worker and friend of Landau Korets was dismissed from UFTI and arrested by the NKVD, Professor Landau stopped work at UFTI in protest, and Gorsky, although he did not make a public protest, in fact held an 'Italian' strike for several days, showed up at work, but did nothing, and wholly agreed with Landau. Gorsky joined the collective protest made to Narkomtjazhprom against Korets's dismissal. Gorsky also took part in giving help to the family of the arrested Korets by collecting funds from 'common friends'.

At the time of the purge of the party in 1934 Gorsky had a relation with the commission member Averum (as I was told later, Averum was arrested as a Trotskyist) and provided her with false information on members of the party with the aim of compromising them. Then Gorsky declared to me in the presence of his wife: "Trotsky gave the correct slogan that you need to sandpaper communists and now they ask us, non-party people, to give the sand and so we did."

In 1936 Gorsky, Shubnikov and others again organised a strike at Kharkov State University in protest against Professor Landau's dismissal, who was accused of adopting a mocking attitude to the question of the training of cadres. Thanks to Comrade Zatonsky's intervention this incident was settled. All participants of the 'strike' made a statement acknowledging their mistake. Akhiezer, one of the participants of the 'strike', subsequently declared to me that Gorsky gave him directions to play a double deal, i.e. to admit that the collective protest was an error, but in essence not to back down from one's convictions in this matter. Gorsky, Landau and Shubnikov systematically carried out work directed at frustrating qualified members of the Ukrainian Communist Party and the Komsomol. Thus, Landau got rid of his aspirant-Komsomol member Pjatigorsky because the latter acted as a witness of the prosecution in the court case against Korets. Landau refused to give an explanation to Pjatigorsky on the reasons of his expulsion.

Gorsky forced the young specialist Braude to leave UFTI just because Braude was a very able person. Braude, when away from Gorsky, became a good specialist on ultra-short waves. The party member Borushko, sent to work with Gorsky, was exposed to all kinds of insults, was discredited and forced to leave, whereby Borushko openly declared that Gorsky was an anti-Soviet element, did not like to increase the number of cadres and did not care about cadres.

And finally in 1936 I was personally sent to work with Gorsky. For a year Gorsky did not allow me to use the X-ray equipment, charged me to do metal work, rejected it, spreading rumours about me being unsuitable for scientific work etc., not giving me the possibility to specialize. In all kinds of ways he impeded the completion of my diploma work and, when this work was done 'excellently', gave it a negative assessment. The Komsomol member Shkapenko and the Ukrainian Communist Party member Strelnikov can confirm how Gorsky insulted me.

Once at a meeting against Gorsky's training method of cadres the member of the Communist Party Gusak appeared. The next day after this meeting Gorsky refused to work with me and declared: "because of you they are still criticising me".

For two years Gorsky conducted a struggle against the Ukrainian Communist Party member Strelnikov, who was the first in the Soviet Union to construct a powerful X-ray tube. Gorsky spread rumours that Strelnikov was a dimwit, an untalented person. In 1936 Gorsky, having used Strelnikov's blueprints, wrote and published a paper, issuing it as his own, on the work of Strelnikov's X-ray tube. This plagiarism was unmasked and published in the journal "Uspekhi Fizicheskikh Nauk" in 1936. As a result Strelnikov was forced to leave UFTI.

I can still add the following anti-Soviet statements by Gorsky.

In 1936, after the execution of the counter-revolutionaries Kamenev and Zinoviev, Gorsky said to me: "... Have they really shot Kamenev and Zinoviev? After all they are cadres of Lenin." When I wanted to explain the actual situation of the matter, he refused to listen and declared: "... don't you read semiliterate things to me, I am myself politically literate...". As regards the Spanish events Gorsky said to me: "Stalin is afraid to send 1000 planes and tanks to Spain. The revolution there will die." I explained to him that this would lead to war, to which Gorsky remarked: "You are afraid of war, we all will go and make war, let's have war sooner".

**Question**: What does Gorsky work on at UFTI?

**Answer**: For 11 years Gorsky has studied special alloys—copper, gold. Nowadays the study of these alloys cannot be considered a primary subject, since the X-ray laboratory has very expensive equipment, which is used in solving scientific-technical problems for industry, while alloys of copper and gold have only a very narrow theoretical interest.

**Question**: Do you know about the relations between Landau and Gorsky?

**Answer**: Landau and Gorsky are friends. They still know each other from Leningrad.

All truthfully written in my own words, and read by me　　　　　*Shavlo*

Interrogation held by the Security Officer, Sergeant of State Security　　　　　*Vesëlyj*

# The interrogation of Vladimir Gej in Leningrad dated 17–18 May 1937.

## Protocol of Interrogation
### *17–18 May 1937*

I, Second Lieutenant Reznikov, Security Officer of the third section of the State Security Administration, have interrogated

1. Name: Gej
2. First name and patronymic: Vladimir Venjaminovich
3. Date of birth: 1900
4. Place of birth: Pskov
5. Address: Leningrad, Lesnaja-Ananevskaja Street 40, apartment 6
6. Nationality: Russian, USSR
7. Passport no:
8. Place of work, function: scientific worker, Scientific Research Institute no. 9
9. Social origin: agent of county bank
10. Social position: office worker
11. Family composition: wife Gej, Marija Jakovlevna
12. Education: higher, finished Physico-Mechanical Institute in Leningrad in 1930

...

**Question**: What do you know about the existence of a counter-revolutionary sabotage organisation at UFTI?

**Answer**: Based on the information available to me and on personal observations during five years of uninterrupted work at UFTI I believe that at UFTI there existed and exist two politically oriented counter-revolutionary groups which at a certain stage of their sabotage activity carried out joint counter-revolutionary work.

**Question**: What is the composition of this counter-revolutionary sabotage group at UFTI?

**Answer**: I believe that the counter-revolutionary sabotage groups indicated by me consist of the following persons:

1. Landau, Lev Davidovich—formerly the scientific head of the theory group at UFTI—currently a scientific collaborator of the Institute of Physical Problems at the Academy of Sciences in Moscow.
2. Korets, whose name and patronymic I do not remember—formerly a scientific collaborator of the UFTI theory department.
3. Obreimov, Ivan Vasilevich—scientific head of the crystal laboratory at UFTI.
4. Shubnikov, Lev Vasilevich—scientific head of the cryogenic laboratory at UFTI.

5.    Trapeznikova, Olga Nikolaevna—wife of Shubnikov, scientific collaborator of the cryogenic laboratory.
6.    Rozenkevich, Lev Viktorovich—scientific collaborator at UFTI.

This group is headed by L.D. Landau.
The other counter-revolutionary sabotage group consists of foreign specialists who arrived at various times in the USSR for work at UFTI, namely:

1.    Weissberg, Alexander Semënovich—the former head of OSGO, an Austrian subject.
2.    Houtermans, Fritz Ottovich—German citizen, scientific head at UFTI.
3.    Ruhemann, Barbara—German, German or English citizen, scientific collaborator of the UFTI cryogenic laboratory.
4.    Ruhemann, Martin Zigfridovich—German, English citizen, scientific collaborator of the UFTI cryogenic laboratory.
5.    Schlesinger, Charlotte—German, German citizen, arrived from abroad in the USSR together with Houtermans under the guise of being his relative.
6.    Tisza, whose name and patronymic I do not remember—Hungarian citizen, scientific collaborator of the UFTI theoretical department.

This counter-revolutionary group is led by A. S. Weissberg.
Lejpunsky, Aleksandr Ilich—Academician, currently director of UFTI—is closely connected with both counter-revolutionary groups and protects and encourages both Landau's counter-revolutionary group and Weissberg's group.

**Question**: What is their political platform and what is the political orientation of the counter-revolutionary groups you mentioned?
**Answer**: Judging from the political convictions and views of the groups of Landau and Weissberg, I believe that the Landau group is counter-revolutionary Trotskyist and the Weissberg group is a counter-revolutionary group of the 'rightist' deviation.
The above-mentioned Lejpunsky—Academician, is 'rightist' according to information available to me, which I will explain in more detail.
In 1935 after party and government had charged the institute with a number of important tasks, these two counter-revolutionary sabotage groups at UFTI, consisting of persons that are closely connected with each other, acted jointly in actively developing sabotage work at UFTI.
**Question**: Which tasks did party and government charge UFTI with in 1935?
**Answer**: Early in 1935 the Labour and Defence Council of Narkomtjazhprom charged UFTI for the first time on a large scale with a number of new tasks, extremely important for the defence of the country. Among them were, as far as I remember, the following defence-related subjects:

1.    For the laboratory of Professor Slutskin: the creation and application of powerful generators of ultra-short waves;
2.    As a personal task for Rjabinin of the laboratory of Professor Shubnikov: the development of oxygen vaporizers, devices important for high-altitude aviation, as well as hydrogen vaporizers[6];
3.    The construction of a machine for liquefaction according to the Kapitsa principle[7];

---

[6]See Chap. 7 where Alexander Weissberg is quoted with saying that Davidovich charged Rjabinin with defence-related work and in doing so passed over Shubnikov, who was Rjabinin's boss.
[7]This was actually also on Shubnikov's list of research topics (item 13). See Chap. 8.

4.          The construction of a Van de Graaff generator of 5–7 million volts and a column for it. Professors Sinelnikov and Valter had to run this generator.

All these questions had either significance for defence or great future significance for the national economy and because of their importance are especially secret.

For securing the utmost secrecy of the research work carried out in connection with these tasks by order of Narkomtjazhprom on the basis of resolutions of the USSR government the following was proposed to the institute:

1.          To review the entire staff of the institute;
2.          To introduce a number of restrictions and measures for securing the nondisclosure of work carried out at UFTI.

These measures had already been introduced somewhat earlier at the Leningrad Electrophysical and the State Physico-Technical Institutes and had become widely known at UFTI and in particular to the participants of the counter-revolutionary groups of Landau and Weissberg from personal correspondence with collaborators of the Leningrad institutes. At the moment of the start of the implementation of these measures at UFTI the groups of Landau and Weissberg consolidated and started to show strong opposition to the implementation of these measures.

**Question**: Which general aims united these two groups of Landau and Weissberg at UFTI?
**Answer**: The main aim unifying the counter-revolutionary groups of Landau and Weissberg was:

1.          Disruption of the government tasks obtained by UFTI on defence-related topics in accordance with the policy of the management of the institute;
2.          Directing the work at UFTI in the future towards purely theoretical research, and no practical activity on applied and defence-related topics, i.e. to isolate the institute from defence topics.

**Question**: How did the counter-revolutionary sabotage activity of the groups of Landau and Weissberg on the disruption of defence work at UFTI manifest itself concretely?
**Answer**: The concrete counter-revolutionary activity of the groups of Landau and Weissberg manifested itself in the following[8]:

1.          The construction of a machine for the liquefaction of hydrogen according to the Kapitsa principle. Shubnikov, a member of the counter-revolutionary group, should have constructed this machine already in 1935;
2.          The vacuum column for the Van de Graaff generator was not constructed at all; the generator itself did not achieve the planned 5–7 million volts (but only 4 million volts). Moreover, the research with this device proposed by government resolutions was not carried out either;
3.          The idea of Rjabinin and others on non-flammable gasoline, which has great technical and defence-related significance, was shelved.

The defence-related problems on ultra-short waves studied by Professor Slutskin and his laboratory, but also the topics of Rjabinin on vaporizers and tasks related to them were carried out in time and partly over-fulfilled, in spite of being frustrated by participants of the Landau-Weissberg counter-revolutionary group and the actual expulsion of Rjabinin from the institute.

---

[8]It should be noted that Landau opposed some of the proposals made by Davidovich on scientific grounds, for instance the use of liquid methane as aviation fuel, which he thought was a waste of precious resources. His opposition was interpreted as sabotage, but it became clear later that he was right.

There is no doubt that there was significant sabotage at UFTI; I can however not give any more detailed evidence as I left the institute at the end of 1935.

**Question**: How did the counter-revolutionary sabotage group of Landau and Weissberg work?

**Answer**: Analysing the situation at the institute in the period 1934 and 1935 and summarizing the information available to me, I believe that the counter-revolutionary groups of Landau and Weissberg united on the basis of their political convictions and their main task of disrupting defence-related work. For this purpose they used:

1. All legal possibilities, such as:
   (a) Appeals to the highest state and party bodies. Presenting applications to the Scientific-research sector of Narkomtjazhprom and the Central Committee of the Communist Party under the guise of being dissatisfied with the management of the institute. Sessions of the scientific council, meetings and wall posters were also used as a platform, but also the newspaper '*Izvestija*', which printed Landau's article at the very moment that the struggle started, which had a definite political significance.
   (b) The use of the authority of prominent party workers of the central apparatus (Modest Rubinshtejn,[9] member of the Central Control Commission of the Communist Party). He was an acquaintance of Weissberg and was used for presenting a letter of Landau's counter-revolutionary group to the Central Committee of the Communist Party, which I know from Korets's letter, attached to this protocol, but also from information obtained from Comrade Gorokhov, deputy head of the scientific department of the Central Committee of the Communist Party.
   (c) Ultimatums given by Shubnikov and Landau to the management demanding Rjabinin's dismissal. In this respect it is characteristic that the demand originated from Shubnikov and the ultimatum was supported by Landau. A similar demand in the form of an ultimatum was given by Landau demanding that an apartment be granted to Rozenkevich, a member of the counter-revolutionary group.
   (d) The participants of the Landau counter-revolutionary group practiced the clearly counter-revolutionary method of strikes, which they held in 1934 at the Kharkov Machine Building Institute and in 1936 at the university.
   (e) Soviet inclined scientific collaborators expelled from the institute after harassment were widely persecuted by the group. Concretely persecuted at their place of work were Kizilbash[10] in Tbilisi, and Bunimovich[11] in Leningrad, specifically as a result of letters sent by Obreimov.

2. Illegal methods, such as:
   (a) Sending encrypted letters. In this respect I can mention a letter of Korets to his second wife in Leningrad, to Andronnikov and Elefejter in Tbilisi, and I believe a letter from Landau to Kikoin in Leningrad. Below I will tell about the nature of the connection between these people.

---

[9]See the quotation from *Conspiracy of Silence* (Ref. [3], p. 161) in Chap. 7 for Weissberg's very positive assessment of this Rubinstejn. It shows how accurate Weissberg's book and how well-informed Gej is.

[10]Boris Borisovich Kizilbash, a Georgian pioneer in cosmic-ray physics working in Tbilisi from 1935 to 1938. "Boris Borisovich Kizilbash was an excellent physicist, a great expert and first-rate experimenter", according to Elevter Andronikashvili. I have not been able to find out which great injustice was done to Kizilbash, but perhaps he wanted to come to UFTI at some point and was not accepted.

[11]Possibly Vladimir Iosifovich Bunimovich (1904–1981), a radio-technical physicist.

(b) The organised harassment of scientific workers who were engaged in defence-related problems. Professor Slutskin and his collaborators underwent such harassment from Obreimov, Landau and others, the scientific worker Rjabinin from Shubnikov and Landau. Rjabinin was eventually forced to leave the institute. Soviet inclined scientific workers who were in opposition to the counter-revolutionary sabotage groups (Pjatigorsky, Kan, Kizilbash, Bunimovich and others) were similarly harassed. Pjatigorsky underwent especially unrestrained harassment on the part of Landau. He acted as a prosecution witness in the case of Korets, who was arrested by the NKVD, and he was chased out of the institute by Landau sanctioned by Lejpunsky, the director of UFTI.

Simultaneously with the harassment of Soviet inclined scientific workers and people who carried out defence work, the participants of the Landau-Weissberg counter-revolutionary sabotage groups besieged the management of the institute with demands for admitting foreign specialists to work at the institute and for calling them from abroad. Especially active in this respect was Weissberg, who demanded that Kashiev (Bulgaria), Shuftan (Germany) and others whom I do not remember were invited from abroad. Under pressure from Landau in 1934 I invited the Hungarian national Tisza to work at the institute.

Characteristic in this question is the solidarity of the counter-revolutionary group with directives of Lejpunsky. During his stay abroad the latter systematically demanded in letters to me (they are added) invitations from UFTI to foreign specialists for work in the USSR. In this respect it is important to note that not one of the foreign specialists proposed by Lejpunsky were of any value to the Union and completely unknown in the scientific world (with the exception of Lange).

**Question**: What do you know about the participants in the counter-revolutionary groups at UFTI mentioned by you?

**Answer**: The following is known to me about the participants of the combined counter-revolutionary groups mentioned by me:

1. Landau, Lev Davidovich, is the son of a wrecker arrested by the organs of the NKVD.[12] According to his political convictions he is a Trotskyist. In 1933 he showed his Trotskyist attitude publicly in the House of Scientists in Kharkov by eulogizing the enemy of the people Trotsky and slandering Comrade Stalin, "putting himself in the service of Trotsky". In questions of science Landau assumes eclectic positions consisting of a mixture of idealism and mechanism and has propagated these theories among Soviet students at Kharkov State University and the Kharkov Electrotechnical Institute (KhETI). When Landau was unmasked as an idealist by the students, and the management of the university demanded an explanation from Landau, he organised a strike in which Shubnikov, Gorsky, Lifshits, Akhiezer, Brilliantov and other UFTI scientific workers who also lectured at the university participated.

   A similar strike was organised by Landau in 1933–1934 at the Kharkov Mechanical-Machine Building Institute (KhMMI), in which Landau, Obreimov, Sinelnikov and others whom I do not remember participated. It is characteristic that the director Efimov of the KhMMI, after having obtained statements from the strikers, called a typist while the strikers were present and started to dictate to her a telegram to Petrovsky of the State University on the strike at the institute. The decisiveness of the director forced Landau, Obreimov and the

---

[12]A particularly vicious remark; a punch below the belt as it had nothing to do with Gej or with UFTI and Landau's father had long since been released.

others to back down by which the strike was averted. Yet this group subsequently left the KhMMI.[13]

Landau is an anti-Soviet person. Still connected with Leningrad as a close friend of Ivanenko, a counter-revolutionary, sent away from Leningrad by the organs of the NKVD after the murder of Kirov (Ivanenko is the son of an editor of "Kievljanin"[14]), and of Gamow a physicist who failed to return to the USSR from abroad.

Landau is one of the authors of the provocative fictitious order which in 1934 was hung up at UFTI to mock Slutskin's laboratory. In Leningrad Landau was excluded from the trade union because of political hooliganism,[15] and in Kharkov in 1937 because of the strike at the State University.

Hiding behind lush phrases on "pure science" and expressing scorn for everything applied, Landau spoke at scientific conferences displaying this attitude and thus carried out extensive sabotage work directed at the disorganisation of Soviet physics.

At UFTI Landau also propagated these counter-revolutionary 'theories' of his, essentially aimed at the disruption of applied and defence-related work. For this purpose he widely used the participants of the counter-revolutionary group and, in particular, Korets, and finally Landau was the organizer of the harassment of the candidate member of the party Strelnikov at the time of the party purge in 1934, of Rjabinin in 1936, and of Pjatigorsky in 1936 who were expelled from UFTI.

In the matter of the disruption of defence work, which UFTI was charged with by the party and government in 1935, Landau played a decisive role together with Weissberg, masking his counter-revolutionary work by 'quarrels' and discontent with the management of the institute.

2.     Weissberg, Aleksandr Semënovich, Austrian citizen. Arrived in the USSR in 1931 at Obreimov's invitation, to whom Weissberg in turn recommended Houtermans. His wife Eva Stricker[16] arrived in the USSR together with Weissberg. She was arrested by the NKVD in Leningrad in 1936.

Weissberg pretended to be a member of a Communist sister party, however he never became a member of the Bolshevik Communist Party. According to his

---

[13]Not much seems to be known about Landau's (and Obreimov's) activities at this institute. Landau was connected with this institute from 1933 to 1937, where he was head of theoretical physics. Kompaneets, the first student of Landau who passed the theoretical minimum exam, was a graduate of this institute.

[14]A conservative Russian newspaper, published in Kiev from 1864–1919; folded in December 1919, as the Red Army stepped into Kiev.

[15]He probably means the prank with Gamov, Ivanenko and others in connection with the article by Boris Gessen on the ether (see Chap. 9).

[16]Eva Stricker (1906-2011) went to Russia in 1932 as a visitor, declaring herself the fiancée of Alex Weissberg in order to obtain a visa. Remained there as a designer, worked at the Lomonosov Manufactory designing dinnerware. Married Weissberg in 1933. In 1935 she moved to Moscow to serve as Artistic Director of the China and Glass Industry; invited to design perfume bottles for the perfume trust, headed by Polina Zhemchuzhina, wife of Molotov; she was arrested (on 17 May 1936) before she could complete this commission. Accused of anti-Soviet agitation (in her ceramic designs Zionist stars and swastikas would allegedly be visible) and expelled from the Soviet Union in 1938. Rehabilitated in 1990. Her arrest and 15 months imprisonment, accused of plotting against the life of Stalin, forms the basis for *Darkness at Noon* by her life-long friend Arthur Koestler. See http://www.evazeisel.org/files-to-update/EvaStrickerZeiselChronologyFeb2012.pdf and Ref. [4], pp. 500–501.

political convictions and views he was a rightist and Weissberg propagated his counter-revolutionary rightist views. With me personally Weissberg had several talks on political topics, from which I remember his rightist views on the question of the uprising of the Austrian proletariat in 1934[17] against fascism, slandering the Austrian working class and its communist party, claiming that the uprising should inevitably be suppressed.

On questions of internal politics of the USSR, on questions of Soviet price policy and the planning of the socialist economic system, Weissberg also expressed his rightist views, demanding the introduction of a single gold currency in the USSR. The rebuff given by me to Weissberg in these two conversations forced him to stop expressing his rightist theories in front of me.

Weissberg has an unusually wide circle of acquaintances among foreign scientific specialists, but also among prominent personalities in the USSR. From the latter I know Ratajchak, Pushin, Norkin and Tamm, with whom Weissberg repeatedly had talks and conversations on his work at OSGO.

In the light of the fact that Weissberg is the leader of a counter-revolutionary sabotage group of foreign specialists at UFTI these connections of his deserve special attention. Weissberg, a master in fraudulent tricks, after his arrival smuggled typewriters and his own car into the USSR (selling the car later at a price which was almost four times higher).

It also deserves attention that Weissberg, pretending to be a specialist in aerodynamics, willingly took up the construction of the deep-cooling station, which has great significance for defence, i.e. did not work in his specialty, and in general as a foreign specialist is not of any value.

3.      Obreimov, Ivan Vasilevich, son of a nobleman, has a brother abroad who emigrated with the Whites. Participant of Landau's counter-revolutionary group. Jesuitical double-faced person, skilfully covered by the mask of a Soviet person. Tried to get into the party. Obreimov is the author of the notorious counter-revolutionary theory of 'communicating doors', according to which young Soviet specialists should stay no longer than 1–2 years at UFTI, making place for others, while there is only a place at the institute for 'specially selected', highly-qualified indigenous physicists.

Such a 'theory' in the hands of Obreimov was in essence directed at chasing away from the institute 'disagreeable' young Soviet specialists and selecting 'his own' people at the institute. Precisely this theory of Obreimov may explain the exclusive clogging existing at the institute to the present day. In 1935 the district committee of the Ukrainian Communist Party carried out an inspection of the institute as a result of which a very large number of people had to be dismissed. From the very founding of the laboratory of Professor Slutskin (defence-related work) Obreimov systematically conducted a harassment campaign against him, trying to discredit the scientific name of Professor Slutskin and his collaborators. At Obreimov's apartment, meetings of the counter-revolutionary sabotage group were organised, at which questions of the struggle against the management of the

---

[17]The Austrian Civil War (*Österreichischer Bürgerkrieg*), also known as the February Uprising (*Februarkämpfe*), is a term sometimes used for a few days of skirmishes between socialist and conservative-fascist forces between 12 February and 16 February 1934. Several hundred people (including paramilitaries, members of the security forces and civilians) died in the conflict; more than a thousand were wounded. The authorities tried and executed nine Schutzbund (paramilitaries organized by the Social Democrats) leaders under martial law. In addition, more than 1,500 people were arrested. Leading socialist politicians, such as Otto Bauer, were forced into exile. See *Wikipedia*.

institute were discussed, but in fact against the special work carried out at the institute.

At the very start of the struggle of this counter-revolutionary sabotage group Obreimov, appointed after the first inspection as deputy director in charge of scientific matters, conducted a cunning manoeuvre of treachery and did not sign the second appeal to the Central Committee of the Communist Party. He came out with a cunning paper against the excessive theoretical bias of Landau, after having directed sharp papers against Professor Slutskin. In this light I consider it necessary to mention the following case known to me which I cannot explain. In the days of mourning after the villainous murder of Comrade Kirov, at Comrade Kravchenko's place (my deputy) Obreimov provocatively had a probing conversation with me on whether there was someone at UFTI connected with the participants of the Leningrad counter-revolutionary organisation, and whether Gorsky had been mentioned in this respect, his favourite pupil and promoted worker, whom Obreimov and Landau had always supported. At the time I informed the organs of the NKVD about this.

4.      Shubnikov, Lev Vasilevich, nobleman.[18] A person with an anti-Soviet attitude, who often has counter-revolutionary talks with the people around him. Shubnikov has great strike experience in the USSR. Already in 1921 at Leningrad University Shubnikov posted a fictitious provocative order on ceasing work, thereby disrupting the work in a number of groups.[19] Shubnikov himself told Rjabinin about this. Landau spoke about this as follows: "Of all Shubnikov's pranks this one whereby one person dismisses the entire university is really first class."

In 1934–1937 Shubnikov took part in the strike in Kharkov at the Machine-Building Institute and the State University together with Landau. Shubnikov is an active participant of Landau's counter-revolutionary sabotage group at UFTI. Clandestine meetings took also place at his apartment.

In the struggle against defence-related work at UFTI Shubnikov harassed with special consistency the specialist worker Rjabinin, ultimately demanding the latter's departure from the institute. Shubnikov did not carry out defence-related work assigned to him.

In 1921 under strange circumstances Shubnikov escaped abroad on a yacht, lived for some time in Finland, subsequently in Germany and on his return he told that his yacht had been carried away to the sea on the wind.[20]

At the height of the struggle at UFTI in 1935 Shubnikov declared in an official conversation with Davidovich that there were two camps at the institute—a revolutionary one, standing behind science, and including the entire Landau-Weissberg group, and a counter-revolutionary group which is against science and included the party organisation and the Soviet part of the scientific workers.

Shubnikov had especially close and friendly relations with the entire group of foreign specialists of Weissberg and others.

---

[18]How Gej got this idea is not clear.

[19]Nothing seems to be known about this. Landau was not yet at Leningrad University in 1921, so he can only have knowledge about this from hearsay.

[20]It is very unlikely that Shubnikov said anything like this, as they were sailing on Lake Ladoga, not on the sea. Gej was born in 1900 and therefore he and Shubnikov must have attended the same classes at university. At the time he must have heard about this sailing trip. It is striking that Gej has nothing to say about Shubnikov's stay in Leiden.

5.  Korets, the son of a disfranchised person, who hid his social origin, for which he was twice expelled from the Komsomol. Korets was a very close friend of Landau and an active participant of this counter-revolutionary group. In 1935 he was arrested by the organs of the NKVD. Upon Korets's arrest Landau's group reacted very energetically, appealing against the arrest up to the Supreme Court. As a result of the active interference Korets was freed.

    From an enciphered and coded letter from Korets, added by me to this protocol, I believe that he is the connecting link between the Kharkov counter-revolutionary group and counter-revolutionary groups existing in other towns.

6.  Rozenkevich, Lev Viktorovich, a nobleman, son of a collegiate assessor.[21] In Leningrad he was close to Ivanenko, Gamow (a non-returnee from abroad) and Landau, connections which were not broken off in Kharkov. He took part in clandestine meetings of the counter-revolutionary sabotage group and in all its sabotage activity. Rozenkevich was an active co-author of the forged provocative order of 1934 at UFTI, which had the aim of scoffing at Professor Slutskin and his laboratory.

7.  Trapeznikova, Olga Nikolaevna, noblewoman (wife of Shubnikov), living with him abroad. Took part in counter-revolutionary meetings and signed the first letter of the counter-revolutionary group addressed to the Scientific-Research Sector of Narkomtjazhprom. Judging from Korets's letter Trapeznikova played the role of the connecting link between them and Shubnikov, through whom the latter also operated.

8.  Houtermans Fritz, German subject, a very close friend of Weissberg. Arrived in the USSR in 1935 from England at Lejpunsky's invitation. He pretended to be a member of the German Communist Party and hinted at having carried out clandestine party work abroad. Abroad he worked at many places and has unusually extensive connections among foreign and Soviet physicists. Took an active part in the activity of the Landau-Weissberg counter-revolutionary group. When at the institute a security and permit system was introduced upon the proposal of the government, he cynically scoffed at these measures, hung his permit on his back and showed it to the guard by turning his back to him. He agitated against special work at the institute, claiming that the secrecy introduced in connection with defence work would result in stupid and senseless work being carried out under this cover.

9.  Ruhemann, Barbara, former wife of a naval officer, the English physicist Blackett,[22] subsequently of Houtermans[23] and finally of Martin Ruhemann, who was under her strong influence. The two Ruhemanns arrived in the USSR on the recommendation of their long-time acquaintance Weissberg. Barbara Ruhemann displayed special activity in the counter-revolutionary sabotage activity at UFTI.[24] Together with Martin Ruhemann she participated in counter-revolutionary meetings at the

---

[21]*Kollezhsky asessor*: from 1717 to 1917 a civil rank; corresponding to the military rank of major; from 1722 occupying place VIII in the Table of Ranks, which in total had fourteen grades, the lowest being collegiate registrar (*Kollezhsky registrator*).

[22]Patrick Blackett (1897–1974), a lieutenant in the Navy in World War I, experimental physicist, winner of the 1948 Nobel Prize. Blackett spent some time in 1924–1925 in Göttingen where he may have met Barbara, but since he married Costanza Bayon in 1924 any relationship with Barbara must have been very brief.

[23]Houtermans was never married to Barbara; he married for the first time to Charlotte Riefenstahl in 1930. Gej probably wants to portray Barbara Ruhemann here as a depraved woman.

[24]Actually she was the greatest Stalinist among the foreign scientists at UFTI and, according to Weissberg (Ref. [3], p. 71ff), even after her return to England she remained faithful to her Stalinist views. She also believed that Weissberg was indeed a counter-revolutionary.

apartments of Weissberg, Shubnikov and others and signed both appeals addressed to the Scientific-Research Sector of Narkomtjazhprom and the Central Committee of the Bolshevik party.

10.  Ruhemann, Martin, agitated against defence work assigned by government and party to UFTI, and claimed that he knew from authoritative sources that the line of the party, directed precisely at the scientific workers of the institute and the imposition of defence work by the management of the institute, distracted the institute from its proper tasks. Martin Ruhemann took an active part in the harassment of Rjabinin, making common cause with Shubnikov. When Shubnikov forced Rjabinin to leave UFTI, Ruhemann forbade him to take part in his laboratory in work that he earlier himself had proposed to Rjabinin.

11.  Schlesinger, Charlotte, German and a German subject. Arrived in the USSR together with Houtermans under the guise of being his relative, was closely connected with Korets. Did not work at UFTI. I knew her little. After a short time she left Kharkov, apparently, for Kiev.

12.  Tisza, a Hungarian national. Arrived in the USSR in 1934 on an urgent requirement, invited by Landau. Tisza was closely connected with the entire group of foreigners, especially with Weissberg. I cannot say anything concretely on his participation in counter-revolutionary sabotage work at UFTI.

**Question**: Name the workers at UFTI known to you who supported the Landau-Weissberg counter-revolutionary group.

**Answer**: The Landau-Weissberg counter-revolutionary sabotage group enjoyed support from the following workers at UFTI:

1.  Korets, scientific collaborator of Landau, took part in a strike together with Landau;

2.  Brilliantov, also took part in the strike;

3.  Shtekkel,[25] foreign specialist, arrived together with Ruhemann, scientific collaborator.

4.  Lifshits, scientific collaborator of Landau, participant in the strike.

5.  Akhiezer, scientific collaborator of Landau, participant in the strike.

6.  Vereshchagin, scientific collaborator of Shubnikov.

7.  Khotkevich, scientific collaborator of Shubnikov, was subject to dismissal according to a decision of the district committee of the Ukrainian Communist Party. It is very suspect that Khotkevich nowadays works in the special laboratory of Shubnikov, organised by him after getting rid of Rjabinin.

The persons mentioned by me have made common cause with the counter-revolutionary sabotage group on a number of issues. About their direct involvement in this group I have no information.

---

[25]Practically nothing is known about this person. His German name is probably F.A. Steckel. There exist a few papers with an author whose name is variously spelled: F.A. Shtekkel and N.M. Tsin, i.e. Natalja Mironovna Tsin, a collaborator from Shubnikov's group (Determination of the Composition Diagram for the Liquid-Gas System Methane-Nitrogen-Hydrogen, *Journal of Chemical Industry (U.S.S.R.)* 16 (1939) 24-28), as well as F. Schtekkel (*J. Tech. Phys. U.S.S.R.* 6 (1936) 137–140) and F. Steckel in *Phys. Z. Sow.* 8 (1935) 337-341. He also worked with Francis Simon (F. Simon, F. Steckel, *Zs. physik. Chem. Bodensteinband* (1931) 737).

**Question**: What was the relation of the director of UFTI Lejpunsky with the counter-revolutionary sabotage groups mentioned by you?

**Answer**: In all his activities Lejpunsky encouraged all participants mentioned by me of the counter-revolutionary sabotage group. Having received repeated signals from a number of party members and non-party members about the presence at UFTI of a counter-revolutionary sabotage group, Leipunsky reacted with a kind of criminal coolness to these signals, considering that they were all 'quarrels'. It is very important that in its appeals to the Scientific-Research Sector of Narkomtjazhprom and to the Central Committee of the Communist Party one of the demands of the counter-revolutionary sabotage group was that Lejpunsky return from abroad and be appointed director of UFTI.

**Question**: What do you know about the relations of the Landau-Weissberg counter-revolutionary sabotage group with other countries and the USSR?

**Answer**: I know about individual connections of the participants of the Landau-Weissberg counter-revolutionary sabotage group with a number of persons abroad and in the USSR, namely:

1.  Connections of Landau: Gamow, scientific collaborator of Leningrad University, fled the USSR in 1934–1935, is in the USA, from letters of Lejpunsky from abroad it is known that he had talks with Gamow about his return; Kikoin, based in Sverdlovsk, is currently temporarily in Kharkov at UFTI, acquaintance of Landau from Leningrad, Kikoin was informed by Landau by letter about the events at UFTI in 1934–1935.

    A common acquaintance of Landau and Kikoin, but also of Weissberg, is Frenkel, Jakov Ilich, scientific collaborator of the Physico-Technical Institute in Leningrad. There exists a special friendship between Frenkel and Kikoin in spite of their difference in age and position. Weissberg turned to Frenkel for assistance in the matter of the release of Eva Stricker, who had been arrested in Leningrad, and especially travelled for this purpose to Leningrad. Frenkel knew Eva Stricker personally.[26] From Frenkel's connections the following deserve attention: Vasilev—former deputy director of the complex of the Physico-Technical Institute in Leningrad, Trotskyist or rightist; Budnitsky—former deputy director of LFTI, Trotskyist or rightist; Glazanov—former head of the second department of Scientific-Research Institute no. 9, Trotskyist or rightist; Rubanovsky—former scientific collaborator of the Institute of Chemical Physics in Leningrad, who feigned drowning, but in essence escaped the cordon; Talmud—director of the Institute of Chemical and Physical Research in Leningrad, who was also connected with Vasilev, Budnitsky and Glazanov.[27]

---

[26]Frenkel's son Viktor does not mention Stricker, nor Weissberg in the biography of his father (*Yakov Ilich Frenkel*, Birkhäuser Verlag, Berlin, 1996); nor does Weissberg mention Frenkel in [3]. Moreover, Stricker was arrested in Moscow (probably because of a denunciation in Leningrad) and taken to Leningrad (Ref. [4], p. 500). Weissberg (Ref. [3], p. 17) probably went both to Moscow and Leningrad to try to secure her release. Gej is making some things up here it seems, probably as he likes (or is asked) to involve Frenkel in the matter.

[27]Of these people only Talmud is mentioned in Viktor Frenkel's book about his father; therefore the latter's connection with these people (apart from Talmud who is mentioned as a family friend) cannot have been very close.

2.      Connections of Korets: Andronnikov, Elefter—scientific collaborator at Tbilisi State University; from Vigdorchik, Kan and Rjabinin I know that Korets informed Andronnikov about the 'events at UFTI'; Nora (I do not know her surname)—Korets's second wife[28] to whom he addressed letters in Leningrad lives in Kharkov on the territory of UFTI in the hostel of scientific workers.

3.      Connections of Houtermans: Lange—professor, works at UFTI, arrived in the USSR on Houtermans' recommendation and Lejpunsky's invitation; Blackett— English physicist, professor at a London college, was in the USSR, stayed often with Houtermans, got acquainted through the latter with Lejpunsky, at Blackett's departure from Kharkov, in spite of active opposition, Houtermans specially flew to Moscow to see him off.

4.      Connections of Martin and Barbara Ruhemann: Krauser—former physicist, now correspondent of the newspaper the 'Manchester Guardian', married to a sister of Barbara Ruhemann, came to the USSR in 1934, stayed with the Ruhemanns; Krauser introduced Obreimov to the English scientific world, which he knew well personally; in Lejpunsky's opinion Krauser is "a propagandist of Soviet science abroad", Lejpunsky recommended, but also supported his invitation and his coming to the USSR.

The protocol was read by me. Truly written in my own hand to which I put my own signature.

*V. Gej*

Second Lieutenant of State Security                      *Reznikov*

---

[28]Korets was married to Elenora Lazarevna Epstein who indeed was Korets's second wife. His first wife was Alexandra Simonova. He met his third wife Serafima Iosifovna Rudova while in a prison camp (see Ref. [5], p. 72 and 425).

# Appendix 3
# Documents of the Cases of Shubnikov, Rozenkevich and Gorsky (Ref. [2], pp. 226–272)

## TELEGRAM

From Kiev telegraphed to Shumsky, Kharkov, NKVD.
In the Katod-Kredo Affair arrest Lev Shubnikov, Pëtr Komarov and Lev Rozenkevich. Arrests have been agreed in Moscow by Mezhlauk, Narkomtjazhprom, and prosecutor Leplevsky. NR 45804 Ivanov.

*Transmitted 24.07.37, 7:26*

[on the telegram the note by Comrade Reznikov, urgent]

## RESOLUTION
### *Kharkov, 5 August 1937*

I, Sergeant Vajsband, Security Officer of the third section of the State Security Administration of the Kharkov District Administration of the NKVD, having considered the material on Shubnikov, Lev Vasilevich, born in 1901 in Leningrad, Russian, citizen of the USSR, non-party, engineer-physicist, working at UFTI, residing at Tchaikovsky Street 16, apartment 19, accused of the crimes envisaged in Articles 54-1 and 54-11 of the UkSSR Penal Code, being manifest from the fact that he is a participant of a counter-revolutionary Trotskyist group and carries out subversive work with the aim of disrupting defence work at UFTI,

### HAVE FOUND:

that his being at large can affect the course of the inquiry and that he can evade the inquiry and the court.
Based on the above and guided by Articles 143, 145 and 156 of the UkSSR Code of Criminal Procedure

### HAVE DECIDED:

1. To take the accused Shubnikov, Lev Vasilevich, in preventive detention in Kharkov prison in order that he will not evade the court and the inquiry.
2. Present this decision to the public prosecutor of the Kharkov region.

Acting Security Officer of the third section of the State Security Administration, Sergeant of State Security

*Vajsband*

© Springer International Publishing AG 2018
L. J. Reinders, *The Life, Science and Times of Lev Vasilevich Shubnikov*,
Springer Biographies, https://doi.org/10.1007/978-3-319-72098-2

Agreed: Deputy Head of the third section of the State Security Administration of the Kharkov District Administration of the NKVD, Captain of State Security

*Tornuev*

Approve: Acting Head of the Ukrainian NKVD of the Kharkov District, Colonel

*Shumsky*


# MEMORANDUM

On Shubnikov, Lev Vasilevich, born in 1901, in Leningrad, Russian, citizen of the USSR, non-party, with higher education, engineer-physicist by profession, works at UFTI as scientific head of the low-temperature laboratory.

Shubnikov is a participant in a counter-revolutionary group of specialists at UFTI, led by Professor Landau.

The counter-revolutionary group at UFTI in which Shubnikov participates sets itself the task of disrupting defence-related research at the institute and of disorganising all work.

With a view to the active participation in the group of foreign German specialists who engaged in espionage activity and were closely connected with Shubnikov, there is also reason to suspect his participation in espionage activity.

In 1921, while being a student at a Leningrad institute, under the guise of sailing on Lake Ladoga, Shubnikov fled to Finland, from where he moved to Berlin, where he lived for almost two years, after which he returned to the USSR. Shubnikov lived in Leningrad, was closely connected with the more reactionary part of LFTI—Landau, Ivanenko and others.[29] In the period of the murder of Kirov the latter was arrested and exiled because of counter-revolutionary activity.

Shubnikov arrived in Kharkov from Leningrad together with Professor Landau.[30]

In April 1936[31] we secretly took into custody and interrogated the member of the counter-revolutionary group Rozenkevich, who confirmed the presence at UFTI of a counter-revolutionary group led by Professor Landau.

Among the other participants of the group he mentioned himself and Shubnikov, while he declared that the start of this counter-revolutionary group went back to 1930 to LFTI, where it was organised by Ivanenko, Frenkel and Professor Gamov, who failed to return from abroad.

Later Rozenkevich retracted parts of his statements, after having confirmed the existence of a counter-revolutionary group at LFTI.

In his first statements Rozenkevich gave evidence of conspiratorial meetings of the UFTI counter-revolutionary group held at the apartments of Shubnikov and Landau.

The repudiation by Rozenkevich of part of his statements on UFTI is of no decisive significance for the case, i.e. similar information, except for the information on LFTI, we also have from intelligence. It is quite clear that Rozenkevich retracted his statement on the instruction of Shubnikov and Landau, with whom he undoubtedly shared the outcome of his visit to the NKVD.

Shubnikov and the entire counter-revolutionary group set as their task the disruption of work of significance for defence and therefore organised a quarrel against the UFTI

---

[29]This is at least partly incorrect. Shubnikov may have known Landau from Petrograd or may have heard about him, but they only got to know each other well in Leiden, where they became friends. Shubnikov enrolled at Petrograd University in September 1918. Landau was born in 1908 and arrived in Petrograd as a student in 1924. At that time Shubnikov was at LFTI since September 1923. Landau came to LFTI in 1926, the year in which Shubnikov left for Leiden.

[30]This is also incorrect. Shubnikov arrived in Kharkov in 1930 and Landau in 1932.

[31]This must probably be April 1937.

director, party member Davidovich who firmly kept raising the question of special topics, compromised him and sought his resignation from the job.

Apart from this, the participants of the group in which Shubnikov took part chased from the institute scientific workers who engaged in defence-related research (Rjabinin, Kravchenko, Pjatigorsky and others), and disrupted the work. In this way research was thwarted (oil as an explosive, devices for high-altitude flights, non-flammable fuel for balloons, etc.).

All this work, which is of enormous significance for the defence of the country and the development of which was already in progress, was thwarted by Shubnikov.

The group achieved this position when at UFTI no proper attention was paid to work of a defence-related nature.

Shubnikov twice placed results of his research of defence-related significance in foreign journals.

Shubnikov, Landau and others announced in practice a 'strike' in protest against the order to perform public work.

While teaching at Kharkov State University, Landau and Shubnikov allowed a number of anti-Soviet attacks against dialectical materialism as a science, on which basis they were asked to criticize their mistakes. Landau and Shubnikov categorically refused to acknowledge their mistakes, and as a protest laid down their work together with a number of UFTI collaborators teaching at Kharkov University.

Similar strikes were also for other reasons called by Landau and Shubnikov at UFTI.

Taking into account the fact that the presence of a counter-revolutionary group at UFTI is currently obvious, we consider it necessary to arrest Shubnikov for developing the case.

Acting Head of the first branch of the third section of the State Security Administration, Second Lieutenant of State Security                                          *Reznikov*

Agreed: Deputy Head of the third section of the State Security Administration, Captain of State Security                                                             *Tornuev*

Confirmed: Acting Head of Kharkov District Administration of the NKVD, Colonel
*Shumsky*

Decision on the basis of the memorandum: arrest of Shubnikov approved, District Prosecutor Leonov, 5.08.1937.

## Protocol of search
### *1937, 6 August*

I, Vajsband, collaborator of the Kharkov District Administration of the State Security Administration of the USSR NKVD on the basis of order no. 32 of the Kharkov District Administration of the State Security Administration of the UkSSR NKVD, carried out a search at citizen Shubnikov, Lev Vasilevich, residing in Kharkov at Tchaikovsky Street, house no. 16, apartment no. 19.

When carrying out the search the following persons were present: Pevnyi B.N., Panfilova Z.P., Trapeznikova L.V., Trapeznikova O.N., wife of the accused.

In accordance with the orders obtained, citizen Shubnikov, Lev Vasilevich, was detained. The following items were confiscated for presentation at the Kharkov District Administration of the State Security Administration of the NKVD:

| 1. | Passport of L.V. Shubnikov | 1 | no. 747242 |
|---|---|---|---|
| 2. | Trade-union card | 1 | no. 169647 |
| 3. | Army leave-card | 1 | |
| 4. | Various photographs | 16 | |
| 5. | Various personal letters | 22 | |
| 6. | Pocket notebooks | 3 | |
| 7. | Correspondence with Wiersma | 16 pages | |
| 8. | Plan of Berlin | 1 | |
| 9. | Plan of Leiden | 1 | |
| 10. | Certificate of Shubnikov | 1 | no. 9 |
| 11. | Various correspondence | 15 | |
| 12. | Correspondence between Trapeznikova and Shubnikov from 1926–27 | | |

No complaints were received.
Signed:
For the searched                                                                 *Trapeznikova*
Representative of the house administration                                        *Pevnyi*

## Questionnaire of the arrested person

| 1. | Name: Shubnikov |
|---|---|
| 2. | First name and patronymic: Lev Vasilevich |
| 3. | Date of birth: 29 September 1901 |
| 4. | Place of birth: Leningrad |
| 5. | Address: Kharkov, Tchaikovsky Street, house 16, apartment 9[32] |
| 6. | Profession and speciality: engineer-physicist (low temperatures) |
| 7. | Place of work and function: UFTI, scientific head of the laboratory of low temperatures |
| 8. | Passport issued: in the third district of the Kharkov militia no. 747246, series XK |
| 9. | Social origin: son of office worker |
| 10. | Social position: |

|   | a. | Before the revolution: studied |
|---|---|---|
|   | b. | After the revolution: studied and worked |

| 11. | Education: higher, engineer-physicist |
|---|---|
| 12. | Party card: non-party |
| 13. | Nationality: Russian |
| 14. | Category of military registration: officer staff with special assignment |
| 15. | Service in white and other armies: no |
| 16. | Subjected to any repression under Soviet power: no |
| 17. | Family composition: |

|   | 1. | Wife—Trapeznikova Olga Nikolaevna, born in 1901, collaborator at UFTI |
|---|---|---|
|   | 2. | Father Vasily Vasilevich, 55 years old, bookkeeper |
|   | 3. | Mother Ljubov Sergeevna, 55 years old, housewife |
|   | 4. | Brother Kirill Vasilevich, 22 years old, student |
|   | 5. | Sister Ljudmila, 33 years old, housewife |

*10.VIII.1937*

---

[32]Above it says 19.

# Protocol of interrogation
## *7 August 1937*

I, Skralivetsky,[33] collaborator of the third section, interrogated as the accused

1.      Name: Shubnikov
2.      First name and patronymic: Lev Vasilevich
3.      Date of birth: 29 September 1901
4.      Place of birth: Leningrad

........................................................................

The accused testified

**Question**: Material of the inquiry of the State Security Service has shown you to be a participant in a counter-revolutionary-espionage-subversive organisation. The inquiry requires from you exhaustive truthful testimony.
**Answer**: I deny being guilty of participation in a counter-revolutionary-espionage-subversive organisation. I do not wish to give any testimony, since I do not consider myself to be guilty of this.
**Question**: In connection with the wish expressed by you to give testimony, the inquiry allowed you to set out such testimony below.
**Answer**: I will answer in the inquiry only to questions that will be put to me.

# To the investigator Comrade V.I. Skralivetsky of the Kharkov District Administration of the NKVD from the prisoner L.V. Shubnikov

## Declaration

With this statement I admit my guilt before Soviet power and, if I am given the opportunity, I want in future through honest and conscientious work to make up for my guilt. Moved by this desire, I declare that I am a member of a Trotskyist sabotage group working within the walls of UFTI. I promise to testify honestly and exhaustively about my activities as well as about the activities of others. I hope that my voluntary confession will reduce the heavy punishment a little and allow me to return to my work, which in our country is a matter of valour, honour and heroism.

*7 August 1937*
*signature*

## Personal testimony of L.V. Shubnikov
*8.8.1937*

Supplementary to the statement I gave about my participation in a Trotskyist sabotage organisation I will give below exhaustive and detailed testimony.

The history of counter-revolutionary activity at UFTI is divided into two periods: a so-called 'quiet' period from 1932 to 1934 and a period of active formation of an organisation from 1934 to the present day.

---

[33]This NKVD collaborator V.I. Skralivetsky is known to have used 'illegal methods of interrogation.' See *Voskresnye chtenija* No 2 January 2012, p. 34 ff, in which another case is described as dealt with by this man.

(1)        A Trotskyist organisation at UFTI was formed as a result of persons with an anti-Soviet attitude appearing in the apparatus and among the scientific workers. Because of this attitude certain political convictions and certain harmful views on the goals and tasks of physics in our country took shape.

(2)        The composition of the national Trotskyist group was: Shubnikov, Leipunsky, Obreimov, Korets, Landau, Rozenkevich, Gorsky, Brilliantov, and Usov. Moreover, a double-dealer position was taken by Garber (engineer of Obreimov). The local Trotskyist group was headed by Leipunsky and Korets, who were the ideological leaders of the group. The active conductors of the group were I, Shubnikov, and Landau.

In parallel with the local Trotskyist group there existed another counter-revolutionary group primarily consisting of foreigners led by the OSGO engineer Weissberg.

As far as I know, M. and B. Ruhemann, F. and Ch. Houtermans and Tisza were members of the second counter-revolutionary group led by Weissberg.

At important moments for the country and the institute a change occurred in the views and practical activity of the group (I will especially give detailed statements on the unification of these groups and on the subversive work jointly carried out by them). The close connection between these groups was established by Leipunsky and Weissberg, who kept close contact as regards political activity.

(3)        Main landmarks of our subversive activity:

The participants of both groups set the following goals, which were achieved to a significant extent:

(a)  The development at the institute of theoretical work at the expense of work of a technical and defence-related nature.
(b)  Unwillingness to carry out work of a technical and defence-related nature.
(c)  An irrational development of the institute was conducted by Leipunsky who had decided that a colossal amount of resources were to be spent on high-voltage installations which could not lead to practical results.
d)  Incorrect formulation of the question of the training of young collaborators and post-graduate students.
(e)  The grandiose construction of OSGO, on such as scale that the premises and equipment cannot be fully used in the coming years.
(f)  Slow pace of construction of high-voltage building and OSGO, carried through by Leipunsky and Weissberg.

I undertake to give additional exhaustive statements on the essence of the above but also on any question put to me by the inquiry.
*8.8.1937*                                    *signature*


## Additional statements by L.V. Shubnikov

Supplementary to my statement of 8 August 1937 I declare that after the organised flight abroad, led by the organiser of the escape Merezhkovsky, I was recruited for espionage activity by I. Dessler under the following circumstances.

Having no means of existence I engaged in inventions with the material support of Dessler. My inventions failed. Then I proposed to continue the work on obtaining transparent quartz from sand I had worked on in Leningrad at the Optical Institute. I did not succeed in obtaining any results, but my proposal and his communication that he worked as a German spy made Dessler want to recruit me. When I left Germany for the Soviet Union he demanded from me that I inform him about work in physics that could have applied value. This I did during my trip to Holland in the autumn of 1926. At that time I informed him about a method for obtaining large crystals of salt from molten states to produce fine-grained emulsions. I was in Holland until the end of 1930. After my return to the Soviet Union I started to work at UFTI with the aim of organising a low-temperature laboratory there. However the former director Obreimov opposed in every possible way that I took this matter in hand, and until the autumn of 1931 I was in fact without work at the institute. I tried to obtain a *komandirovka* abroad with the purpose of remaining abroad. This was refused to me. In the summer of 1931 the institute charged me with the formation of a laboratory of low temperatures, which I started after the holidays. At that time I did not have any connection with Dessler. In 1934 Weissberg told me that my former connection was known to him, proposed to draw a conclusion from this and from that time he obtained information from me. Moreover, in 1936 he proposed that I take the post of director or scientific head of OSGO, very strongly insisting on this, and promised, in case I would agree, to get an order from on high.

Qualified German personnel at OSGO should guarantee that any results obtained would not be applied in the USSR and be passed on to German Industry.

*11.8.1937*

## Official position of all participants of the group[34]

(1)  *Ivan Vasilevich Obreimov* has been working at UFTI as scientific head of the solid-state laboratory from the moment of the founding of the institute, i.e. from 1930. Moreover in the period from 1930 to 1933 he was also the director of the institute. Until then he worked as the scientific head of the solid-state laboratory at the Leningrad Physico-Technical Institute (LFTI) from the moment of the founding of the institute, i.e. from 1923 through to 1928 when he was charged by Academician A.F. Ioffe with the organisation of UFTI. The actual organisation of UFTI, both as regards projects and as regards selecting staff. In the period 1926–1928 he combined this with being assistant director of LFTI.

(2)  *Aleksander Ilich Lejpunsky* works at LFTI as scientific head of the atomic nucleus laboratory, has also been heading the work of all subdivisions and groups of this large laboratory from the moment of the formation of the laboratory, i.e. approximately from 1932. He was also director of the institute in the period 1934–1935 and from 1936 to the present day. Until 1932 he was the head of the laboratory of elementary processes. Until his arrival in Kharkov in 1930 he worked at LFTI as a physicist in the laboratory of N.N. Semënov, approximately from 1924. He was invited to UFTI by Obreimov.

(3)  *Lev Vasilevich Shubnikov* worked at UFTI as head of the low-temperature laboratory from the moment of the founding of the laboratory, i.e. from 1931 until the day of his arrest in 1937. Before that, starting from 1923, he worked as a physicist at LFTI, in the solid-state laboratory, led by I.V. Obreimov. In the period 1935–1936 he held the chair of professor of solid-state physics at Kharkov University. He was invited to work at UFTI by I.V. Obreimov.

---

[34]It is not a priori clear that this document was written by Shubnikov, as it is mostly written in the third person also when it involves Shubnikov himself. Only in a few places the first person is used.

(4)     *Lev Davidovich Landau* worked at UFTI as scientific head of the theoretical department from 1934 to March 1937, when he went to work at the Institute of Physical Problems of the Academy of Sciences, led by P.L. Kapitsa. Until 1934 he worked at LFTI in the same capacity. In 1934–1935 he also gave lectures at the physics-mechanics faculty of the Machine-Building Institute and occupied the chair of theoretical physics at Kharkov University in 1936 and 1937. He was invited to work at UFTI by A.I. Lejpunsky.

(5)     *Mikhail Korets* worked at UFTI in 1935. I do not know the place of his earlier work, I only know that he was in Sverdlovsk and Leningrad and had a connection with the teaching of physics at physics-mathematics faculties. At UFTI he was an engineer, at the same time leading problem classes with students at Kharkov University at the physics-mathematics faculty. He was invited to work at UFTI by Lejpunsky on Landau's proposal.

(6)     *Vadim Sergeevich Gorsky* has been working at UFTI from its formation, i.e. from 1930, first as a physicist in the laboratory of I.V. Obreimov, and from 1937 as scientific head of the Roentgen Laboratory. At the same time he led problem classes with students at the physics-mathematics faculty of the Kharkov Machine-Building institute approximately in 1933–1935 and with students of the physics-mathematics faculty of Kharkov University in 1935–1936, and was a professor at the university. Until his arrival in Kharkov he worked at LFTI as a physicist in the solid-state laboratory, led by Obreimov. He was invited to UFTI by Obreimov.

(7)     *Lev Viktorovich Rozenkevich* has been working at UFTI from the moment of its formation, first as a theorist, but subsequently from 1935 he switched to experimental work on the study of the atomic nucleus. Lately he worked as scientific head. Parallel to his work at UFTI he had a number of other jobs, of which the following are known to me: teaching at the physics-mathematics faculty of the Machine-Building Institute approximately in 1932–1934, teaching at Kharkov University in 1934–36 and work as deputy editor of the journal 'Soviet physics'. Until his transfer to Kharkov he worked at LFTI under the direction of Professor Rozhansky. He was invited to work at UFTI by Obreimov.

(8)     *Aleksandr Semënovich Weissberg,* German, worked at UFTI in 1931 with Obreimov, in 1932 at the low-temperature laboratory and from 1933 he directed the work on the preparation of the OSGO project, and subsequently also its construction. I do not know his previous place of work, he arrived in Kharkov directly from Germany at the invitation of the former director Obreimov on a recommendation from Lejpunsky.

(9)     *Martin Zigfridovich Ruhemann,* English citizen with German background, working at UFTI since 1932, first at the low-temperature laboratory, and from 1934 as head of the OSGO laboratory, temporarily staying in an UFTI apartment. Before that he worked at the low-temperature laboratory with Professor Simon in Berlin. After Professor Simon moved to Breslau he worked in Stuttgart with Professor Ewald. Invited to work at UFTI by former director Obreimov on a proposal of Weissberg and with my support.

(10)    *Barbara Frantsevna Ruhemann,* English citizen, originally German, working at UFTI since 1932 in the low-temperature laboratory as a physicist. Before that she was abroad, studying and living with her husband. Invited by Obreimov.

(11)    *Fritz Houtermans,* German citizen, working at UFTI since 1935 as scientific head in a subgroup of the atomic nucleus laboratory, led by Lejpunsky. Before that he was in England and worked as a physicist in the laboratory of a radio-gramophone factory. Invited for work at UFTI by Lejpunsky.

(12)      *Charlotte Houtermans,* German citizen, worked on the editorial staff of the
          journal 'Soviet Physics' in 1935 and 1936. Invited together with her husband.

The resources and staff made available during 7 years at the Ukrainian Physico-Technical
Institute had to give the country, apart from physical discoveries, also a number of real
applications of physics for the needs of industry and the defence of the country. And what
in fact do we have? The result of the wrecking activity of the counter-revolutionary
organisation at UFTI is clearly visible.

The main counter-revolutionary sabotage in the field of physics was that in the entire
country the leading scientific cadres led the struggle for the disruption of physical-technical
problems and related defence work under the counter-revolutionary slogan of the incompatibility of scientific work with applied work. It is therefore no accident that the foreigners,
when they arrived in the USSR and observed that such a policy was pursued by the Russian
physicists, joined in the general counter-revolutionary choir and took part in weakening the
defence capabilities of the country.

The Ukrainian Physico-Technical Institute, led by the counter-revolutionary Lejpunsky,
had in its midst an organised counter-revolutionary organisation which carried out the
practical wrecking activity set out below.

In order to ensure that their illegal undermining work had the desired effect the participants
of the organisation Landau, Korets, Rozenkevich, Obreimov deliberately created impossible conditions for the realization of extremely important defence work, by harassing and
isolating workers who pursued defence subjects, and, if successful, by sabotaging the
completion of work which could be useful for strengthening defence.

The whole essence of the sabotage carried out at UFTI consisted of assigning only remote,
promising value to the solution of problems of the institute for industry and for the defence
of the country and, covering this up, by deliberately not providing any possibilities for the
development of immediately applicable work of a defence and technical value.

Who led the counter-revolutionary work of the participants of the group and in which
sector?

*Obreimov.* Obreimov opposed the development of theoretical and defence-related work, he
bullied young specialists and chased them from the Institute, both in his own laboratory in
the case of Strelnikov, Braude, Vereshchagin, and also in other parts of the institute in the
case of Bunimovich, Kizilbash and others. As former director he sabotaged the development of research work of the Institute, based on the theory that scientific and applied
research work are incompatible. For several years after his dismissal as director he carried
out sabotage at the institute. As a result, in the seven years of its existence the laboratory led
by Obreimov has not accomplished a single remarkable scientific achievement, and
absolutely nothing was done in the area of technical and defence work. He deliberately
delayed the formation of the low-temperature laboratory as an important establishment of
technical, defence-related and scientific work.[35] He created at the institute an atmosphere of
quarrels, spreading clearly erroneous information and gossip, engaged in provocative work
in order to create estrangement and mutual hostility among the staff. He actively opposed
the training of young physicists within the walls of the institute.

*Lejpunsky.* He led the institute as a wrecker, directing its development in a direction which
was less effective for the country, in the first instance by opposing technical and
defence-related work. He liquidated a laboratory that had carried out important technical
work, e.g. the laboratory of ion transformations. He chased staff from the institute and
deliberately failed to create the conditions for the selection and training of new physicists
and post-graduate students, as a result of which the institute did not train a single qualified

---

[35]This sounds as personal criticism of Obreimov by Shubnikov, who was idle for the best part of a
year before he could start setting up his laboratory (Chap. 6).

physicist in the last years, using staff that had been obtained at one time from other institutes of the Union, mainly from LFTI. He carried out sabotage by spending resources assigned to the institute for wrecking purposes. Here I refer to multimillion amounts of money for the construction of high-voltage installations that nobody needed, which, after they had been constructed, did not turn out to work and did not give any results, were subsequently dismantled in order to again build in its place an installation nobody needed. As a result, in more than five years of work in investigating the atomic nucleus the institute has literally achieved no result whatsoever. The same wrecking activity was also carried out by Lejpunsky jointly with Weissberg in connection with the Deep-Cooling Research Station (OSGO) by endlessly prolonging the period of construction of the station, increasing the scale of the construction beyond the limits of necessity, which resulted in the construction period being dragged out and the construction becoming more expensive. For this purpose at one time the chief of construction was Zabalsky, a person who already had failed with more than one construction in the Union. He supported and realized jointly with Weissberg a sabotage plan for selecting and training staff for the OSGO laboratory. For this purpose the main leading positions in the OSGO laboratory should be distributed among foreigners in order to ensure delay in the application of technical achievements of the station in the Union and to provide the information to foreign countries. Lejpunsky deliberately frustrated the training of Soviet cadres and failed to timely prepare housing and workplaces for OSGO staff at UFTI.

*Landau.* Opposed technical and defence work at the institute, propagating the idea of the incompatibility of scientific and applied work. He harassed young specialists: Pjatigorsky, Bunimovich and Rjabinin. Together with Rozenkevich he was the author of the well-known April Fool's day order, in which scientific degrees were deliberately incorrectly assigned to the collaborators of the institute, whereby people who engaged in technical and defence work were 'awarded' the lowest degrees. He held counter-revolutionary anti-Soviet conversations within the institute, criticising the decisions of the party and the government and the higher organs of power, leading the fight against the application of dialectical materialism in physics. He took part in the counter-revolutionary anti-Soviet action of a group of people: Shubnikov, Gorsky, Brilliantov, Lifshits, Akhiezer, Kompaneets, Kikoin in the form of a protest strike against Landau's dismissal from the university.

*Rozenkevich.* Opposed technical and defence work at the institute, propagating the idea of the incompatibility of scientific and applied work. Together with Landau he was the author of the well-known April Fool's day order. He was sharply anti-Soviet and tried to obtain a *komandirovka* abroad with the purpose of staying abroad. He held counter-revolutionary anti-Soviet conversations within the institute, criticising the decisions of the party and the government and the higher organs of power, expressing himself against the application of dialectical materialism in physics.

*Korets.* Opposed technical and defence work at the institute, propagating the idea of the incompatibility of scientific and applied work. For this purpose he spread a whole series of theories within the institute on the expediency of assigning defence work to a new establishment of UFTI and dividing the institute into two parts: one for defence work and the other for scientific work, which would imply the exclusion of the entire old UFTI building from technical and defence subjects.

Korets is the type of an inveterate counter-revolutionary, who engaged in anti-Soviet propaganda within the institute, directed against the decisions of party and government. The activity of Korets was rapidly stopped by his arrest. In court I, Shubnikov, appeared in defence of Korets.

*Gorsky, Brilliantov* opposed technical and defence work. They spread counter-revolutionary anti-Soviet propaganda within the institute, criticising the decisions of the party, government and the higher organs of power, expressing themselves against the application of dialectical materialism in physics. They were participants of counter-revolutionary anti-Soviet action in the form of the organised strike in protest against L.D. Landau's dismissal from Kharkov University at the end of 1936.

*Weissberg.* Conducted counter-revolutionary sabotage activity on the orders of German intelligence both at OSGO and at UFTI. As regards the construction of OSGO he tried to prolong the period of completion of the construction, increased the necessary size of the construction, which led to an increase of the construction time and to unwise spending of resources. According to Weissberg's wrecking plan the scientific staff should mainly consist of foreign specialists which would make it possible to withhold the achievements of OSGO from Soviet industry and transfer them abroad. In order to justify the hiring of foreign specialists, he publicly suggested the idea of the necessity to attract already trained qualified cadres, since Soviet cadres could not be trained in a short period of time. All this shows Weissberg's wish to hire for work at OSGO such people as Weisselberg, Shtramer and Taunis, who were not at all qualified specialists. The training of Soviet cadres was deliberately made to fail, since the number of people accepted for training did not at all secure the work of OSGO, and the abilities for qualification of most people accepted were very low.

At UFTI he engaged in wrecking activity by opposing technical and defence work, propagating the idea of the incompatibility of scientific and applied work.

Weissberg, being the consummate type of a counter-revolutionary, often expressed his negative attitude towards Soviet power and engaged in propaganda against the line of government and party.

*M. Ruhemann* is closely connected with A.S. Weissberg as regards wrecking activity, he worked with him on the construction, the training of staff and the research subjects of the future Deep-Cooling Research Station. I do not know if Ruhemann and Weissberg knew each other from abroad, but here they were very close both personally and as regards work. This explains Ruhemann's rather strong defence of Weissberg, which went as far as accusing the entire party organisation of sabotage in a well-known letter.[36]

He conducted anti-Soviet propaganda at UFTI, directed against the government and the party. He believes that it is not possible to work honestly in the Soviet Union, hence after Weissberg's arrest he decided to completely stop this work and leave for England.

*B. Ruhemann* conducted anti-Soviet propaganda at UFTI, directed against the government and the party. She believes that it is not possible to work honestly in the Soviet Union, hence after Weissberg's arrest she decided to completely stop this work and leave for England. Opposed technical and defence-related work at UFTI by spreading the idea of the inferiority and ignorance of the people engaged in defence topics.

*F. Houtermans.* Good friend of Weissberg, still when abroad, where they together engaged in party work. On Houtermans' recommendation, communicated by him to Lejpunsky and Obreimov, as far as I know, Weissberg was accepted for work at UFTI.[37] After Weissberg's arrest he completely stopped work at UFTI and decided to leave for England. Opposed technical and defence work of the institute and spread propaganda against party and government resolutions.

*Ch. Houtermans* is an anti-Soviet inclined person, but did not express her anti-Soviet views. As far as I know, she had no greater involvement in counter-revolutionary work.

---

[36]Nothing is known about this letter, but Ruhemann was the only one who spoke up for Weissberg at the staff meeting organized by the NKVD to condemn Weissberg, when he had just been arrested as an enemy of the people. Ruhemann was never arrested, but in 1938 was forced to leave the country when the director of the institute was instructed not to renew his contract (Ref. [3], p. 71).

[37]This seems to be completely incorrect. Houtermans only came to UFTI in 1935, Weissberg already in 1931.

*L.V. Shubnikov* opposed technical and defence work at UFTI, spreading the idea of the incompatibility of scientific work with applied work. Harassed workers who engaged in defence work, dismissed Rjabinin from the low-temperature laboratory while he was carrying out defence-related work on the application of liquid hydrogen as a combustion fuel in airplanes. With the aim of sabotage he spread the opinion on the work of Slutskin's laboratory that they were ignorant and inferior physicists. As an anti-Soviet inclined person, he saw the purpose of physics in the development of purely scientific work, with a value far in the future, and in the laboratory led by him he developed precisely such work, not starting work of technical and defence value. In the founding of OSGO he saw a way of saving the low-temperature laboratory from the necessity of carrying out technical and defence work, i.e. after the formation of OSGO such work should be done there. As a result of my[38] counter-revolutionary wrecking activity the low-temperature laboratory, which contributed a number of discoveries of promising scientific value, did not contribute anything to the economy and the defence of the country. I informed Weissberg about the progress of work carried out by the laboratory. Politically I shared the counter-revolutionary views of the group on resolutions of the party, government and higher powers, and also on the interrelations between Marxist philosophy and physics. I was an active participant of the organised protest strike against L.D. Landau's dismissal from Kharkov University at the end of 1936.

The practical activity of the counter-revolutionary sabotage group consisted of:

(1)      The creation of an atmosphere of quarrelling and mutual mistrust among the scientific workers of UFTI. This task turned out to be easy, since most UFTI collaborators were anti-Soviet people, who were only interested in their own personal ambitions and career possibilities, but not in the interest of the Soviet land. As an example the well-known April Fool's day order can be cited, published by Rozenkevich and Landau, whereby its text offended a number of collaborators, since scientific degrees were deliberately awarded inversely proportional to the duties of the person as a scientist.

This also has to do with the harassment of people working on special subjects. At the institute the opinion was put forward that Slutskin and his collaborators are inferior and ignorant physicists.

Finally, this relates to the expulsion of specialists who worked on special subjects. As an example I can point to my quarrel with Comrade Rjabinin, a person who is good at technical work and completed defence work on the use of liquid hydrogen for aircraft engines. As a result of the quarrel Rjabinin was transferred to another laboratory while I had undoubtedly all the means at my disposal for resolving the conflict in a peaceful manner. The quarrels distracted the attention of the scientific workers from accomplishing work and lowered the pace of fulfilling the plan.

(2)      The closure of a laboratory that had an applied purpose. In 1936 the laboratory of ion transformations, which was of great interest for modern electronics, was closed. This laboratory was transferred to the Kharkov Electromechanical Factory. The reason for the liquidation of the laboratory was twofold. First, the scientific head of the laboratory E.M. Sinelnikov was extremely seldom in the laboratory, leaving it without supervision. Second, the sabotage group spread the opinion that it was expedient to transfer the laboratory to the factory.

(3)      The transmission of important discoveries of a defence nature abroad through publication in the scientific press. I know that this year with permission from Lejpunsky Ruhemann sent two papers to the journal 'Khimstroj' on the

---

[38]A few times such words make clear that this is actually Shubnikov's statement.

extraction of helium from natural gases. Insofar as I know, the papers were not secret, but they are of enormous interest for the construction of equipment for the extraction of helium. Other facts I do not know.

(4)     Deliberate delay in the utilization of achievements at the institute.

Scientific achievements were not delayed, communication about them went quickly to the printer. But technical achievements were usually not brought to an end or they were delayed. The reason for this is simple. It had in the first place to do with the wrecking attitude of the management and collaborators of the institute towards technical work, in the second place with the difficulty to hand over a new product for manufacture, i.e. again because of the wrecking attitude. The most significant sabotage took place here in respect of the manufacture of metallic Dewars: the country had an enormous need for vessels for storing liquid nitrogen and other liquid gases; UFTI (Kostenets) developed the technology for preparing Dewars, of a considerably higher quality than foreign ones, and still after almost three years the mass manufacture of Dewars had not yet been organised.

Furthermore Strelnikov's X-ray tube, which was of great interest to factories, industrial laboratories and scientific research institutes, also did not go into production. It is true though that in this case the tube was not constructed to full perfection.

Finally, the wrecking attitude applies to a whole series of minor technical inventions, needed by other laboratories of the Union. In most cases these inventions were simply lost within the UFTI laboratory.

(5)     Sabotage in the training of cadres, in particular for OSGO.

In this matter Weissberg and Lejpunsky played a leading role. Weissberg argued everywhere that it was necessary to attract foreigners for the leading cadres of OSGO, since it was not possible to train Soviet cadres in a short period of time. In the first place it was necessary to attract Germans, since the German cooling industry was the most perfect. These conversations were meant to convince the public opinion of the necessity to attract foreigners. In fact the attraction of foreigners had to serve the goals of espionage, since Weissberg did not intend to attract such great specialists for the work, but people like Weisselberg, Shtramer, Tausik, i.e. people, who were hardly better trained than Soviet students who had completed a higher technical school. Lejpunsky supported Weissberg's point of view. The younger staff at OSGO should according to Weissberg consist of Soviet specialists. The set-up of their training was extremely poor. With sabotage in mind Lejpunsky opposed the hiring of new people in every possible way, not wishing to provide regular places although the workers had to work on UFTI topics; for a long time he did not provide housing although at the time a number of rooms of the demolished high-voltage installations stood empty.

(6)     Grandiose scale of deception of party and government, expressed in the well-known telegram announcing the splitting of the lithium nucleus.

English physicists had done this work a few months earlier and published it modestly in a scientific journal; the showmen of UFTI made a global affair out of it. The organisers of this affair were Latyshev, Sinelnikov, Obreimov and, if I am not mistaken, also Shepelev, who at the time was secretary of the party committee. Lejpunsky was at the time on holiday in the south.

(7)     Strikes, disruption of work and other anti-Soviet actions.

As far as I know, the first strike was organised by the UFTI collaborator Kan when reading lectures at the physics-mechanics faculty of the Machine-Building Institute. Soon after this there was the strike of Brilliantov, Rjabinin and Kan (it is possible that the names are not completely correct) demanding that rooms be provided below the dining-hall of the hostel. Subsequently Gorsky went on strike at the Machine-Building Institute (no precise information). Finally, the biggest anti-Soviet action in the form of a strike was at the end of 1936 at the physics-mathematics faculty of the university. Seven people: Shubnikov, Gorsky, Brilliantov, Lifshits, Akhiezer, Pomeranchuk and Kikoin simultaneously declared to leave the university as a protest against Landau's dismissal.

I proposed to Gorsky and Brilliantov to draw up a statement, the others wrote their statements independently from me, but of course we agreed with each other. After having made the statement it was twice proposed to me to take it back, but I persisted in my anti-Soviet conduct and did not wish to take into account the norms of behaviour in Soviet society. Only much later did I realise my conduct and repented. I showed this unwillingness to take into account the rules of Soviet society also in conversations with the director. If I did not succeed in securing something for the work of the laboratory, I threatened to leave the institute, being conscious of the fact that nobody could replace me at work.

Finally, a large number of anti-Soviet and counter-revolutionary speeches could be heard from Landau and Korets. These speeches sharply expressed a negative attitude towards applying dialectical materialism in science, criticism of resolutions of party and government, criticism of the development of science in the USSR, and criticism of the development of economic life of the country.

(8)     It can be stated that UFTI possesses more possibilities for work than any other institute of the Union and hardly any institute abroad. The low-temperature laboratory is one of the largest in the world and in size comparable to the Leiden laboratory. The institute has the best cadres in the Union, it can be said that 1/3 of all qualified physicists are assembled at UFTI. In respect of resources and support the institute never experienced a shortage. With such colossal possibilities it can also be demanded from the institute to deliver work accordingly. The name of the Physico-Technical Institute itself already shows that the institute should engage at the same time in the development of both technical and physical problems. Furthermore, one of the main tasks of the institute is to be a school for training qualified physicists for factory laboratories and other institutes of the Union, where the level of knowledge of the collaborators is significantly lower than at UFTI. It should be noted that industrial production facilities have difficulty mastering a significant part of new inventions simply because there are no qualified engineer-physicists at the production facilities. The institute has the possibility to train at least 10 qualified physicists per year, but in fact does not train a single one. The institute has done enough as regards scientific research and practically nothing as regards technical research. This is due to the activity of the sabotage organisation and a result of the fact that no political-educational work was done among the young people and that they only lived for interests cut off from the interests of the country. In this respect it is characteristic that at UFTI, more than in any other research institute where this phenomenon can also be seen, an atmosphere of 'Soviet bohemianism' was developed, which entailed that the scientific worker is a 'free creator', not bound by any obligations in his work, an aesthete and eccentric. Although as far as research is concerned the institute has accomplished a number of achievements, also here sabotage played its vile role and without it the achievements would have been many times more.

This applies in particular to the atomic nucleus laboratory, on which the most resources have been spent, where an endless number of different high-voltage installations has been built, but not completed and the installations demolished, while without doubt it can be maintained that in the case that the work had been carried out in practice results would have been obtained. It is clear that to this day this laboratory has achieved the least of all both technically and scientifically with a large degree of advertising and showmanship.

The wrecking organisation has caused the greatest harm to OSGO whose construction has not been completed to the current day, and the possibility of using the beautiful laboratory in the near future is infinitesimally small due to the lack of a sufficient number of qualified physicists.

Weissberg and Lejpunsky mainly carried out the recruiting into the wrecking organisation.

(9)        Recruitment into the counter-revolutionary group of UFTI.

Weissberg was invited for work at UFTI by Obreimov and Lejpunsky. The latter obtained a recommendation about Weissberg from Houtermans who knew Weissberg from party work abroad. However Lejpunsky was attracted to the counter-revolutionary group by Weissberg, as he often spoke about the fact that it was essential to influence Lejpunsky, a weak and double-dealing person, in order that he would follow the 'correct' line at the institute. Lejpunsky was undoubtedly strongly influenced by Weissberg, as was Obreimov.

Weissberg attracted the foreigners, M. and B. Ruhemann, and F. and Ch. Houtermans into the counter-revolutionary group in 1935, at the time of the struggle with former director Davidovich. This fact I know from Weissberg who told me that he recruited foreigners and gave them instruction to fight against the director, demoralize the scientific collaborators of the institute and impede the increase of technical and defence-related topics. Weissberg was closely connected with the foreigners and enjoyed exceptionally great authority and influence among them. Ruhemann and F. Houtermans took some part in the struggle with the former director Davidovich and subsequently engaged in the struggle against technical and defence-related topics at UFTI.

I recruited Landau into the counter-revolutionary group in 1935 at the time of the struggle with the former director Davidovich.[39] I gave the task to demoralize the scientific collaborators of the institute and impede the increase of technical and defence-related topics. Landau took an active part in the struggle with former director Davidovich, he opposed technical and defence-related work of the institute and continued the struggle afterwards. Moreover, at the end of 1936 I recruited Gorsky and Brilliantov into the counter-revolutionary group. I told them about the goals of the organisation, namely about the necessity to carry on the struggle against technical and defence-related work. They agreed to take part in this work. Moreover, at the end of 1936 they were given instruction to take part in a collective strike at Kharkov University. They took active part in this strike.

From Landau I know that he recruited Korets for counter-revolutionary work in the struggle against technical and defence-related work of the institute. I know from the actions of Korets that he actively carried out this instruction.

I do not know who recruited Rozenkevich for counter-revolutionary work. I assume that Landau recruited him. Rozenkevich was strongly under Landau's influence.

...................................................................................

My flight abroad in 1921 was conditioned by my negative attitude towards the Soviet system. I, the son of a low civil servant, was educated at a bourgeois gymnasium. The arrival of the February revolution was friendly received in my age group and by my

---

[39]This is a peculiar statement as there are many other statements accusing Landau of having been the leader of a contra-revolutionary group already in Leningrad.

teachers, but the arrival of the October revolution was met with hostility. An even more hostile attitude towards Soviet power I encountered in the autumn of 1919[40] from the students and professors of Leningrad University, when I had enrolled at the physics-mathematics faculty. The people around me and my own lack of interest in the victory of the proletarian revolution determined my political character. I therefore was attracted to the idea of going abroad, although I did not have any means whatsoever for achieving this, did not know anything and could not yet work.

The preparations for the flight on the yacht were made without my participation. I joined at the moment of the departure of the yacht from the yacht club. However I soon learned about the real aims of the trip. These aims were to take Merezhkovsky and his wife with their luggage across the Soviet border. Because of my negative attitude towards the Soviet system, I found it impossible to leave at the last stopping place of the yacht at Shlisselburg on the Soviet shore and report to the authorities.

The trip was organised by Merezhkovsky who used for this purpose the captain of the yacht club Nagorny[41] (or Zagorny, I do not now remember the exact name) as an intermediary. According to Nagorny's initial plan the yacht should take Merezhkovsky to the eastern shore of Lake Ladoga, to a place close to the Finnish border, from where Merezhkovsky would continue independently. Except for Merezhkovsky and his wife, Nargorny and a woman, whose name I do not remember and who had a relationship with Nagorny, there were two sailors on the yacht, the student Ilja Dessler, who insofar as I remember studied at the Mining Institute, was twenty-seven years old and a Jew (his brother was a businessman in Helsinki), an old Estonian and I.

The initial plan of the trip did not work out, since Nagorny apparently did not know that the eastern shore of Lake Ladoga is shallow and not suitable for landing. Therefore already on Lake Ladoga Nagorny decided that he would steer the yacht directly to the steep northern shore of the Finnish part of Lake Ladoga. Since I had an anti-Soviet disposition, I did not object to the decision of the captain.

At first, upon arrival in Finland, we were arrested and subsequently sent to Mikkeli, except for Merezhkovsky, his wife and Dessler, who used their acquaintances and managed to take up residence in Helsinki. In Helsinki he wanted to set up his own chemical laboratory on a patented matter and proposed that I work with him for bread and board. I agreed since I had no other choice. In Helsinki Merezhkovsky failed to get the laboratory organised and after 5–6 months he decided to move to Berlin. I moved to Berlin together with Merezhkovsky in January 1922. Dessler had arrived there somewhat earlier than me. The remaining participants of the flight stayed in Finland, and I lost touch of them.

Merezhkovsky proposed that I busy myself in the laboratory with developing day-light developer for photographic plates. He had obtained (bought or stolen, I do not know) the recipe for this developer from Kononov a collaborator of the Photo Institute[42] in Leningrad. My work did not lead to any result, and in the autumn of 1922 Merezhkovsky sacked me. Literally dying of a severe flue and of hunger I turned for help to Dessler. He was interested in my work on the developer and decided to give me money to live on and to carry out experiments. After four months of work the results were still poor, and the work with the developer was stopped. Dessler demanded from me another invention and I decided to continue with the work on obtaining transparent quartz from sand which I had also worked on at the Optical Institute in Leningrad. From this neither any invention resulted after several months of work.

---

[40]Other sources state that Shubnikov enrolled in the autumn of 1918.

[41]Should be Nagornov, as the Finnish archives show.

[42]Probably the Higher Institute of Photography and Photo-technique, founded in 1918, from 1924 the State Photo Technikum and from 1931 the Leningrad Institute of Cinema.

At the time I had already submitted a petition for the restoration of my citizenship and my return. I should note that this decision was suggested to me by de collaborators of the Optical Institute Arkhangelsky and Grebenshchikov, who saw my return to the mother country as the only correct path of my development.

After my unsuccessful inventions Dessler found me a job at a bank. This was an establishment of the type of a currency exchange office, of which many had been set up in Berlin in the time of inflation. The bank's name was 'Berliner Credit- und Handelsbank'. Its owners were some Germans and White immigrants. Of the latter I remember the name Dubinsky, who at the time of my departure to the USSR proposed that I inform him by letter about the prospects of the development of NEP and on the possibilities of opening a branch of the bank in Moscow.

Before my departure to the USSR Dessler came to me and demanded that when in the USSR I fulfil a commission for him. He warned me that in case of refusal he would report about what I had told him about the work of the Optical Institute and that I even had continued the work with quartz I started in Leningrad. He required from me that I inform him about work in physics which could have practical applications.

In the autumn of 1923 I returned to Leningrad and started to work at the Leningrad Physico-Technical Institute in the laboratory headed by Obreimov. The political situation at the institute was on average significantly more Soviet than at the Physics Institute of the university. However, after evaluation it should be recognized that it was also anti-Soviet and that most people working at the institute were also anti-Soviet people. This also applied to my scientific supervisor Obreimov and to the collaborator Strelkov, who worked with him (now works with Kapitsa). Understandably this environment did not enable me to reconsider my political views, and I remained alien to Soviet power.

Upon finishing the institute in 1926 a foreign *komandirovka* was proposed to me. Ioffe made this proposal directly to me with the help of the Dutch physicist Ehrenfest, whom I had got to know at the time of his stay (if I am not mistaken in 1926) at the physico-technical institute. I obtained a *komandirovka* for a year to Leiden (Holland), where I had to work as an assistant of the low-temperature laboratory at a salary of 180 guilders (150 roubles).

I went to Holland via Berlin where I again met Dessler. There I told him about a series of work that could have practical applications. He told me that such activity is not what he requires and that upon my return to the Soviet Union he would force me through his people to work differently. I remained in Holland until the end of 1930, i.e. for four years. On the basis of an understanding with former UFTI director Obreimov, after my return to the USSR I should work at Kharkov on the foundation of a low-temperature laboratory. Obreimov did not allow me to do this work, as he wanted the large imported equipment to be used for conducting his own petty topic. For a long period I remained completely without work, which made me realize that if the situation here were to continue longer, it would cause enormous damage to my development as a scientist, and since I still was a person with an anti-Soviet attitude, I decided that the best way out of this would be to return to Holland, where before my departure to the USSR I was offered very good working conditions. I started to make efforts to obtain a *komandirovka* and a passport for foreign travel, intending to stay abroad permanently in case my position at UFTI would not change. By that time, against the wish of the director, the party committee proposed that I get busy with the organisation of the laboratory. I started with this matter after the holidays of 1931. By the end of 1933 the laboratory had been set up, in scope and possibilities the equal of the Leiden low-temperature laboratory. My idea of setting up a special technical laboratory for low temperatures, the experimental station, dates from 1932. At this laboratory experiments on the separation of gases could be carried out on a semi-industrial scale. But as a person with an anti-Soviet attitude, I was not interested in work of a technical nature, and therefore decided to transfer the practical realization of this idea to someone else, so that I could be busy with scientific questions only. The former director Obreimov transferred construction and budget questions to Weissberg, and I transferred the training of cadres, getting acquainted with technical questions and the preparation of research topics to Ruhemann, a collaborator of the laboratory I headed. Ruhemann's group was soon made into an independent laboratory by Lejpunsky and Ruhemann was appointed as its scientific head.

My acquaintance with Weissberg dates from 1932, when he arrived from abroad. Around 1934 we knew each other rather well and my anti-Soviet attitude was clear to Weissberg. Once he told me that my sinful connection with German intelligence was known to him, as well as my anti-Soviet attitude and that I therefore should get involved with him in general counter-revolutionary work.

He instructed me that within the institute I should oppose carrying out technical and defence work and should spread the idea of the incompatibility of scientific and applied work.

I agreed to carry out this task and fulfilled it along the following lines:

(1) Develop the work of the low-temperature laboratory led by me only along the lines of scientific work to the detriment of technical and defence-related work.
(2) Harassment of specialists involved in special topics, whereby I removed Rjabinin, who had carried out defence work, from the laboratory.

## Protocol of interrogation
### (type written)

Shubnikov, Lev Vasilevich, born in 1901 in Leningrad, nationality Russian, citizen of the USSR, non-party, trained as a physicist-engineer, professor, scientific collaborator of the low-temperature laboratory of the Ukrainian Physico-Technical Institute in Kharkov.

*13–14 August 1937*

**Question**: Material of the state security service shows that you are a participant of a counter-revolutionary organisation. Do you confess to this?

**Answer**: Yes, I do. I was a participant in a counter-revolutionary organisation from 1932. I was recruited into the organisation by the foreign specialist Alexander Semënovich Weissberg, an Austrian national, who had arrived from abroad and whom I worked with at the Ukrainian Physico-Technical Institute. Before my recruitment into the organisation by Weissberg there existed at the institute a group of people who were hostile towards Soviet power, towards the measures and politics of the Communist Party without any definite political colouring. Soon after Weissberg's arrival from abroad in one of a number of frank discussions with me he told me that he had connections with foreign Trotskyist organisations and that he considered the situation at the institute sufficiently suited for carrying out organised counter-revolutionary activity, and proposed that I take part in setting up a counter-revolutionary Trotskyist organisation. Subsequently, Weissberg told me that he was an agent of German intelligence in Ukraine and proposed to me to carry out jointly with him tasks for German intelligence on diversion and espionage. Without any hesitation I accepted Weissberg's proposal to participate in setting up a Trotskyist organisation and to carry out tasks for German intelligence on diversion and espionage, since at the time I already was an agent of German intelligence, having been recruited already in Berlin in 1922, about which I informed Weissberg.

**Question**: Elaborate both on your escape from the USSR and your recruitment by Dessler.

**Answer**: The reason for my flight abroad in 1921 was my hostile attitude towards Soviet power. My anti-Soviet convictions had been strengthened by the counter-revolutionary students amongst whom I found myself in that period. I had the intention to move abroad and found a suitable occasion for this. As a member of the yacht club in Leningrad I learned about a trip of a yacht on Lake Ladoga. Taking advantage of this I joined a group of people, who were unknown to me and turned out to be: a nephew of the writer Merezhkovsky, a chemist by profession, who departed together with his wife, a woman unknown to me, two sailors, the captain of the yacht Nagorny and Dessler, about whom I have given evidence above. Already on the yacht I learned from Merezhkovsky and Dessler about their plan to escape abroad[43] and decided then to also carry out my intention to move abroad. Upon arrival in Finland, all of us, except Merezhkovsky and Dessler, who had relatives in Helsinki, were arrested and sent to Vyborg[44] prison, and subsequently deported to the town of Mikkeli. Merezhkovsky and Dessler stayed in residence in Helsinki.[45] After half a year I was set free, returned to Helsinki and accepted Merezhkovsky's proposal to work for him in a laboratory on developer.[46] In 1922 I moved to Berlin with Merezhkovsky. Dessler had arrived there slightly earlier. After a number of failures with experiments in Merezhkovsky's lab he fired me, as a consequence of which I found myself without any means of existence. I again contacted Dessler and turned to him for help. Being interested in my work on developer, Dessler offered me money for carrying out some experiments. After a number of failures with my inventions paid for by Dessler, he refused any further assistance and in 1923 he helped me to get a job at the 'Berlin Credit- und Handelsbank', run by Germans.

In Germany I accidentally met Arkhangelsky and Grebenshchikov, two collaborators of the Optical Institute in Leningrad who were there on *komandirovka* and convinced me that it was expedient for me to return to the USSR. I therefore made an application at the Soviet embassy in Berlin for returning to the USSR. I told Dessler about my meeting with Arkhangelsky and Grebenshchikov, as well as about their advice to return to the motherland. After Dessler had convinced himself of my firm intention to return to the USSR, he approved it. Using the help he had given me before and also knowing my anti-Soviet attitude, Dessler proposed that I take part in espionage activity for Germany, of which he was an agent. I accepted Dessler's proposal, agreeing to carry out his instructions upon my return to the USSR. Just before my departure from Berlin I obtained from Dessler instruction to provide through him information to the German secret service about important discoveries in physics, in particular about the work on quartz that I had started at the time in Leningrad. Dessler drew my special attention to the necessity of informing him on physical-technical work, which could have a practical application both in industry and for defence. I accepted Dessler's instructions.

One of the shareholders of the bank in Berlin, a certain Dubinsky, who had emigrated after the October revolution, also made me a proposal about espionage activity for Germany after my return to the USSR. Since I had already been recruited by Dessler, I rejected Dubinsky's proposal. At the end of 1923 after my return to Leningrad I started to work at the Physico-Technical Institute, in the laboratory led at the time by Obreimov, later appointed director of the Ukrainian Physico-Technical Institute.

**Question**: With whom in the USSR did Dessler bring you into contact as regards espionage activity?

---

[43]From statements in the Finnish archives it is known that Dessler only came on the boat in Shlisselburg. They had departed from Petrograd a couple of days earlier without Dessler. See Chap. 2. So, it is unlikely that Dessler was one of the persons organising the escape, as in that case he would certainly have shown up when they departed from Petrograd.

[44]Since 1940 part of Russia as a result of the Winter War (Peace of Moscow).

[45]This is contradicted by a document in the Finnish archives, which states that Dessler (and Shubnikov) were resident foreigners in Mikkeli until January 1922. See Chap. 2.

[46]Within half a year of his arrival in Finland Shubnikov had already moved to Germany. The time scale suggested here is not correct.

**Answer**: Dessler had promised to communicate with me in the USSR through a person I trusted with whom I should stay in contact in respect of espionage work; however, until 1926, i.e. until my *komandirovka* abroad, nobody from Dessler showed up. Therefore, in 1926 after having obtained a *komandirovka* abroad from LFTI to Holland, I contacted Dessler when passing through Berlin and gave him some information about the USSR, which I had at my disposal. The information communicated by me to Dessler concerned work carried out at LFTI which had a practical application in industry and for defence of the country. I stayed in Holland until the end of 1930.

**Question**: Did you go to Germany during your *komandirovka* abroad?

**Answer**: Yes. Once in 1927 I was called by Obreimov who was on *komandirovka* in respect of questions connected with the purchase of equipment for the Ukrainian Physico-Technical Institute that was being organised in Kharkov.[47] A second time I was in Germany during the holidays, in the summer of 1929, making a tourist trip on the Rhine.

**Question**: Which tasks did you get from Weissberg?

**Answer**: Apart from recruiting persons with a hostile attitude towards Soviet power into our organisation at UFTI and carrying out sabotage in the area of my work, low temperatures, Weissberg proposed that I collect and communicate to him information on work of significance for defence and industry in the field of physics. Weissberg also pointed out to me the necessity of creating impossible conditions for normal work of specialists engaged in defence-related topics, through harassment, provocations and quarrels. I fulfilled Weissberg's tasks during the entire period of my participation in the counter-revolutionary organisation, up to my arrest. In 1936 with the aim of ensuring the sabotage and disruption of the work in finishing the construction of the Deep-Cooling Research Station (OSGO), Weissberg proposed that I try to get myself appointed as director of this station, while he declared to me that, if I did not succeed in obtaining this appointment locally, he would do this through his connections higher up. Weissberg also told me that for fully guaranteeing the wrecking of the work of OSGO he would also secure from Lejpunsky, the director of the institute, that the station would be manned by Germans.

**Question**: Who, apart from you, became part of the counter-revolutionary organisation at UFTI?

**Answer**: Apart from me, participants of the counter-revolutionary organisation at UFTI were:

1.  Lejpunsky, Aleksandr Ilich, director of the institute, working at this institute also as scientific director of the atomic nucleus laboratory.
2.  Weissberg, Aleksandr Semënovich, Austrian subject, former head of the construction of OSGO, arrested half a year ago by the organs of the NKVD.
3.  Landau, Lev Davidovich, head of the theory department of the institute, now moved to the Institute of Physical Problems of the USSR Academy of Sciences.
4.  Obreimov, Ivan Vasilevich, former director of UFTI, now head of the solid state laboratory.
5.  Rozenkevich, Lev Viktorovich, scientific head of one of the subgroups of the atomic-nucleus laboratory.
6.  Korets, Mikhail, engineer-physicist, worked together with Landau, is now in Moscow.
7.  Gorsky, Vadim Sergeevich, scientific head of the UFTI X-ray laboratory.

---

[47]In 1927 it had not yet been decided to set up an institute in Kharkov. Perhaps Obreimov was in Germany during that time to purchase equipment for the lab Ioffe intended to set up in Leningrad. Shubnikov does not mention here that he had to go to Germany every six months to have his visa renewed as Trapeznikova claims.

8.      Brilliantov, Nikolaj Alekseevich, engineer at the solid state laboratory, led by Obreimov. From April 1937 he moved to work in Moscow at the Institute of Physical Problems of the USSR Academy of Sciences.

9.      Ruhemann, Martin Zigfridovich, foreign specialist, English subject, scientific head of OSGO.

10.    Ruhemann, Barbara Frantsevna, wife of Ruhemann, German, English subject, engineer at the low-temperature laboratory.

11.    Houtermans, Fritz Frantsevich,[48] foreign specialist, German subject, scientific head of one of the subgroups of the atomic-nucleus laboratory.

12.    Tisza, foreign specialist, engineer of the UFTI theory department, Hungarian.

Apart from the direct participants of the Trotskyist organisation at UFTI I knew a number of counter-revolutionarily inclined persons, who had expressed their anti-Soviet convictions in conversations with me and their displeasure with the politics of party and government. These are:

Parusov, Aleksandr Ivanovich,[49] engineer of the low-temperature laboratory, Charlotte Houtermans, German, wife of Fritz Houtermans, Fomin, Valentin Petrovich, engineer of a subgroup of the atomic-nucleus laboratory, who returned from Germany to the USSR in 1931. According to Fomin, in Germany he was the chairman of the union of Soviet students.

**Question**: How do you know about the participation of those mentioned above by you in the counter-revolutionary organisation at UFTI?

**Answer**: Directly from Weissberg, who recruited me as a member of the organisation, and subsequently also for sabotage activity for Germany. I became aware of the involvement of Lejpunsky and Obreimov in the above-mentioned organisation. Weissberg told me about this at the end of 1934. Moreover, Obreimov as a long-time acquaintance from LFTI confirmed this to me. From Weissberg I also heard that the foreign specialists Houtermans, Martin Ruhemann, Barbara Ruhemann and Tisza were recruited by him into the organisation. On Weissberg's instruction I personally recruited the following persons into the organisation: Landau, Lev Davidovich, in 1935, Gorsky, Vadim Sergeevich, and Brilliantov, Nikolaj Alekseevich, early in 1936. From Landau I know that he recruited into the organisation Rozenkevich, Lev Viktorovich, and Korets, Mikhail, whom he attracted for work at UFTI from Sverdlovsk.

**Question**: How was recruitment into the organisation carried out?

**Answer**: The recruiting of members of the organisation carried out by Weissberg, myself, and Landau, was preceded by first approaching people selected for sabotage and subsequently properly influencing them in a counter-revolutionary spirit. On Weissberg's instruction I should in particular recruit people into the organisation who had a hostile attitude towards Soviet power. On this basis I approached the young specialists of the institute, Vadim Sergeevich Gorsky and Nikolaj Alekseevich Brilliantov. The first was the son of a nobleman and the second of a teacher. They were approached by getting their support for the idea of the incompatibility of applied work with work of a purely scientific nature. In the process of discussing these, at first glance, purely scientific problems I satisfied myself of their negative attitude towards the party in respect of the question of collectivization, the liquidation of the kulaks as a class, etc. They also expressed dissatisfaction with their position, as well as with the position of the intelligentsia, of people with a socially alien background who in their opinion were oppressed under the dictatorship of the proletariat. For successfully recruiting them into the organisation I, on my part, supported

---

[48]Should be Ottovich. Houtermans' father was called Otto.
[49]Nothing is known about this person.

their anti-Soviet attitude, drawing their attention to the cruel punishment of dissidents by Soviet power, in particular I pointed at the repression that took place in connection with the murder of Kirov, the mass expulsion of intelligentsia from Moscow, Leningrad and other towns. Finally after being convinced that I had influenced them successfully, I proposed at another time to Gorsky and Brilliantov to take part in sabotage activity in our counter-revolutionary organisation. They agreed to my proposal and were included in the practical execution of sabotage tasks. I achieved the recruitment of Gorsky and Brilliantov in 1936. I told Weissberg about having recruited them into the organisation. Later in the same year 1936 Gorsky and Brilliantov, participants of the organisation, were recruited by me to participate in the strike in connection with Landau's dismissal for counter-revolutionary speeches at Kharkov University.

**Question**: How was the activity of the counter-revolutionary organisation at UFTI set up and how did it develop?

**Answer**: The history of the counter-revolutionary activity of our organisation at UFTI is divided into two periods, into a so-called quiet period from 1932 through 1934 and a period of active formation of an organisation from 1936 until my arrest. The so-called quiet period of the counter-revolutionary activity of our organisation, in particular, of Weissberg, myself and Rozenkevich, and later from 1936 also of Landau was characterized by the fact that through counter-revolutionary influence on cadres in the form of a struggle for pure science we created impossible conditions for the development of both scientific and practical activity of the institute. The participants of our organisation, Landau and the engineer-physicist Korets who worked with him, inspired by Weissberg, were the main organizers of the disruption of technical and defence-related work at the institute. Landau and Korets and jointly with them Rozenkevich and me led the struggle against technical and defence-related work of the institute. Through quarrels that were artificially created by us and various provocations we carried out counter-revolutionary work directed at destroying the unity of the collective by preventing the part of the institute that did not share our views from working successfully. By various dirty tricks and harassment we drove the young specialists Pjatigorsky, Bunimovich, Rjabinin, Strelnikov and others from the institute. Aiming to cause a directly provocative blow on the specialists engaged in special topics (the laboratory led by Professor Slutskin), Landau, Rozenkevich and I published by prior agreement the provocative order of 1 April 1934 which contained for the collaborators abusive content that for a long period of time would unsettle them in the normal course of work. With the aim of misinforming higher organisations, as well as party organisations, in the course of 1936 I, Weissberg, Obreimov, Rozenkevich and Landau sent out letters falsely setting out the state of affairs at the institute to the scientific-research sector of Narkomtjazhprom and the central committee of the party, aiming at the removal from the institute of people unpleasant to us, who pursued the policy of introducing technical and defence-related work at the institute. At the end of 1936 I and the other participants of our organisation, Gorsky and Brilliantov, allured a number of other scientific collaborators of the institute into a strike organised by us in protest against Landau's dismissal from Kharkov University, who had proclaimed openly counter-revolutionary concepts against the application of dialectical-materialism in physics. In 1935 the participant of our organisation Korets was called to account in a criminal case for open counter-revolutionary agitation within the institute, directed at the disruption of defence-related work. In defence of Korets, Landau and I gave false statements to the court, preventing by this a possible failure of the counter-revolutionary activity we carried out at the institute. The foreigners Houtermans and Martin and Barbara Ruhemann, participants of our organisation, coordinated to a large extent by Weissberg, took part in the organised disruption of defence-related work, signed false information about the state of the institute and, finally, took part in harassing people who did not share our position.

In 1936 our organisation became significantly more active. The participants of the organisation gathered at illegal meetings at the apartments of Weissberg, Houtermans, Ruhemann, Obreimov, and sometimes mine or in Landau's office. At these meetings we discussed various political questions, in the process of which the content became increasingly more clearly Trotskyist, both with the head of the organisation Weissberg, and with the other participants of the organisation. Weissberg, Landau and Korets expressed more actively than the other participants of the organisation bitterness about the existing regime in the USSR, spoke sharply slanderous and insulting words at the address of the leader of the party and the head of Soviet power. In one of such counter-revolutionary conversations soon after the murder of Kirov Landau expressed his satisfaction about the murder, considering Kirov guilty of the extermination of Trotskyists in Leningrad, as a consequence of which many people had suffered, in particular from the intelligentsia, both party and non-party people. Weissberg, Landau and Korets hoped for a change of the existing order in the USSR, were indignant about the cruel oppression by Soviet power in connection with the murder of Kirov, and developed before us, the participants of the organisation, the idea of strengthening the struggle against Soviet power. The following people took part in illegal meetings and discussions at UFTI: Weissberg, I, Shubnikov, Rozenkevich, Obreimov, Korets, Landau, Martin Ruhemann, Barbara Ruhemann and Houtermans.

**Question**: Set out the practical sabotage activity carried out by you and other participants of the counter-revolutionary organisation at UFTI.

**Answer**: The practical sabotage activity carried out by participants of the counter-revolutionary organisation at UFTI came down to the following:

**In the field of research of the atomic nucleus**.

Sabotage in this field in terms of the material damage caused to the republic, estimated roughly at no less than 10 million roubles, is the most appreciable. In 1936 this laboratory had equipment in the form of a high-voltage installation with a voltage of 1 million volts, which made it possible to carry out work in investigating the atomic nucleus. Based on the mere repetition of experiments on splitting the atomic nucleus of lithium, which had already been achieved at that time in Cambridge, Lejpunsky, as head of the atomic-nucleus laboratory and director of the institute in that period, and Obreimov sent to the leaders of party and government a telegram with false content on an outstanding achievement of UFTI, and obtained in this way large allocations for the large-scale construction of high-voltage installations, which did not contribute anything useful to the institute. Lejpunsky, Obreimov and the scientific collaborators Sinelnikov and Valter working under their direction demolished the current installation with a voltage of about 1 million volts, and subsequently constructed an impulse generator with a voltage of almost 2 million volts, which was also demolished and instead a Van de Graaff installation was set up with a voltage of almost 2 million volts, which also gave no results and was supposedly sold. Not limited to this, an installation colossal in size with a voltage of 6 million volts was constructed for which a special building was built. All this together did not only fail to provide any results in the field of research on the atomic nucleus, but on the contrary hampered the scientific and research work in this field for many years. Directing their attention at the construction and reconstruction of large high-voltage installations, Lejpunsky and Obreimov did not carry out any research work for five years and did not train any specialists for conducting further work.

While sabotage leading to the disruption of applied work on ion transformations was carried out under the pretext that it was inexpedient to further work out earlier achievements at the institute, the very important sabotage in respect of manufacturing metallic Dewars consisted in Lejpunsky's failure to pass on the results obtained to industry.

With the aim of deliberately disrupting applied work carried out successfully on the construction of ion transformers, which is of great interest to modern electronics, the laboratory for this was liquidated by Lejpunsky in 1936 under the pretext that it was expedient to transfer this work to the Kharkov Electromechanical Factory. In fact this branch of work was liquidated instead of developing perspectives for such work. Knowing full well that the country has a great need for Dewars for storing liquid nitrogen and other liquid gases, developed by the collaborator Kostenets of the institute, and having obtained much higher quality than abroad, Lejpunsky for three years did not pass on the developed technological production processes for exploitation and mass production in industry.

In 1934 the UFTI engineer Strelnikov developed an X-ray tube with a power ten times larger than achieved before. Strelnikov's invention should have been widely used in industry for eliminating defects in metals and for the X-ray investigation of structures. Following a clear sabotage policy in physics to disrupt work that has any technical application, Obreimov and Lejpunsky, in order not to give Strelnikov's invention to industry, dismissed him from the institute under the pretext that it was not possible to continue applied work at the institute.

Direct sabotage in the field of low temperatures was committed by me, Shubnikov, and the foreign specialists, the Germans Martin and Barbara Ruhemann. We committed sabotage in this most important field with significance for defence under the direction of one of the most active participants of our organisation, connected with German intelligence, the foreign specialist Weissberg, with Lejpunsky being fully aware of this.

The UFTI low-temperature laboratory, one of the largest in the world and comparable in size to the Leiden laboratory, should, apart from making discoveries in physics, have researched practical applications of low temperatures for separating coke gas with the aim of separating the nitrogen-hydrogen mixture and helium, which is found in negligible quantities in the country. In fact nothing was done on this, including that a number of large technical problems, which could have had an effect on the national economy, were not solved. At the same time Rjabinin who carried out successful work on the application of liquid hydrogen as fuel for aviation engines was first chased from the laboratory, and subsequently from the institute through harassment and artificial quarrels by me, Weissberg, Landau and Rozenkevich. After Rjabinin's departure from the institute the work he had partly completed was not further developed. Driven to despair Rjabinin beat up Landau.

### In respect of the construction of OSGO.

While the wrecking in the field of atomic nuclei consisted of the completely unnecessary construction and reconstruction of large high-voltage Van de Graaff installations and suchlike, sabotage in respect of the construction of OSGO, which was greatly needed for industry and defence, was committed in a completely different direction. OSGO was exclusively constructed for technical and defence-related purposes, mainly questions of the liquefaction of gases and the separation of gas mixtures by the method of deep cooling. Both as regards the investigation of physical processes, the measurement of physical constants needed for equipment work, and also as regards the development of new equipment designs on the basis of laboratory research. The solution of such important problems was in the hands of foreign specialists, Weissberg, an agent of foreign intelligence, who in turn attracted the foreign specialists Martin and Barbara Ruhemann, who were participants of our counter-revolutionary organisation.

The sabotage, both in the construction of OSGO and in the creation of unfavourable conditions for its exploitation, was carried out by Weissberg on the direct instruction of the German secret service, about which Weissberg spoke to me when giving instructions on espionage activity. As a result, Weissberg and other participants of our organisation, Ruhemann, Lejpunsky and Obreimov, succeeded in successfully delaying the construction of the research station by an extra two years; instead of the usual two years the construction continued for four years.

Weissberg did not limit himself to this and, in order to ensure that there would be further sabotage and diversion in the process of setting up OSGO, he urgently insisted on inviting foreign specialists, Germans, as management staff for organising the practical activity of the station, excluding in this way our national cadres. OSGO, constructed with a large delay, on a larger scale than initially planned, cannot be used in the coming years in the direction needed for the country.

Results of research work carried out by the foreign OSGO specialist Martin Ruhemann on the separation of natural gases, containing helium, and calculations showing how to correctly construct equipment for separating pure helium were published, with wrecking purposes, in journals of the chemical industry.

This fact as well as the completion of equipment of OSGO by foreign specialists is a direct fulfilment of tasks of the German secret service through Weissberg, its trusted agent in Kharkov. In conclusion I consider it necessary to sum up the sabotage committed by the counter-revolutionary organisation at UFTI. During the entire period of its existence the institute did not yield a single invention or proposal useful for industry and for the defence of the country, in spite of the fact that unlimited resources and opportunities for carrying out scientific research and technical work were granted by the country. Moreover, in individual cases when on the initiative of a number of co-workers technical and defence-related work was proposed, such initiative was strangled by participants of the counter-revolutionary organisation, and the co-workers were chased from the institute by creating an atmosphere of quarrels and harassment. Laboratories carrying out useful technical work were closed. The fight against technical and defence work was carried out under the slogan of the incompatibility of scientific and applied technical work. That this slogan is only a cover for sabotage is clear from the fact that as regards scientific work the institute has performed insignificantly little compared to the means that were spent on it. Colossal resources allocated by the country were spent wastefully (sabotage-like) for the set-up of installations nobody needed and which subsequently did not work and, when no results were forthcoming, were demolished. This sabotage activity was carried out on a large scale at the atomic-nucleus laboratory, which failed to yield any results after having swallowed up 10 million roubles.

The low-temperature laboratory led by me had all possibilities for research work that is extremely important for industry and for the defence of the country in the field of the liquefaction of gases and the separation of gas mixtures by the method of deep cooling. The laboratory did nothing in this direction, having as a blind person switched over exclusively to scientific work.

The construction of OSGO, whose aim was to carry out technical and applied work, was with wrecking intentions given over to Weissberg, an agent of the German secret service, who in every possible way prolonged the construction time, and in the exploitation period intended to fill the leading positions with foreigners for facilitating espionage-subversive work.

In the course of the further inquiry and with the aim of eliminating the aftereffects of the sabotage I will set forth all its considerations which can be used in a direction useful for the country.

Concluding my statements on my participation in the counter-revolutionary organisation and on carrying out sabotage work I also consider it necessary to communicate another fact of my guilt to the inquiry.

When I returned from Holland to the USSR in 1930 and failed to find adequate conditions from my point of view for my further development, I intended to obtain a *komandirovka* abroad in order not to return to the USSR.

The answers written in my own words are true and have been read through by me, which I confirm with my signature

*(signature)*

Interrogated by the collaborator of the Kharkov District Administration of the NKVD

*Skralivetsky*

Correct: Security Officer of the third section of the State Security administration, Sergeant of State Security

*Semechkin*

# Resolution

*Kharkov, 23.VIII.1937*

I, Skralivetsky, Security Officer of the third section of the State Security Administration of the Kharkov District Administration of the UkSSR NKVD, having considered the material of the inquiry in respect of citizen L.V. Shubnikov accused of the crimes envisaged in Articles 54-11, 54-6 and 57-7 of the UkSSR Penal Code, have found that the inquiry has established that L.V. Shubnikov as an agent of the secret service of a foreign state carried out espionage work on the instruction of a representative of this secret service, the foreign subject A.S. Weissberg. On Weissberg's proposal Shubnikov took part in the formation of a counter-revolutionary Trotskyist sabotage organisation at UFTI. He recruited a number of specialists in this organisation.

Based on what has been set out in Article 126 and guided by Article 127 of the UkSSR Code of Criminal Procedure

## HAVE DECIDED:

To call Lev Vasilevich Shubnikov to account as the accused, after having presented him with the charges according to Articles 54-11, 54-6 and 57-7 of the UkSSR Penal Code, and to provide a copy of this to the Kharkov military prosecutor.

Security Officer of the third section                                    *Skralivetsky*
Agreed: Head of the third section                                            *Fisher*
Confirmed: Acting Head of the political administration,
Captain of State Security                                                *Rejkhman*
The resolution was communicated to me on 23 August 1937            *Shubnikov*

# Protocol of interrogation

*27 August 1937*

I, Skralivetsky, Security Officer of the third section of the Directorate of State Security of the Kharkov District Administration of the UkSSR NKVD, have interrogated the defendant Lev Vasilevich Shubnikov.

**Question**: Do you admit to being guilty of the charges presented to you according to the resolution of 23/VIII/37 about your participation in a counter-revolutionary Trotskyist sabotage organisation at UFTI and that you have carried out espionage tasks as an agent of the German secret service?

**Answer**: Yes, I fully admit to being guilty of the charges presented to me.

Written truthfully in my own words, which I confirm with my signature.

                                                                      (*signature*)
Interrogated by the Security Officer                                    *Skralivetsky*

The following document is the concluding indictment in the joint cases of Shubnikov, Rozenkevich and Gorsky. Apart from this there are also separate concluding indictments for each of the defendants.

## Concluding indictment (Ref. [2], pp. 269–271)

In inquiry no. 47894 on the charges against Shubnikov L.V., Rozenkevich L.V., and Gorsky V.S.

The third section of the State Security Administration, Kharkov District Administration of the NKVD has at its disposal documents setting out that the scientific collaborators of UFTI Shubnikov, Rozenkevich and Gorsky engaged in espionage and sabotage activity. On this basis the aforementioned people were arrested and called to account.

The inquiry carried out in this matter has established that:

Shubnikov, having a hostile attitude towards Soviet power, fled in 1921 together with other people on a yacht from Leningrad across Lake Ladoga to Finland and from there moved to Germany. In 1923, while living in Berlin, he was recruited for espionage work by the agent Dessler of the German secret service and in the same year sent to the USSR with espionage instructions.

In the following years Shubnikov while travelling on foreign *komandirovki*[50] systematically[51] informed Dessler on discoveries in physics in the USSR, which had practical application for industry and for the defence of the country. In 1932 he was recruited for a second time for espionage and subversion by the agent Weissberg (arrested) of the German secret service, to whom Shubnikov proposed to hand over espionage information. At the same time Shubnikov was recruited by Weissberg into a counter-revolutionary Trotskyist sabotage group existing at UFTI. Shubnikov personally recruited the UFTI specialists Gorsky (arrested) and Brilliantov (not arrested) into the Trotskyist organisation. Through provocations and quarrels he harassed UFTI specialists engaging in defence topics. He carried out sabotage work at the laboratory of low temperatures, led by him, which had the utmost significance for industry and the defence of the country.

The scientific collaborator Rozenkevich was a participant of this Trotskyist sabotage organisation. From 1928 to 1930 Rozenkevich was a participant of the counter-revolutionary group existing at LFTI and a member of the counter-revolutionary organisation at Leningrad University.

In 1930, after having arrived for work at UFTI, Rozenkevich entered in the same year the Trotskyist-sabotage organisation existing at UFTI, of which he was a participant until the day of his arrest. The practical sabotage activity of Rozenkevich at UFTI on instruction of the organisation was mainly directed at the total disruption of special subjects of significance to defence.

Side by side with Shubnikov and Gorsky he took an active part in the harassment of scientific workers who engaged in the practical solution of defence-related subjects and did not share the counter-revolutionary views of the participants in the organisation, through quarrels that were artificially created by them and by creating conditions that made it impossible to organise the work of the institute.

---

[50]As far as known Shubnikov did not go on a *komandirovka* abroad in the years before he went to Leiden.

[51]Shubnikov had only stated, confessed or admitted, or whatever word is appropriate here, that he had met Dessler once in 1926 while on his way to Leiden, which can hardly be called 'systematically'.

He participated in illegal meetings of the counter-revolutionary organisation at which the trials of the participants of the discovered Trotskyist-Zinovievite and parallel anti-Soviet centres were discussed. At these meeting the participants openly urged to be prepared to take revenge on the Soviet government for repression. Together with Shubnikov and Gorsky Rozenkevich was the organiser of strikes arranged by members of the counter-revolutionary organisation at UFTI in the form of open demonstrations against the party organisation of the institute.

The scientific worker Gorsky was a participant of the same Trotskyist sabotage organisation. In 1936 Gorsky was recruited by Shubnikov into the counter-revolutionary Trotskyist sabotage organisation at UFTI. On Shubnikov's instruction Gorsky took an active part in the harassment of scientific collaborators who engaged in the practical solution of defence-related topics and did not share the counter-revolutionary views of the participants of the organisation. On instructions of the organisation Gorsky sabotaged work of an applied nature of the senior UFTI engineer Strelnikov. This work advanced the technical problem of X-ray cinematography, by means of powerful tubes, and has important significance for metallurgy.

Together with Shubnikov and Rozenkevich he was an active participant and organiser of strikes, openly arranged by the counter-revolutionary organisation.

Shubnikov is exposed as a spy and saboteur by his own confessions and the statements of Rozenkevich and of the witnesses Gej, Shavlo and others.

Rozenkevich is exposed as a member of the counter-revolutionary sabotage organisation by his own confessions, the statements of Shubnikov and of the witnesses Gej, Shavlo and others.

Gorsky is exposed as a member of the sabotage organisation by the statements of Shubnikov and Rozenkevich, by a confrontation with them and by the statements of the witnesses Gej, Shavlo and others. He himself has only admitted counter-revolutionary activity.

Based on the above the following charges are brought:

1. against Shubnikov, Lev Vasilevich, born in 1901, native of Leningrad, Russian, citizen of the USSR, non-party, worked up to his arrest as scientific head of the low-temperature laboratory at UFTI, that he engaged in espionage and sabotage activity for Germany; that he was an active participant of a counter-revolutionary Trotskyist sabotage organisation, i.e. the crimes envisaged in Articles 54-6, 54-9, 54-11 and 54-7 of the UkSSR Penal Code.

2. against Rozenkevich, Lev Viktorovich, born in 1905, native of Leningrad, Russian, citizen of the USSR, son of a non-hereditary nobleman, non-party, worked up to his arrest as scientific head of the radioactive-measurement laboratory at UFTI, that he was an active participant of a counter-revolutionary Trotskyist sabotage organisation at UFTI and that he carried out together with other participants sabotage work on the disruption of defence-related tasks, i.e. the crimes envisaged in Articles 54-11, 54-10 and 54-7 of the UkSSR Penal Code.

3. against Gorsky, Vadim Sergeevich, born in 1905, native of Gatchina, Leningrad region, son of a nobleman, Russian, citizen of the USSR, non-party, worked up to his arrest as scientific head of the X-ray laboratory at UFTI, that he was an active participant of a counter-revolutionary Trotskyist sabotage organisation at UFTI and that together with other participants of the organisation he carried out sabotage in disrupting the work of the institute, i.e. the crimes envisaged in Articles 54-11, 54-10, paragraph 1, and 54-7 of the UkSSR Penal Code.

## HAVE DECIDED

To submit case no. 9411 on the charges against Shubnikov Lev Vasilevich, Rozenkevich Lev Viktorovich, and Gorsky Vadim Sergeevich in accordance with Articles 54-6, 54-11, 54-10, paragraph 1, and 54-7 of the UkSSR Penal Code for instruction to the Main Directorate of State Security[52] of the USSR NKVD according to order no. 00485 of the USSR NKVD.

Security Officer of the third section of the State Security Administration, Sergeant of State Security,                                                                                     *Vajsband*

Agreed: Deputy Head of the third section of the State Security Administration, Captain                                                                                                    *Tornuev*

Confirmed: Acting District Procurator                                                       *Leonov.*

## Concluding indictment

Shubnikov, Lev Vasilevich, born in 1901, native of Leningrad, Russian, citizen of the USSR, non-party. Arrested on 5 August 1937. Worked up to his arrest as scientific head of the low-temperature laboratory at UFTI. To refer to category 1.[53]

Shubnikov, having a hostile attitude towards Soviet power, fled in 1921 together with other people on a yacht from Leningrad across Lake Ladoga to Finland and from there moved to Germany. In 1923, while living in Berlin, he was recruited for espionage work by the agent Dessler of the German secret service and in the same year sent to the USSR with espionage instructions.

In the following years Shubnikov while travelling on foreign *komandirovki* systematically informed Dessler on discoveries in physics in the USSR, which had practical application for industry and for the defence of the country.

In 1932 he was recruited for a second time for espionage and subversion by the agent Weissberg (arrested) of the German secret service, to whom Shubnikov proposed to hand over espionage information. At the same time Shubnikov was recruited by Weissberg in a counter-revolutionary Trotskyist sabotage group existing at UFTI. Shubnikov personally recruited the UFTI specialist Gorsky (arrested), Landau and Brilliantov (not arrested) into the Trotskyist organisation.

Through provocations and quarrels he harassed UFTI specialists engaging in defence topics. He carried out sabotage work at the laboratory of low temperatures, led by him, which had the utmost significance for industry and the defence of the country.

Shubnikov is exposed as a spy and saboteur by his own confessions, the statements of Rozenkevich (arrested) and the witness statements of Gej and Shavlo.

Confessed: spy, saboteur, wrecker.

Deputy Head of the Kharkov District Administration of the NKVD,
Captain of State Security                                                                     *Rejkhman*
Acting District Procurator                                                                     *Leonov*
Correct:                                                                                        *Reznikov*

---

[52]*Glavnoe upravlenie gosudarstvennoj bezopasnosti* (GUGB) was the name of the Soviet intelligence service and secret police from July 1934 to April 1943.

[53]Meaning to be condemned to death.

## Extract

From protocol no. 13 of the resolution of the People's Commissariat on Internal Affairs of the USSR, General Commissar of State Security Comrade Yezhov and the Procurator of the Union Comrade Vyshinsky of 28 October 1937.
Heard: the materials on the accused presented by the administration of the USSR NKVD for the Kharkov District based on order N 00439 of 25.07.1937 of the USSR NKVD.
Decided: Shubnikov Lev Vasilevich

### is to be shot

The USSR People's Commissar of Internal Affairs, General Commissar of State Security Yezhov, USSR Procurator Vyshinsky.
Correct: Security Officer, third section of the State Security Administration, Lieutenant of State Security                                                                             *Reshetnev*

                                                                             *Top secret*

## Certificate

The decision of the NKVD Commission and the USSR procurator of 28 October 1937 in relation to
                          *Shubnikov Lev Vasilevich*
was carried out on 10 November 1937.
Head of the second section of the third sector of department "A" of the USSR MGB[54]
                                                                             *Kamensky*
Certificate written on 26.11. 1953.

# Documents pertaining to Rozenkevich

## MEMORANDUM

on Rozenkevich, Lev Viktorovich, born in 1905 in Leningrad, son of a non-hereditary nobleman, Russian, citizen of the USSSR, with higher education, professor of physics, married, scientific head of the atomic nucleus laboratory at UFTI.
According to the information at our disposal L.V. Rozenkevich is a member of a counter-revolutionary group of scientific collaborators of UFTI, headed by Professors Landau and Shubnikov which has set as its goal the disruption of defence-related research and the general disorganisation of work at UFTI.
On 11 April 1936[55] Rozenkevich was secretly detained and interrogated by us.
In the interrogation Rozenkevich confirmed the presence at UFTI of a counter-revolutionary group led by Landau and also confirmed our information on the aims and tasks of this group.
Rozenkevich indicated that he was recruited into the counter-revolutionary group for the first time in 1930/1931 while working at the Leningrad FTI by FTI collaborators Ivanenko, Frenkel and Landau.

---

[54]*Ministerstvo gosudarstvennoj bezopasnosti* (Ministry of State Security) was the name of the Soviet intelligence agency from 1946 to 1953.
[55]This must probably be 1937.

Rozenkevich indicated that at his recruitment into the counter-revolutionary group Ivanenko and Landau directly set him the task to struggle against Soviet power for the restoration of lost rights.

After having arrived in Kharkov at UFTI together with Landau,[56] Rozenkevich agreed with the latter to continue participating in counter-revolutionary activity, while Landau, as in Leningrad, pointed out the necessity to first of all direct his activity at the disruption of defence-related work.

A few months later Rozenkevich retracted parts of his statements, after having confirmed the existence of a counter-revolutionary group in Leningrad which he and Landau entered, but denying the anti-Soviet nature of the group at UFTI.

It is completely clear that Rozenkevich 'corrected' his statements under the influence of Landau and others, whom he told about his summons to the NKVD.

Rozenkevich's denial of part of his statements is not of great significance for the case, since in the first interrogations he confirmed intelligence information we already had.

There is no doubt about the close and direct relation of Rozenkevich with Landau and others, and in view of his former relationship with Landau in respect of counter-revolutionary activity in Leningrad, there is every reason not to pay attention to his denial of parts of his statements.

Both according to intelligence information and to information of Rozenkevich himself he repeatedly participated in counter-revolutionary meetings at the apartments of Landau and Shubnikov.

On this basis we consider it necessary for the development of the case to arrest Rozenkevich.

Acting Head of the first branch of the third section of the State Security Administration, Second Lieutenant of State Security                                                                    *Reznikov*

Agreed: Deputy Head of the third section of the State Security Administration, Captain of State Security                                                                                                      *Tornuev*

Confirmed: Acting Head of Kharkov District Administration of the NKVD, Colonel                                                                                                                            *Shumsky*

Resolution: arrest of Rozenkevich sanctioned by the District Prosecutor Leonov on 5.08.

# Order no. 28

*Issued on 5.08.1937*
*Valid for 2 days*

Second Lieutenant, Comrade Reznikov, collaborator of the Kharkov District Administration of the USSR NKVD, is charged with carrying out the search and arrest of Rozenkevich, Lev Viktorovich, residing in Kharkov in Tchaikovsky Street, house 16, apartment 15.

All organs of Soviet power and citizens of the USSR are required to provide legal assistance to the bearer of the order in the performance of his mission.

Deputy Head of the Kharkov district administration of the NKVD, Colonel

*Shumsky*

Head of the eighth section of the State Security Administration, Lieutenant of State Security

*Signature (illegible)*

---

[56]Incorrect, Rozenkevich arrived at UFTI in 1930, two years before Landau.

## Protocol of interrogation (Ref. [2], p. 252–258)

Rozenkevich, Lev Viktorovich, born in 1905 in Leningrad, up to his arrest residing in Kharkov at Tchaikovsky Street 16, Russian, citizen of the USSR, professor, scientific head of the UFTI laboratory of radioactive measurements, son of a non-hereditary nobleman, official of the State Duma, higher education, non-party.

*12–13 August 1937*

**Question**: The material of the inquiry shows that you are a participant of a counter-revolutionary sabotage organisation. Do you admit this?

**Answer**: Yes, I do.

**Question**: Describe the organisation of which you were a participant.

**Answer**: At the moment of my arrest I was a participant of a counter-revolutionary sabotage organisation at UFTI. Until the end of 1932 the political nature of this organisation was profoundly anti-Soviet, subsequently, when it was headed by the engineer Aleksander Semënovich Weissberg, who had arrived from abroad, the political character of the organisation clearly became Trotskyist.[57] Personally until my arrival in Ukraine (in 1930) I was a participant of a counter-revolutionary group that had been active at LFTI where I was sent at the end of 1927.

**Question**: Give details of your counter-revolutionary activity.

**Answer**: The environment in which I found myself when I was invited to LFTI as a post-graduate student was clearly counter-revolutionary.

Until joining this institute I was only busy with science, was a post-graduate student and was susceptible to developing a hostile attitude to Soviet power.

At the time LFTI was almost the only institution supplying young cadres in physics and technology. Apart from this, LFTI served as a centre for training cadres that were counter-revolutionarily inclined and were subsequently sent all over the Union to places where physico-technical institutes were organised (Ukraine, Urals, Siberia).

The basis for cultivating counter-revolutionarily inclined cadres at LFTI was the presence of a leadership that 'humiliated and offended' the October socialist revolution. The young people at LFTI also consisted to a large extent of people who were socially and ideologically close to this category.

My closest friend at the institute was the post-graduate student D.D. Ivanenko. As I grew closer to Ivanenko he expressed to me his counter-revolutionary views and attitudes. Under that influence I, who before was apolitical, started to respond to Ivanenko's ideological treatment, and later shared his counter-revolutionary focus.

In one of the conversations with me in 1928 Ivanenko developed his counter-revolutionary views and advanced the idea of the necessity of active struggle against Soviet power, he made me understand that at LFTI there existed an organisation in the form of a counter-revolutionary group into which I was later recruited by him.

**Question**: Elaborate on the circumstances of your recruitment into the counter-revolutionary group at LFTI.

**Answer**: The basis for my recruitment into the counter-revolutionary group at LFTI was the constant ideological influence on me of the LFTI post-graduate students D.D. Ivanenko, and Georgy Antonovich Gamov and Professor Jakov Ilich Frenkel of the institute. The ideological influence of the people mentioned above on me and, in particular, of Ivanenko, the son of a Kharkov black-hundred nobleman, manifested itself in the fact that in many conversations with me each of them in his own way argued to me that the existing political

---

[57]Here Rozenkevich does not seem to follow the script. He suddenly is a member of Weissberg's group. Landau's group no longer enjoys a separate existence. Moreover Weissberg's group was supposed to be 'rightist', not Trotskyist.

regime of the dictatorship of the proletariat was based on a cruel, boorish suppression of the rights of the defeated class and the intelligentsia.

Ivanenko told me that he experienced various forms of oppression, that this not only concerned him, but a significant part of talented people, who under Soviet power were deprived of opportunities for developing their talents because of their social origin and the discrepancy between their political views and the politics of Soviet power and the party.

Ivanenko repeatedly demonstrated to me the necessity of the organised struggle for changing the existing political order in the country. Sharing Ivanenko's counter-revolutionary convictions, I agreed to take part in the counter-revolutionary group at LFTI.

**Question**: Who, apart from you and Ivanenko, were part of the LFTI counter-revolutionary group?

**Answer**: After I was persuaded to become part of the counter-revolutionary group, Ivanenko told me that apart from him the following persons were participants of the counter-revolutionary group at LFTI: the post-graduate student Georgy Antonovich Gamov, who in 1934 obtained a *komandirovka* to Denmark and refused to return to the USSR; the post-graduate student of Leningrad University Lev Davidovich Landau and the professor of the polytechnic institute, combining this with a job at LFTI, Jakov Ilich Frenkel, who as became known to me later led the counter-revolutionary group at LFTI.

**Question**: Was the counter-revolutionary group in the LFTI theory department a local group or was it part of the underground?

**Answer**: The counter-revolutionary group of theoretical physicists at LFTI was only a part of a branched network of similar cells both in the LFTI apparatus itself and at the university.

The first stage in the preparation of counter-revolutionary young people was the physico-mechanical faculty of the Leningrad Polytechnic Institute. Here socially alien youth enjoyed primary attention of the scientific leadership. The counter-revolutionary upbringing of the young people was ensured by the counter-revolutionary activity of the leading staff of the physico-mechanical faculty at LFTI in the persons of e.g. former baron Frederiks, his close friend Professor Bursian, A.F. Valter, Frenkel, Ivanenko, Bronshtejn and others.

At that time a monarchist spirit reigned both at the Leningrad Physico-Technical Institute and at the physico-mechanical faculty of the university. On this basis young cadres were trained. In 1929 Ivanenko told me that he recruited young people at the institutes of higher education in Leningrad. He did not mention to me the people he had recruited.

The political attitude of the young people connected with Ivanenko and others is evident, if only, from the following. In 1933 I was in Leningrad and invited to an evening at Nina Gavrilovna Prusakova, the daughter of a professor at the Forestry Academy, held at the apartment of her father on the territory of the Forestry Academy. Apart from me, Ivanenko and his wife, the collaborator of the Leningrad Physico-Technical Institute, M.P. Bronshtejn, a fellow student of Prusakova, Natalja Iosifovna Nikitinskaja, and several students were present. At this evening V.A. Miljukov's speech on the radio was discussed in very sympathetic terms, subsequently after supper Prusakova proposed a toast of 'heil Hitler', and we all welcomed this.

**Question**: Where are Ivanenko and other participants of the counter-revolutionary group in Leningrad at present and where do they work?

**Answer**: In 1935 Ivanenko, together with the collaborators Nelidov[58] and Zhuze[59] of LFTI and many others were arrested in connection with the murder of Kirov and subsequently

---

[58]Ivan Jurevich Nelidov (1905–1936?). Graduated from the physics faculty of Leningrad State University in 1929. Known for his work in solid state physics. Arrested on 1 March 1935 and sent to Gorki a few days later. Arrested for the second time in March 1936 and probably shot in August 1936. (http://www.ihst.ru/projects/sohist/repress/lfti/nelidov.htm.)

[59]Vladimir Pantelejmonovich Zhuze (1904–1993). Worked from 1931 at LFTI, arrested on 2 March 1935 for spreading "false rumours about the murder of Kirov". On 26 March 1935 sent into exile to Saratov for five years with his family as a "socially dangerous element". Worked as a lecturer at Saratov University. Returned to LFTI after the war. (http://www.ihst.ru/projects/sohist/repress/lfti/juze.htm.)

exiled. Ivanenko is currently in Tomsk where he works at the university. Professor Frenkel and Bronshtejn still work at LFTI, Gamov, as already mentioned above, refused to return to the USSR from abroad.

In order to complete my statements on the Leningrad period, I should report on my guilt. In 1930 in one of his candid conversations with me Ivanenko confessed to me that if he were to obtain a *komandirovka* abroad he would not return. I shared Ivanenko's intention in that respect hoping to obtain a *komandirovka* abroad and remain there. In the case of obtaining a foreign *komandirovka* I had conceived a plan for fleeing abroad. I intended to illegally move to Turkey and subsequently go to England.

**Question**: At the start of the interrogation you mentioned that, apart from taking part in the counter-revolutionary group in Leningrad, you were part of a Trotskyist organisation at UFTI. Elaborate your statement on this.

**Answer**: In 1930 I was sent on *komandirovka* to UFTI together with a number of other people from Leningrad. Apart from me and Ivanenko, as participants of the Leningrad counter-revolutionary group, Shubnikov, who was on *komandirovka* abroad, joined later.[60] Thus from the very beginning of its existence there was a counter-revolutionary group at UFTI.

The period of the first one and a half to two years (until Weissberg's arrival at the institute) can be called a relatively 'quiet' period of our counter-revolutionary activity. The activity of our counter-revolutionary group, to which I also belonged, consisted of the creation of conditions that would make it impossible to organise and develop the work of the institute, this under various pretexts and under the guise of quarrels that were artificially created by us.

From Weissberg's arrival at UFTI in 1931 the activity of the counter-revolutionary group was revived and was constantly organisationally strengthened. It also took shape politically as a Trotskyist organisation.

The foreign specialist Weissberg, appointed in the function of director of OSGO, was invited from abroad by the former director I.V. Obreimov. According to Obreimov he met Weissberg in 1931 in Berlin. Weissberg is of no value as a specialist. Weissberg's rapid assimilation at UFTI gives me reason to claim that Obreimov had informed Weisberg when still abroad on the state of affairs at the institute and on the political inclination of the cadres. It is therefore not accidental that upon his arrival at UFTI Weissberg soon approached me, Shubnikov and Ivanenko, who at the time formed the first counter-revolutionary cell at the institute.

In conversations with me, Shubnikov and Ivanenko, Weissberg talked very much about political questions. He criticized the line of the party and government on the struggle with the kulaks, told about political life abroad, about, from his point of view, possible directions of the politics of the party and government of the USSR, about strengthening and activating in this way our counter-revolutionary attitudes and gradually channelling them into critical ones. Weissberg talked much and often about political topics, authoritatively and categorically showed the incorrectness of the measures of the Central Committee of the Bolshevik Communist Party and the Soviet government on a number of issues. On almost every question he had a judgement, which he tried to implant in us.

In 1933 the engineer-physicist Korets arrived at Kharkov from Sverdlovsk for work at UFTI and also joined our counter-revolutionary organisation.

Weissberg's Trotskyist influence on the participants of our counter-revolutionary organisation was most pronounced in the days of Kirov's murder and at the explanations in the newspapers of the results of the subsequent court proceedings.

---

[60] So here Rozenkevich seems to testify that in 1930 Shubnikov was a member of the LFTI counter-revolutionary group. If this were true, he should have become a member of such group before he left for Leiden in 1926. This is extremely unlikely, almost impossible.

In conversations with us Weissberg sharply condemned the answer of the party, of the Soviet government and the anger of the people to the murder of Kirov, expressing outrage against the cruelty of the punishment of the participants of the murder, as well as in general against the repression of the Soviet government. Weissberg, and with him Landau and Korets, openly called upon us to be prepared to take revenge on the Soviet government for the repression.

The participants of the group often gathered in the apartments of Weissberg and Shubnikov, and in Landau's office at the institute, and discussed questions of the trials of participants of the discovered Trotskyist-Zinovievite and parallel anti-Soviet centres.

At the meetings Shubnikov, Weissberg, Korets, Martin Ruhemann, Barbara Ruhemann and I were present. Apart from political questions we discussed at these meetings methods of carrying out sabotage, with as its most effective result a disruption of the activity of the institute. We also discussed questions of the necessity of harassing and scientifically isolating people who worked on defence-related tasks and did not share our position.

**Question**: Who, apart from the persons you mentioned, were members of the counter-revolutionary organisation at UFTI?

**Answer**: Apart from the persons mentioned above, the following were also members of the organisation: Fritz Ottovich Houtermans, foreign specialist, Vadim Sergeevich Gorsky, senior engineer of the X-ray laboratory, Valentin Petrovich Fomin, who obtained his education in Berlin in the Soviet period, Nikolaj Sergeevich Brilliantov, engineer of the crystal laboratory.

Weissberg and Landau told me in 1935 about the participation of these people in the organisation. Houtermans, Gorsky, Fomin and Brilliantov attended together with me several meetings of the counter-revolutionary organisation; in personal conversations with me they expressed their counter-revolutionary opinions and together with us they carried out sabotage activity at the institute.

**Question**: Elaborate on the practical sabotage activity of the counter-revolutionary organisation at UFTI.

**Answer**: Before setting out the facts known to me (as a participant of the counter-revolutionary organisation at UFTI) of the sabotage activity carried out, I want to dwell separately on illuminating Lejpunsky's role and activity as director of the institute. The scientific work was planned under Lejpunsky's direct leadership. The planning at UFTI of applied work was always put on the back burner due to Lejpunsky's sabotage arrangements. Everywhere Lejpunsky advanced the idea that the significance of applied work was incompatible with successfully carrying out 'purely scientific work'. This line of Lejpunsky reflected the sabotage line of part of the leading physicists of the USSR, who hid behind the slogan that it was necessary to work out only problems of physics as a pure science and deliberately exclude work on applied physics, which in the near future would make it possible to solve problems of technical and defence-related value faced by the country.

For several years under Lejpunsky's eyes scientists-specialists who engaged in defence-related topics were harassed and openly frustrated by Landau, Shubnikov and Korets. Strikes took place at the institute as a demonstration against the party line of UFTI. Lejpunsky's attitude with respect to foreign specialists and finally the fact that OSGO and the Van de Graaff generator, on the construction of which more than 20 million roubles were spent, had so far not yielded any practical results, places Lejpunsky among the saboteurs.

Lejpunsky could not understand that his position on the question of the advancement and solution of problems of defence and technical significance was a sabotage line. At the institute really ostentatious work of physicists was given great visibility at the expense of abstract work for which applications can only be found after many years. Meanwhile, with the advanced methodology and the highly qualified physicists at the institute it was full well possible to advance and carry out work of a topical defence-related value. Lejpunsky's line of conduct at UFTI gave us total freedom for carrying out our sabotage activity.

I will now go over to setting out the sabotage activity per institute.

### In the field of the splitting of the atomic nucleus.

From the very foundation of the institute work on the atomic nucleus was carried out under Lejpunsky's direction. Impulse generators of the Van de Graaff type were constructed, as well as electric machines that made it possible to obtain voltages of up to two million volts. In addition there were also transformer installations. Up to 1933 all this made it possible to develop research work in any direction, whereby a number of important technical problems could have been solved.

With an installation of the Van de Graaff type (existing at UFTI already in 1933) the Americans Hafstad[61] and Tuve[62] had carried out brilliant work on proton-proton scattering, belonging to the best experimental work carried out in the last years. In 1933 we had at our disposal at UFTI the entire apparatus necessary for organising similar work. The subject of the research carried out by Hafstad and Tuve was known everywhere together with other very interesting subjects.

Instead of this, in order to prolong work on the existing equipment, the following activity was subsequently displayed at UFTI on the atomic nucleus with the use of high voltages. In 1932 Lejpunsky and his group (Latyshev, Sinelnikov and Valter) repeated the experiment on the splitting of the lithium atom carried out at Cambridge. This was not a difficult experiment and was in itself not a discovery. Nevertheless, two reports were sent to Stalin, Molotov and Ordzhonikidze on an allegedly brilliant achievement of Soviet physics. By sending these reports possibilities for obtaining new resources were opened, which were indeed obtained and used on the construction of a high-voltage building at UFTI. The construction of the high-voltage building at UFTI was not needed for scientific work. On the contrary, for its construction and equipment resources were withdrawn from other institutes and for many years all work on the application of high-voltage installations and the development of physics in the USSR was halted.

Up until now not a single work on the atomic nucleus has been carried out at UFTI with the high-voltage installations, and likewise no tasks of an applied nature, including of a defence-related character, have been carried out. The scientific work of the remaining physicists working on the atomic nucleus (Houtermans, Fomin and me) was in another direction; we did not use any high voltages, but radioactive samples. Our work had a purely academic character without any orientation towards possible technical and defence-related output.

Thus, the results of sabotage in the field of the atomic nucleus led to the situation that almost 7 million roubles were spent on the work, practically stymying the development of physics of the atomic nucleus for a number of years and as yet not providing any practical results. In accordance with the general sabotage line highly-qualified scientific workers were completely cut off from solving practical problems.

---

[61]Lawrence R. Hafstad (1904–1993) was an American electrical engineer and physicist notable for his pioneering work on nuclear reactors. In 1939, he created the first nuclear fission reaction in the United States.

[62]Merle Anthony Tuve (1901–1982) was an American geophysicist. With the Van de Graaff accelerators, Tuve, Hafstad, and fellow DTM (Department of Terrestrial Magnetism) scientist Norman Heydenburg conducted proton-proton and proton-neutron scattering experiments in the 1930s which led to the understanding that the nuclear force between nucleons is attractive and identical.

In the absence of sabotage, in this field in the atomic nucleus laboratory a number of questions of colossal technical significance could have been solved. For instance, a number of questions connected with high voltages, with measuring equipment and suchlike. There also exist a number of problems which are suggested in connection with new work on neutrons. For instance, the existence of chains of radioactive elements with artificial raw material may make it possible to construct a nuclear energy accumulator, which might make it possible to build engines for airplanes with an efficiency that immeasurably exceeds the efficiency of gasoline engines. But this was not done out of sabotage considerations.

### In the field of low temperatures.

The sabotage work in the laboratory of low temperatures was realized by Shubnikov, Weissberg, and Martin and Barbara Ruhemann. The laboratory, which should have been reorganised into the deep-cooling experimental station, on the construction of which more than 10 million roubles were spent, could have carried out a colossal volume of useful research on the question of a rational study of coke-oven gases, which are of great significance for the defence industry. This great question should have been solved long ago. However, because of the sabotage activity of the people mentioned above, in particular Weissberg and Shubnikov, the construction of OSGO was delayed for several years. A number of large coke-chemical installations, constructed at UFTI in recent years, could not use a series of data that was expected from OSGO.

The staff of the low-temperature laboratory, led by Shubnikov, a member of the counter-revolutionary organisation, did not engage in solving applied problems, while the cryogenic laboratory had clear access to industry.

Preliminary research on state diagrams of coke-oven gases was published, as a result of sabotage activity of Shubnikov and Lejpunsky, by the foreigner of German descent Ruhemann. This fact, as well as the transfer of management work of OSGO into the hands of foreign saboteurs (Weissberg, Martin and Barbara Ruhemann) and the uncontrolled nature of their work, I see as the straightforward creation of favourable conditions for espionage activity by the foreign specialists who had become members of our counter-revolutionary organisation. Apart from this, by means of provocations and harassment Shubnikov, with the help of the participants of the organisation Landau and Korets, drove the senior scientific worker Rjabinin from the cryogenic laboratory, who engaged in solving practical tasks on the question of obtaining new types of fuel for internal combustion engines. Driven to despair Rjabinin beat up Landau and in connection with this was forced to leave the institute.

### In respect of ion transformations.

By order of Lejpunsky and with the participation of the members of the counter-revolutionary organisation Landau, Korets, myself and Shubnikov, the laboratory of ion transformations was radically eradicated. A number of scientific workers, Zhelekhovsky, Pomazanov, E.M. Sinelnikov, Bunimovich, was fired. The photo-effect laboratory was also demolished, from which Kan was fired. As a result, from the research topics of UFTI a number of problems, which were very important for the application of physics in electro-technology, was erased. The transferred workers moved to the Kharkov Electromechanical Factory (except Bunimovich who left for Leningrad) and, lacking the necessary support, their work was gradually reduced to zero.

### In the area of X-ray research.

The applied character of the laboratory of Strelnikov was abolished and Strelnikov was gotten rid of by the efforts of the members Gorsky, Landau and Korets of the counter-revolutionary group. Thus, the conditions for a normal development of Strelnikov's work, which deserved special attention in view of its value for metallurgy, were spoiled. Strelnikov developed a method of X-ray cinematography with the help of a tube with a rotating anticathode.

### As regards theoretical physics.

Landau, the scientific head of the laboratory of theoretical physics, banned every possibility of providing help in technical questions. At every attempt to obtain advice on a technical question Landau invariably answered that he cannot and does not want to think about applied problems. Moreover, in his own work on questions which could have a technical application he always took away the basis which could have been of use in technical applications (e.g. the question of the properties of an ion and electron gas in a plasma, whose solution, giving practical results, would have helped the development of e.g. the technology of ultrashort waves, which is of enormous significance for defence).

In conclusion I want to declare the following to the inquiry:

1.  Our counter-revolutionary organisation engaged in sabotage, creating the appearance of great work on the basis of insignificant and purely academic research.
2.  The participants of the UFTI sabotage organisation and the director of the institute consciously suppressed the initiative to carry out applied work, subject to use for defence.
3.  Conditions were deliberately created in which the position of an institution that is useless for the country should be preserved for the future (cf. third five-year plan).
4.  The atmosphere of academism was also spread outside UFTI by the members of the counter-revolutionary group in institutions of higher education, taking away the initiative from young specialists to carry out applied work.
5.  A similar situation exists throughout the entire network of physico-technical institutes, which only by a misunderstanding carry the name physico-technical. Physics in the entire Soviet Union should be thoroughly reviewed and actually put in the service of the country, for which it can and should provide brilliant work.

The answers to the questions posed to me have been written down truthfully in my own words and have been read by me                                                    *Rozenkevich*
Interrogated: Collaborator of the third section of the UkNKVD, Kharkov District
                                                                                          *Skralivetsky*

### Resolution
*Kharkov, 23.VIII.1937*

I, Skralivetsky, Security Officer of the third section of the State Security Administration of the Kharkov District Administration of the UkSSR NKVD, having considered the material of the inquiry on Lev Viktorovich Rozenkevich, accused of the crimes envisaged in Articles 54-11, 54-10 and 54-7 of the UkSSR Penal Code,

### HAVE FOUND

that the inquiry has established that L.V. Rozenkevich is an active participant of a counter-revolutionary Trotskyist sabotage organisation at UFTI. Together with other participants of the organisation he carried out sabotage work by disrupting defence-related tasks of the institute.

Based on Article 126 and guided by Article 127 of the UkSSR Code of Criminal Procedure

## HAVE DECIDED:

To call Lev Viktorovich Rozenkevich to account as the accused according to Articles 54-11, 54-10 and 54-7 of the UkSSR Penal Code and to provide a copy of this to the Kharkov military prosecutor.

| | |
|---|---|
| Security Officer of the third section | *Skralivetsky* |
| Agreed: Head of the third section | *Fisher* |

Confirmed: Acting Head of the political administration Captain of State Security

*Rejkhman*

## Protocol of interrogation
### *26 August 1937*

I, Security Officer... Skralivetsky, have interrogated the accused Lev Viktorovich Rozenkevich

**Question**: Do you admit to being guilty of the charges presented to you according to the resolution explained to you on 23 August 1937 about your participation in a counter-revolutionary Trotskyist sabotage organisation at UFTI?

**Answer**: Yes, I fully admit to being guilty of the charges presented to me.

Written truly in my own words, which I confirm with my signature.

| | |
|---|---|
| *Signature* | *Rozenkevich* |
| *Signature* | *Skralivetsky* |

## Concluding indictment

Rozenkevich, Lev Viktorovich, born in 1905, native of Leningrad, Russian, citizen of the USSR, son of a non-hereditary nobleman, non-party. Arrested on 5 August 1937. Up to his arrest he worked as scientific head of the radioactive-measurement laboratory at UFTI. Inquiry no. 9411. To refer to category 1.

From 1928 to 1930 Rozenkevich was a participant of the counter-revolutionary group existing at LFTI and a member of the counter-revolutionary organisation at Leningrad University.

In 1930, after having arrived for work at UFTI, Rozenkevich entered in the same year the Trotskyist-sabotage organisation existing there, of which he was a member until the day of his arrest.

The practical sabotage activity of Rozenkevich at UFTI, on instruction of the organisation, was mainly directed at the total disruption of work on special subjects of significance for defence.

He took an active part in the harassment of scientific workers who engaged in the practical solution of defence-related subjects and did not share the counter-revolutionary views of the participants in the organisation. Through quarrels that were artificially created by him he created conditions which made it impossible to organise the work of the institute.

He participated in illegal meetings of the counter-revolutionary organisation at which the trials of the participants of the discovered Trotskyist-Zinovievite and parallel anti-Soviet centres were discussed. At these meetings the participants openly urged to be prepared to take revenge on the Soviet government for repression. He was the organiser of strikes arranged by members of the counter-revolutionary organisation at UFTI in the form of open demonstrations against the party organisation of the institute.

He intended to escape abroad or, in case he obtained a *komandirovka*, to remain abroad. He appeared to have been recruited twice.

He is exposed as a member of the counter-revolutionary sabotage organisation by his own confessions and by the statements of Shubnikov (arrested) and of the witnesses Gej, Shavlo and others.

Confessed: participant in a counter-revolutionary sabotage group.

Deputy Head of the Kharkov District Administration of the NKVD,
Captain of State Security

*Rejkhman*

Acting District Procurator                                                                   *Leonov*

There also exists a document signed by Yezhov and Vyshinsky identical to the one reproduced above for Shubnikov condemning Rozenkevich to death, and a certificate of 1953 stating that the sentence has been carried out:

*Top secret*

## Certificate

The decision of the NKVD Commission and the USSR procurator of 28 October 1937 in relation to

*Rozenkevich Lev Viktorovich*

was carried out on 9 November 1937.

Head of the second section of the third sector of department "A" of the USSR MGB

*Kamensky*

Certificate written on 26.11. 1953.

# Documents pertaining to Gorsky

## Information
### *on Gorsky, Vadim Sergeevich*

Gorsky, Vadim Sergeevich, born in 1905 in Gatchina, Leningrad oblast, son of a nobleman, Russian, citizen of the USSR, non-party, scientific worker at UFTI, residing at the address Tchaikovsky Street 16.

According to evidence from the arrested persons Rozenkevich and Shubnikov, Gorsky is an active participant of a counter-revolutionary sabotage organisation at UFTI.

On instruction of the counter-revolutionary organisation Gorsky actively carried out sabotage work, for instance, he ruined and destroyed the work of the UFTI scientific collaborator Strelnikov (later banished from UFTI), which has great significance for metallurgy. Strelnikov developed a method of X-ray cinematography.

At secret meetings of the counter-revolutionary group Gorsky showed a terrorist state of mind. On the basis of the above Gorsky, Vadim Sergeevich, is subject to arrest in order to clarify his counter-revolutionary activity.

Security officer of the third section, Sergeant of State Security                          *Vajsband*
Agreed: Head of the third section, Senior Lieutenant                                        *Fisher*
Confirmed: Deputy Head of the Kharkov District Administration of the NKVD,
Captain of State Security                                                                   *Rejkhman*

# Order no. 72/4

*Issued 21.09.1937*

*Valid for two days*

Junior Lieutenant of State Security, Comrade Solovchenko, collaborator of the Kharkov district administration of the USSR NKVD, is charged with carrying out the search and arrest of citizen Gorsky, Vadim Sergeevich, residing in Kharkov at Tchaikovsky Street 16, apartment 1.

All organs of Soviet power and citizens of the USSR are required to provide legal assistance to the bearer of the order in the performance of his mission.

Deputy Head of the Kharkov District Administration of the NKVD,

Captain of State Security                                              *Rejkhman*

Head of the second branch of the State Security Administration of the Kharkov District Administration, Senior Lieutenant of State Security                        *Chernov*

## Questionnaire of the arrested person

1.  Name: Gorsky
2.  First name and patronymic: Vadim Sergeevich
3.  Date of birth: 1 May 1905
4.  Place of birth: Gatchina, Leningrad oblast
5.  Address: Kharkov, Tchaikovsky Street, house 16, apartment 1
6.  Profession and speciality: scientific worker
7.  Place of work and function: UFTI, scientific head of the X-ray laboratory
8.  Passport:
9.  Social origin: from the family of a priest[63]
10. Social position: office worker

    (a)  Lived with his father
    (b)  Studied

11. Education: higher, finished the Leningrad Polytechnic Institute in 1928
12. Party membership: automatically expelled from the Komsomol in 1928
13. Nationality: Russian
14. Category of military registration: reserve of first rank
15. Service in white and other armies: no
16. Subjected to any repression under Soviet power: in 1923 in Gatchina was arrested together with an acquaintance who helped to drive a sled which turned out to contain weapons.
17. Family composition:

    (a)  Wife—Danilevskaja Nina Vasilevna, 25 years old, scientific worker at UFTI
    (b)  Daughter Ada, 3 years old, in Leningrad
    (c)  Son from first wife Evgenija, 7 years old
    (d)  Father Sergej Pavlovich Gorsky lives in Gatchina

---

[63]The nobleman has turned out to be a priest.

## Protocol of interrogation
### 24 September 1937

(…)

**Question**: You are accused of being a participant of a counter-revolutionary Trotskyist sabotage organisation, on whose instruction you, working at UFTI, carried out sabotage activity. You are offered to give evidence.

**Answer**: I did not know anything about the existence of a counter-revolutionary organisation at UFTI. Nobody has recruited me into this organisation, I never engaged in sabotage activity, nor do I admit to being guilty of this.

**Question**: The material of the inquiry has fully unmasked you as a participant of the said counter-revolutionary organisation and as a saboteur. I warn you about the necessity to give truthful testimony to the inquiry.

**Answer**: I only testify truthfully.

The answers in the protocol have been written down truthfully in my own words and have been read by me.

        24.09.1937                               *Gorsky*

The interrogation has been carried out by the Security Officer of the third section,
Sergeant of State Security                                *Vajsband*


## Protocol of confrontation

Of the accused Gorsky, Vadim Sergeevich, with Shubnikov, Lev Vasilevich, on 23 September 1937.

After identification the accused were asked the following questions.

**Question** (*Shubnikov*): Do you confirm your testimony that you recruited Gorsky who is sitting in front of you into the counter-revolutionary Trotskyist organisation existing at UFTI?

**Answer**: Yes, I do.

**Question** (*Shubnikov*): When and under which circumstances was Gorsky recruited by you into the counter-revolutionary organisation?

**Answer**: On Weissberg's instruction I had to attract into the organisation people who were hostilely inclined towards Soviet power. On that basis I approached the young specialist of the institute, Gorsky, who is currently sitting in front of me. Gorsky was approached by getting his support for the standpoint that the performance of applied work was incompatible with work of a purely scientific nature. In the process of discussions of these, at first glance, purely scientific problems I convinced myself of Gorsky's negative attitude to the policies of the party as regards questions of collectivization, the liquidation of the kulaks as a class, etc. Gorsky also expressed dissatisfaction about his position, as a member of the intelligentsia, the son of a nobleman, of being oppressed, in his opinion, under the conditions of the dictatorship of the proletariat. For a successful recruitment of Gorsky into the organisation I on my part supported his anti-Soviet attitude, by pointing out the cruelty of the punishment of dissidents by Soviet power, in particular, I pointed at the repression which had taken place in connection with Kirov's murder. After I had finally satisfied myself of having successfully influenced Gorsky, early in December 1936, after having retired with him into my office, I proposed to him to take part in sabotage activity in our counter-revolutionary organisation. Gorsky agreed to my proposal and was included in the practical performance of sabotage tasks. I informed Weissberg about the recruitment of Gorsky in the organisation.

**Question** (*Gorsky*): Do you confirm the statement of Shubnikov in which he points you out as a participant of a Trotskyist counter-revolutionary organisation?

**Answer**: No, I do not. Shubnikov did not make such proposal to me.

**Question** (*Shubnikov*): Do you confirm that you recruited Gorsky into the counter-revolutionary Trotskyist organisation existing at UFTI in spite of the fact that he, Gorsky, denies this?

**Answer**: Yes, I confirm that Gorsky was recruited by me into the counter-revolutionary Trotskyist sabotage organisation at UFTI and that on my proposal he carried out sabotage work at UFTI.

The confrontation was performed by the Security Officer of the third section, Sergeant

*Vajsband*

## Protocol of confrontation

Of the accused Gorsky, Vadim Sergeevich, with Rozenkevich, Lev Viktorovich, on 25 September 1937.

After identification the accused were asked the following questions.

**Question** (*Rozenkevich*): Do you confirm your testimony that Gorsky who is sitting in front of you is known to you as a participant of the counter-revolutionary Trotskyist organisation existing at UFTI?

**Answer**: Yes, I do.

**Question** (*Rozenkevich*): Under which circumstances did it become known to you that Gorsky was a member of this counter-revolutionary organisation?

**Answer**: In 1936 Weissberg and Landau told me about Gorsky's participation in the counter-revolutionary organisation. Gorsky took part in a number of campaigns, arranged by the organisation, e.g. in 1936 in December in the collective walk-out from the university together with Shubnikov and Landau, in the harassment of the senior engineer Strelnikov, who was taken off the technical problem of X-ray cinematography, which is of great significance for the metallurgical industry. On the question of the walk-out from the university a meeting took place between some of the members of the counter-revolutionary organisation, at which meeting Gorsky also took part.

**Question** (*Gorsky*): Do you confirm the statement of Rozenkevich in which he points you out as a participant of a counter-revolutionary Trotskyist organisation?

**Answer**: The statements of Rozenkevich are partially true, namely on 26.12.1936 Shubnikov indeed called me and Brilliantov into his office and told us about Landau's dismissal by the rector of the university Neforosnyj, depicting this as a provocation on the part of the rector. However, I was not a member of a counter-revolutionary Trotskyist organisation and did not know anything about its existence.

*25.09.1937*

Confrontation was carried out by

*Vajsband*

## Protocol of interrogation
### 26 September 1937

Of the arrested person Gorsky, Vadim Sergeevich, on 26 September 1937.

**Question**: You are offered to give evidence of your membership in a counter-revolutionary organisation existing at UFTI.

**Answer**: I did not and do not belong to any counter-revolutionary organisation. However I admit that I was drawn into sabotage activity by the scientific collaborator Shubnikov of UFTI, now known to me as a member of a counter-revolutionary organisation at UFTI.

**Question**: How did your sabotage activity concretely manifest itself?

**Answer**: My sabotage activity manifested itself in the following. On Shubnikov's proposal I took part in a demonstrative presentation of a declaration together with Shubnikov and others on a walk-out from work at the university, which in essence was a strike with an anti-Soviet character. Moreover, we undertook this strike in defence of Landau who had been dismissed from the university because of anti-Soviet speeches, about which I learned later.

At the time of the purge of the party at the end of 1934 I took part in slander with the aim of getting rid of the party communist and valuable scientific worker Strelnikov. I took part in the harassment of Strelnikov by Shubnikov, Obreimov and Landau. I ruined the introduction or more accurately the development of a tube by Strelnikov which is of great significance for metallurgy and for other fields of industry.

**Question**: You are not telling the truth, in saying that you are not a member of a counter-revolutionary Trotskyist organisation. The inquiry has at its disposal information that you were recruited by Shubnikov and have carried out your sabotage work on Shubnikov's direct instruction. I demand truthful statements.

**Answer**: I was under Shubnikov's influence, however I did not know that he was a member of a counter-revolutionary organisation.

The answers have been written down accurately in my own words and have been read by me.

16.10.1937                                                                                      *Vajsband*

## Concluding indictment

Gorsky, Vadim Sergeevich, born in 1905, native of Gatchina, Leningrad region, son of a nobleman, Russian, citizen of the USSR, non-party. Arrested on 21.09.1937. Up to his arrest he worked as scientific head of the X-ray laboratory at UFTI in Kharkov. Inquiry no. 9411. To refer to category 1.

In 1936 Gorsky was recruited into the counter-revolutionary Trotskyist sabotage organisation at UFTI by Shubnikov, an agent of the German secret service (arrested).

On Shubnikov's instruction he took an active part in the harassment of scientific collaborators who engaged in the practical solution of defence-related topics and did not share the counter-revolutionary views of the participants of the organisation.

On instructions of the organisation Gorsky sabotaged work of an applied nature of the senior UFTI engineer Strelnikov. This work advanced the technical problem of X-ray cinematography, by means of powerful tubes, and has important significance for metallurgy.

He was an active participant and organiser of strikes, openly arranged by the counter-revolutionary organisation at UFTI against the party organisation of the institute.

He is exposed by the statements of the accused Shubnikov and Rozenkevich (arrested).

Only confessed to counter-revolutionary activity.

Partially confessed: participant of a counter-revolutionary Trotskyist sabotage organisation, wrecker.

*15.10.1937*

Deputy Head of the Kharkov District Administration of the NKVD,
Captain of State Security                                                          *Rejkhman*
Acting District Procurator                                                         *Leonov*

There also exists a document signed by Yezhov and Vyshinsky identical to the one reproduced above for Shubnikov and Rozenkevich condemning Gorsky to be shot and a certificate of 1953 stating that the sentence has been carried out:

*Top secret*

# Certificate

Decision of the NKVD Commission and the USSR procurator of 28 October 1937 in relation to

*Gorsky Vadim Sergeevich*

was carried out on 8 November 1937.


Head of the second section of the third sector of department "A" of the USSR MGB

*Kamensky*

Certificate written on 26.11. 1953.

# Appendix 4
# Personal Statements in the Rehabilitation Procedure of Shubnikov, Rozenkevich and Gorsky (Ref. [2], pp. 275–278)

### Abram Naumovich Chernets (1911–1974)[64]

*On Shubnikov, Lev Vasilevich.* I have known Shubnikov as the head of the laboratory of low temperatures of the Physico-Technical Institute of the USSR Academy of Sciences since I came to work at the institute. I seldom directly associated with him, just in separate instances connected with the work at the institute.

Shubnikov was the head and founder of the laboratory and under his leadership until the day of his arrest his group of young scientists, specialists in the field of low temperatures, grew.

In addition, it should be noted that Shubnikov expressed the opinion, the essence of which came down to the fact that only people who have a special talent and undergo special training can engage in scientific work. At the time this was directed at a group of people—graduate students who came from the factory to the institute without sufficient training, and who had not obtained any special education. In one of his notices in the newspaper he even wrote that after all a hare can learn to strike a match, but that there is no sense to this.

In a number of cases which I can now hardly remember he was among a group of scientific workers who acted against the decision of the management and the party organisation. In general, at the time the institute consisted of two camps, one of which included the group of non-party scientists (some of the foreign specialists then working at the institute associated with this group), who behaved in a haughty manner to the other part of the scientific collective and, simply speaking, caused quarrels.

*On Gorsky, Vadim Sergeevich.* Gorsky was a gifted scientific worker. He had done some serious scientific work. In scientific meetings his speeches were rich in content and interesting.

In the collective he did not stand out especially—I have in mind some appearances not connected with scientific work. Gorsky was a person who, as far as I remember, divided all his time between work and his young wife and was not interested in anything else. That is how he appeared to me at the time. I have known him from October 1931 to the day of his arrest through teamwork at the institute. We did not have any work relations and I did not know him outside the institute.

*On Rozenkevich.* Rozenkevich's behaviour at the institute was very modest, even quiet. He first worked in the theory division, but later became convinced that he was not made for that field and switched to the experimental laboratory. He had done some work on neutron physics, which, as far as I remember, was highly valued at the time. As an experimentalist he was purposeful in his work and in a short period became a good experimentalist, which served then to his credit.

---

[64]Worked as an engineer in Slutskin's laboratory at UFTI.

© Springer International Publishing AG 2018
L. J. Reinders, *The Life, Science and Times of Lev Vasilevich Shubnikov*,
Springer Biographies, https://doi.org/10.1007/978-3-319-72098-2

In those years I happened to attend some lectures by Rozenkevich, which he read to second-year students (1932 courses) at the physics-mechanics faculty of the Kharkov Machine-Building Institute, where I studied, combining study with work at the physico-technical institute. As a lecturer Rozenkevich was poor and soon left this work. Apart from encounters at the lectures, and also separate conversations connected with joint work at the Physico-Technical Institute, I did not have any relations with Rozenkevich. I knew him through joint work at the Physico-Technical Institute from October 1931 until the day of his arrest.

### Statement by P.I. Strelnikov:

I did not know and still do not know anything about hostile anti-Soviet activity of the above-mentioned persons. Nor do I know any facts about espionage, subversive and sabotage activity.

*28 July 1956*

### Statement of Landau addressed
### to the military procurator Captain of justice Kabtsov

*15 August 1956*

I knew L.V. Shubnikov, L.V. Rozenkevich, and V.S. Gorsky well, having worked with them for 5 years at UFTI in Kharkov. They are all known to me as talented scientists and persons who were passionately devoted to the cause of socialism. I can say the following about each of them separately.

Lev Vasilevich Shubnikov was undoubtedly one of the greatest physicists working in the field of low temperatures not only with us in the Union, but also on a global scale. Many of his papers are classics nowadays. To speak about his sabotage activity in the field of low-temperature physics is absurd, taking into account that he actually was one of the founders of this field in our country. His passionate patriotism is underlined by the fact that he voluntarily switched his work in Holland for work in Russia. The loss suffered by Soviet physics by the untimely death of L.V. Shubnikov can hardly be overestimated.

Vadim Sergeevich Gorsky was the greatest specialist we had in the Union on X-ray structure analysis. It suffices to say that up to the present day, in spite of the fact that twenty years have passed, we do not have in this field of physics someone equal to him.

Lev Viktorovich Rozenkevich did not yet succeed to do much in science, since he changed his scientific specialization—he had engaged in nuclear physics only for a few years before his arrest. However in this short period he was singled out as a very talented worker, and undoubtedly might have played an important role in the development of this important field of physics.

*Hero of socialist labour*
*Academician Landau*

### The following statements are from N.A. Brilliantov[65]

I have known Comrade Shubnikov, Lev Vasilevich, for a long period of time (1924–1936). From 1924 to 1929[66] I worked with Comrade Shubnikov in the same laboratory, where he was for me an older comrade both as regards qualification and as regards official position.

---

[65]On him see *Kristallografija* 49 (2004) 972-973. The statements given here do not have a very genuine ring, sound stilted and very much written for the event. They are not very personal and give little useful information.

[66]This must be wrong, can only be from 1924–1926, as from 1926 Shubnikov was in Leiden.

Upon his return from a long foreign *komandirovka* Comrade Shubnikov took a leading position at UFTI (in the post of laboratory head).

Together with a number of young comrades Shubnikov carried out brilliant experiments on the investigation of basic properties of superconducting metals. This work carried out under the fruitful leadership of Comrade Shubnikov made the then young UFTI into the leading physics institute in the Soviet Union.

The high scientific authority of Comrade Shubnikov and his large organizational abilities enabled him in a short period to unite a large young collective that successfully worked in both scientific and technical fields. Still today, 19 years later, there are at Kharkov many scientists who started their work under the leadership of Comrade Shubnikov.

At the moment of his arrest Comrade Shubnikov was already a mature physicist with great scientific authority.

All activity of Comrade Shubnikov, which I knew well, were to the benefit of people and motherland. The arrest of Comrade Shubnikov was a great loss to Soviet science.

*Nikolaj Alekseevich Brilliantov,* Moscow B-261, Borovskoe Road, no. 55, apartment 494.

*9 July 1956*

I have known Comrade Gorsky, Vadim Sergeevich, very well. We studied together at the Leningrad Polytechnic Institute, lived in the same hostel, worked in the same laboratory in adjacent rooms, first at LFTI, than at UFTI in Kharkov.

Still in high school Comrade Gorsky distinguished himself by his talents and splendid progress. The graduate work of Comrade Gorsky on solid solutions of metals was published and from that time onwards it is cited in all monographs as research that lay the basis for a large section of the science of metals.

At UFTI Comrade Gorsky was one of the most talented and promising young scientists. All his activity was directed at the intensive development of Soviet science and industry.

Comrade Gorsky spent much effort and attention to the training of cadres. In his laboratory there always were many graduate and post-graduate students. It is beyond doubt that the activity of Comrade Gorsky was only directed at the benefit of the people and land of Russia.

*9 July 1956*

I knew Comrade Rozenkevich, Lev Viktorovich, through joint work at UFTI in Kharkov from 1929 to 1936.

In those three (*sic*) years Comrade Rozenkevich completed and published a number of valuable theoretical papers. The work of Comrade Rozenkevich nominated him among the most able young theoretical physicists.

Comrade Rozenkevich spent much attention to the training of new cadres. He read lectures on various parts of theoretical physics for the young scientific staff of UFTI. The scientific and social activity of Comrade Rozenkevich during the period from 1926 to 1936 was directed at the benefit of science and of Russia.

*9 July 1956*

*Brilliantov N.A.*

## Statement by K.D. Sinelnikov

*3 July 1956*

1.    I have known Lev Vasilevich Shubnikov, Vadim Sergeevich Gorsky and Lev Viktorovich Rozenkevich very well, first from LFTI and subsequently through UFTI. Lev Vasilevich Shubnikov worked at LFTI in the laboratory of Professor Obreimov, being his student, in the post of junior scientific collaborator until 1927. In 1927 he obtained from Vesenkha a scientific *komandirovka* to Leiden

(Holland), where he worked to August 1930. After his return he worked the first year (1930–1931) in the laboratory of Professor Obreimov at UFTI, and then became the head of the UFTI cryogenic laboratory.

Vadim Sergeevich Gorsky is also a student of Professor Obreimov. After finishing the Leningrad Polytechnic Institute he started to work at LFTI in the laboratory of Professor Obreimov and together with him moved in 1930 to UFTI, where he was a scientific collaborator in Obreimov's laboratory (laboratory of crystals).

Lev Viktorovich Rozenkevich was a student of Professor Frenkel, a theorist. He worked at LFTI from 1927 to 1930 in the function of scientific collaborator of the theory division. In 1930 he moved to UFTI, worked in the theory division of Academician Landau and was deputy editor of the journal *Phys. Zeit. der Sowjetunion*, which was published in Kharkov from 1930 to 1936 inclusive.

2.      I do not know anything about any counter-revolutionary organisations, neither at LFTI, nor at UFTI, nor about the involvement of Shubnikov, Gorsky and Rozenkevich in any political or counter-revolutionary organization.

3.      Neither Shubnikov, not Gorsky, nor Rozenkevich carried out any harassment of Komsomol members and communists. They neither ever acted against collaborators of the institute who engaged in defence-related or applied topics. In those years many discussions were held at UFTI about the direction of the work, there were many arguments and disagreements on the general direction of the scientific activity of the institute, there were many discussions that the institute works poorly because of the presence of a large number of collaborators who in the course of a number of years did not give any scientific production, on the necessity to raise the theoretical level of the employees, on the necessity to regularly 'purge' the institute of unnecessary ballast etc., but I never heard about any 'harassment'.

4.      The aforementioned comrades, like various other employees of UFTI, indeed organized a 'strike', however not at the institute, but at the university, challenging the dismissal of Academician Landau from the university. Professor Landau was dismissed by the rector of the university because of indignation of the students about Professor Landau's mocking attitude towards them. Landau's dismissal by the rector was unlawful, as a professor can only be dismissed by order of the ministry (*narkomat*). A number of UFTI employees who were teaching at the university issued as a form of protest a joint statement about their own departure. The party organization of the university appealed to UFTI for a public investigation in this matter. At a very stormy meeting of UFTI employees the 'strike' was of course discussed. Speaking at this meeting I analysed the reasons that led to Landau's conflict with the students, drew attention to the inadmissible arrogance of a certain part of the scientific collaborators in relation to the students and also pointed out that this arrogance still comes from the old days, when at Leningrad University there indeed existed a group of talented young people (Landau, Ivanenko and others), calling themselves the 'jazz band' which called all other students 'subs', i.e. standing below them. I did never speak about any counter-revolutionary organization. Landau's arrogance is explained by his narcissism known to all, not by any political reasons.[67] All UFTI employees who took part in the 'strike' admitted that the method of 'collective protest' used by them was inadmissible.5.

---

[67]This will explain why Landau and Sinelnikov were no friends, but it must be said that Sinelnikov has a point here.

I do not know anything about any counter-revolutionary activity of Shubnikov, nor do I believe in it, as Shubnikov always kept himself very far from politics. However, it should be noted that the whole matter of a counter-revolutionary (or espionage) organization at UFTI is very unclear (at least to me), since we indeed had at the institute indisputable spies—Fritz Houtermans and Weissberg, and also the obscure person Fomin, who had arrived from Germany, knew Houtermans from Germany and worked with him at UFTI. Fomin made the impression on me of being an abnormal or very frightened person; for instance, he came to me in the laboratory and almost hysterically said that UFTI can "unexpectedly burn down". However much I inquired, he did not add anything further. Quite disturbed by this statement I told the head of the first section Comrade B.N. Pevny and the secretary of the party organization Comrade Zalivadny about it. My anxiety was strengthened when a few days later Fomin tried to commit suicide by drinking hydrochloric acid and jumping from the third floor. When he regained consciousness, he demanded to see Comrade Pevny, after which they put him up in an NKVD hospital and we did not hear anything further about him.

But now, as then, it appears to me that the activity of Houtermans, Weissberg and possibly Fomin was in no way connected with Shubnikov, Gorsky and Rozenkevich, nor with Komarov (head of OSGO organized by our institute) who was arrested and later killed. Apparently, they are all victims of slander and perhaps careerism of some employees of the institute.[68]

6.　　This point I have answered in 4., and I did not speak about the 'jazz band' as about some political or counter-revolutionary organization, and I did not at all connect Shubnikov with it, but Landau.

*12 August 1956*

This deposition has been taken by me and has been written down by Kirill Dmitrievich Sinelnikov, director of UFTI, USSR Academy of Sciences, professor, member of the Academy of Sciences, honoured scientist of the USSR, residing in Kharkov, Tchaikovsky 6, apt. 1.

Senior Security Officer, first section Ukrainian KGB Kharkov District, Senior Lieutenant

*Khoteev*
*14 August 1956*

# Information overview
## of case on Landau, arch. 621925 (Ref. [2], p. 313)

Compiled in connection with the clarification of Landau's statements in respect of the former UFTI employees Shubnikov, Rozenkevich and Gorsky.

Landau, Lev Davidovich, born in 1908, Jew, citizen of the USSR, non-party, before his arrest senior scientific collaborator of the Institute of Physical Problems of the USSR Academy of Sciences, arrested on 28 April 1938 by the organs of the NKVD in Moscow. Statements (copies) of the arrested persons Shubnikov and Rozenkevich in another case, which showed that Landau was involved in joint counter-revolutionary activity at UFTI, served as the ground for his arrest. In the first interrogations Landau pleaded guilty and declared that in 1935 at UFTI under the flag of the struggle for 'pure science' an anti-Soviet group was formed, which apart from him was joined by Rozenkevich, Shubnikov,

---

[68]It is obvious that he could have told more here.

Obreimov, Korets, and Weissberg. The task of this group was the struggle against Soviet power. In connection with this, so Landau declared in the interrogation of 30.08.38, sabotage was carried out at UFTI, mainly, by bringing 'to naught' the practical significance of theoretical work and by harassing Soviet scientific employees who worked on subjects that were topical for the economy and for defence. As a result of the sabotage activity of Shubnikov, Weissberg, Rozenkevich, and Korets, Landau further explained, the work of the atomic nucleus laboratory was completely cut off from the solution of any problems that have practical significance. Landau further declared that early in 1937 he moved to Moscow and continued his anti-Soviet activity. In particular, together with Korets (who was also in Moscow) he composed an anti-Soviet leaflet, which has been added to the file. According to material in the file it had been written by Korets.

In 1939 Landau retracted his 'confessional' statements. On 28 April 1939 by decision of the organs of the USSR NKVD the file in respect of Landau was closed for the following reasons: that he is a major specialist in the field of theoretical physics and can be useful for Soviet science and that Academician Kapitsa expressed the wish to bail him out.

Landau was freed from custody.

In Landau's statements Gorsky does not appear at all.

To the present file the following were added: copies or extracts of statements of L.V. Rozenkevich and L.V. Shubnikov of 13–14 August 1937, of Moisej Abramovich Korets, born in 1908, native of Sevastopol, lecturer in physics at the Moscow Pedagogical Institute, of 15 June 1938 and of Ju.B. Rumer, born in 1901, native of Moscow, senior scientific employee of the Lebedev Physics Institute of the USSR Academy of Sciences, of 4 August 1938. In this respect Korets declared that in 1935 he was recruited by Shubnikov into the anti-Soviet sabotage group, 'took part' in almost all illegal meetings of participants of the group that took place in the apartments of Shubnikov and Landau, at which questions of unrolling 'sabotage acts' were discussed. Rozenkevich and Gorsky do not appear in the statements of Korets.

Rumer declared that the transfer of the participant Landau of the 'anti-Soviet group' from Kharkov to Moscow was provoked by the arrest of participants of the Kharkov anti-Soviet group—Shubnikov, Rozenkevich and others[69] and by the criticism of Landau at meetings of the institute as an accomplice of these people. In these statements Rumer did not mention Gorsky and in respect of Shubnikov and Rozenkevich (about their membership in the anti-Soviet group) he does not give a source for the information.

Survey compiled on 10.07.1956.                                              *Kabtsov*


## Information overview
## of the case on Ivanenko, arch. 86368 (Ref. [2], p. 314)


Compiled in connection with the verification of the case of the prosecution of Shubnikov, Rozenkevich and Gorsky.

Ivanenko, Dmitry Dmitrievich, born in 1904, son of a non-hereditary nobleman, non-party, citizen of the USSR, former scientific employee of the Leningrad Physico-Technical Institute, arrested on 27 February 1935 by the organs of the NKVD in Leningrad.

It is not clear what the basis for the arrest of Ivanenko was. No accusations were brought against him in the inquiry. He was interrogated mainly about his acquaintances living in the USSR and abroad among whom a number of well-known physicists-specialists. Ivanenko

---

[69]Landau had left Kharkov more than half a year before the arrest of Shubnikov and others. When he left nobody had been arrested yet.

has not given any statements on any counter-revolutionary activity of himself or of other people. On the contrary, he has declared that he considers himself utterly standing "on the platform of Soviet power and honestly working on the development of Soviet physics." There is no other evidence in the file.

On 4 March 1935 by decision of the Special Council of the USSR NKVD Ivanenko was sentenced 'as a socially dangerous element' to 3 years of corrective labour camp.

On 30 December 1935 the said decision of the Special Council of the USSR NKVD was changed and Ivanenko was exiled to Tomsk for the remaining time.

*Kabtsov*
*10.07.56*

# References to the Appendices

[1]  B.I. Verkin et al. (1990).
[2]  Ju.V. Pavlenko, Ju.N. Ranjuk and Ju.A. Khramov, *"Delo" UFTI 1935–1938* (The "UFTI" Case 1935–1938) (Feniks, Kiev, 1998).
[3]  A. Weissberg, *Conspiracy of Silence* (Hamish Hamilton, London, 1952).
[4]  Barry McLoughlin and Josef Vogl, … *Ein Paragraf wird sich finden. Gedenkbuch der österreichischen Stalin-Opfer (bis 1945)* (Dokumentationsarchiv des österreichischen Widerstandes, Vienna, 2013).
[5]  M. Shifman, *Physics in a mad World* (World Scientific, 2016).

© Springer International Publishing AG 2018
L. J. Reinders, *The Life, Science and Times of Lev Vasilevich Shubnikov*,
Springer Biographies, https://doi.org/10.1007/978-3-319-72098-2

# List of Shubnikov's Publications

The papers of 1938 and 1939 do not have Shubnikov as a co-author due to the fact that at that time he was officially an 'enemy of the people'.

Apart from paper no. 51 this list is identical to the one published by B.I. Verkin et al. in 1990. Paper no. 51 was discovered by Jury Ranjuk in Kharkov (Amaldi (2012), p. 147).

### 1924

1. I.W. Obreimow, L.W. Schubnikow, Eine Methode zur Herstellung einkristalliger Metalle, *Z. Physik* 25 (1924) 31–36.

### 1927

2. I.W. Obreimow, L.W. Schubnikow, Über eine optische Methode der Untersuchung von plastischen Deformationen in Steinsalz, *Z. Physik* 41 (1927) 907–919.

### 1930

3. L.V. Shubnikov and W.J. de Haas, A new phenomenon in the change of resistance in a magnetic field of single crystals of bismuth, *Nature* 126 (1930) no. 3179, 500.
4. L. Schubnikov, W.J. de Haas, Magnetische Widerstandsvergrösserung in Einkristallen von Wismut bei tiefen Temperaturen, *Leiden Commun.* 207a (1930) 3–6 [*Proceedings Royal Academy of Sciences* 33 (1930) 130–133].
5. L. Schubnikov, Über die Herstellung von Wismuteinkristallen, *Leiden Commun.* 207b (1930) 9–14 [*Proceedings Royal Academy of Sciences* 33 (1930) 327–331].
6. L. Schubnikov, W.J. de Haas, Die Abhängigkeit des elektrischen Widerstandes von Wismutkristallen von der Reinheit des Metalles, *Leiden Commun.* 207c (1930), 17–31 [*Proceedings Royal Academy of Sciences* 33 (1930) 350–362].
7. L. Schubnikov, W.J. de Haas, Neue Erscheinungen bei der Widerstandsänderung von Wismuteinkristallen im Magnetfeld bei der Temperatur von Flüssigem Wasserstoff I, *Leiden Commun.* 207d (1930) 35–53 [*Proceedings Royal Academy of Sciences* 33 (1930) 363–378].
8. L. Schubnikov, W.J. de Haas, Neue Erscheinungen bei der Widerstandsänderung von Wismuteinkristallen im Magnetfeld bei der Temperatur von Flüssigem Wasserstoff II, *Leiden Commun.* 210a (1930) 3–18 [*Proceedings Royal Academy of Sciences* 33 (1930) 418–432].
9. L. Schubnikov, W.J. de Haas, Die Widerstandsänderung von Wismuteinkristallen im Magnetfeld bei der Temperatur von Flüssigem Stickstoff, *Leiden Commun.* 210b (1930) 21–28 [*Proceedings Royal Academy of Sciences* 33 (1930) 433–439].
10. L. Shubnikov, W.L. de Haas, Izgotovlenie i issledovanie kristallov Vismuta (Preparation and investigation of bismuth crystals), *Zhurn. fiz.-chim. o-va, Chast' fiz.* 62 (1930) 530–537.

© Springer International Publishing AG 2018                                                333
L. J. Reinders, *The Life, Science and Times of Lev Vasilevich Shubnikov*,
Springer Biographies, https://doi.org/10.1007/978-3-319-72098-2

**1934**

11. O.N. Trapeznikova, L.V. Shubnikov, Issledovanie uslovij ravnovesija gazoobraznoj i zhidkoj fazy smesi kisloroda i azota (Investigation of the equilibrium conditions of the gaseous and liquid phases of a mixture of oxygen and nitrogen), *Zh. Tekhn. Fiz.* 4 (1934) 949–953.
12. O.N. Trapeznikova, L.W. Schubnikov, Anomaly in the specific heat of ferrous cloride at the Curie point, *Nature* 134 (1934), no. 3384, 378–379.
13. N.S. Rudenko, L.W. Schubnikov, Die Viskosität von flüssigen Stickstoff, Kohlenoxyd, Argon und Sauerstoff in Abhängigkeit von der Temperatur, *Phys. Z. Sow.* 6 (1934) 470–477.
14. N.S. Rudenko, L.V. Shubnikov, Vjazkost' i zavisimost' eë ot temperatury dlja zhidkogo azota, okisi ugleroda, argona i kisloroda (The viscosity and its dependence on the temperature of liquid nitrogen, carbon monoxide, argon and oxygen), *Zh. Exper. Teor. Fiz.* 4 (1934) 1049–1059.
15. J.N. Rjabinin, L.W. Schubnikov, Über die Abhängigkeit der magnetischen Induktion des supraleitenden Blei vom Feld, *Phys. Z. Sow.* 6 (1934) 557–568.
16. L.W. Schubnikov, W.I. Chotkewitch, Spezifische Wärme von supraleitenden Legierungen, *Phys. Z. Sow.* 6 (1934) 605–607.
17. J.N. Rjabinin, L.W. Schubnikow, Verhalten eines Supraleiters im magnetische Feld, *Phys. Z. Sow.* 5 (1934) 641–643.
18. J.N. Rjabinin and L.W. Shubnikov, Dependence of magnetic induction on the magnetic field in superconducting lead, *Nature* 134 (1934), no. 3382, 286–287.

**1935**

19. Ju.N. Rjabinin, L.V. Shubnikov, Zavisimost' magnitnoj induktsii sverkhprovodjashchego svintsa ot polja (Dependence of the magnetic induction of superconducting lead on the field), *Zh. Exper. Teor. Fiz.* 5 (1935) 140–147.
20. G.N. Rjabinin, L.W. Schubnikov, Magnetic induction in a supra–conducting lead crystal, *Nature* 135 (1935), no. 3403, 109.
21. O.N. Trapeznikova, L.W. Schubnikow, Über die Anomalie der spezifischen Wärme von wasserfreiem Eisenchlorid, *Phys. Z. Sow.* 7 (1935) 66–81.
22. O.N. Trapeznikova, L.V. Shubnikov, Anomalija teploemkosti bezvodnogo khloristogo zheleza (The anomaly of the specific heat of anhydrous iron chloride), *Zh. Exper. Teor. Fiz.* 5 (1935) 281–291.
23. O.N. Trapeznikova, L.W. Schubnikow, Über die Anomalie der spezifischen Wärme von wasserfreiem CrCl₃, *Phys. Z. Sow.* 7 (1935) 255–256.
24. N.S. Rudenko, L.W. Schubnikow, Vizkosität des flüssigen Metans und Athylens in Abhängigkeit von der Temperatur, *Phys. Z. Sow.* 8 (1935) 179–184.
25. N.S. Rudenko, L.W. Shubnikow, Vjazkost' zhidkogo metana i etilena v zavisimosti ot temperatury (Viscosity of liquid methane and ethylene as a function of the temperature), *Zh. Exper. Teor. Fiz.* 5 (1935) 826–829.
26. W.J. de Haas, J.W. Blom, L.V. Shubnikov, Über die Widerstandsänderung von Wismuteinkristallen im Magnetfeld bei tiefen Temperaturen, *Physica* 2 (1935) 907–915.
27. J.N. Rjabinin and L.W. Schubnikow, Magnetic properties and critical currents of superconducting alloys, *Phys. Z. Sow.* 7 (1935) 122–125.
28. J.N. Rjabinin and L.W. Schubnikow, Magnetic properties and critical currents of supra-conducting alloys, *Nature* 135 (1935), no. 3415, 581–582.

**1936**

29. O. Trapeznikova, L. Schubnikow, G. Miljutin, Über die Anomalie der spezifischen Wärmen von wasserfreiem CrCl₃, CoCl₂, NiCl₂, *Phys. Z. Sow.* 9 (1936) 237–253.

30. O.N. Trapeznikova, L.V. Shubnikov, G.A. Miljutin, Anomalija teploemkosti bezvodnykh CrCl₃, CoCl₂, NiCl₂ (Anomalous heat capacity of anhydrous CrCl₃, CoCl₂ and NiCl₂), *Zh. Exper. Teor. Fiz.* 6 (1936) 421–432.

31. V. Fomin, F.G. Houtermans, A.I. Leipunsky, L. W. Schubnikow, Slowing down of neutrons in liquid hydrogen, *Phys. Z. Sow.* 9 (1936) 696–698.

32. L.W. Schubnikow, W.I. Chotkewitsch, J.D. Schepelew, J.N. Rjabinin, Magnetische Eigenschaften supraleitender Metalle und Legierungen, *Phys. Z. Sow.* 10 (1936) 165–169.

33. L.W. Schubnikow, W.I. Chotkewitsch, J.D. Schepelew, J.N. Rjabinin, Magnetische Eigenschaften supraleitender Metalle und Legierungen, *Phys. Z. Sow. Sondernummer, Arbeiten auf dem Gebiete tiefer Temperaturen* (1936) 39–66.

34. L.W. Schubnikow, Das Kältelaboratorium, *Phys. Z. Sow. Sondernummer* (1936) 1–5.

35. O. Trapeznikova, L. Schubnikow, Anomale spezifische Wärmen der wasserfreien Salze FeCl₂, CrCl₃, CoCl₂ und NiCl₂, *Phys. Z. Sow. Sondernummer* (1936) 6–21.

36. N.S. Rudenko, L.W. Schubnikow, Die Viskosität von verflüssigen Gasen, *Phys. Z. Sow. Sondernummer* (1936) 83–90.

37. L.F. Wereschtschagin, L.W. Schubnikow, B.G. Lasarew, Die magnetische Suszeptibilität von metallischen Cer, *Phys. Z. Sow. Sondernummer* (1936) 107–110.

38. O. Trapeznikova, L. Schubnikow, Anomale spezifische Wärmen der wasserfreien Salze FeCl₂, CrCl₃, CoCl₂ NiCl₂ und MnCl₂, *Sonderdruck aus den Berrichten des VII Kaltenkongresses* (1936) 1–14.

39. L. Schubnikow, A.K Kikoin, Optische Untersuchung an flüssigem Helium II, *Phys. Z. Sow.* 10 (1936) 119–120.

40. B.G. Lasarew, L.W. Schubnikow, Uber das magnetische Moment des Protons, *Phys. Z. Sow.* 10 (1936) 117–119.

41. V. Fomin, F.G. Houtermans, I.W. Kurtschatov, A. Leipunsky, L.W. Schubnikow, G. Shchepkin, Über die Absorption thermischer Neutronen in Silber bei niedrigen Temperaturen, *Phys. Z. Sow.* 10 (1936) 103–105.

42. L.W. Schubnikow, W.I. Chotkewitsch, Kritische Werte des Feldes und des Stromes fur die Supraleitfähigkeit des Zinns, *Phys. Z. Sow.* 10 (1936) 131–241.

43. L. Schubnikow, Destruction of superconductivity by electrical current and magnetic field, *Nature* 138 (1936), no. 3491, 545–546.

44. V. Fomin, F.G. Houtermans, I.W. Kurtschatov, A.I. Leipunsky, L.W. Schubnikow and G. Shtshepkin, Absorption of thermal neutrons in silver at low temperatures, *Nature* 138 (1936), no. 3486, 326–327.

45. V. Fomin, F.G. Houtermans, A.I. Leipunsky, L.B. Rusinov and L.W. Schubnikow, Neutron absorption of boron and cadmium at low temperatures, *Nature* 138 (1936), no. 3490, 505.

46. L.V. Shubnikov and V.I. Khotkevich, Kriticheskie znachenija polja i toka dlja sverkhprovodjashchego olova (Critical values of field and current for superconducting tin), *Zh. Tekhn. Fiz.* 6 (1936) 1937–1943.

47. L.F. Wereschtschagin, L.W. Schubnikow, B.G. Lasarew, Über die magnetische Suszeptibilität des metallischen Cer und Praseodym, *Phys. Z. Sow.* 10 (1936) 618–624.

48. A.K. Kikoin and L.W. Schubnikow, Optical experiments on liquid helium II, *Nature* 138 (1936), no. 3493, 641.

49. L.W. Schubnikow and N.E. Alexeyevsky, Transition curve for the destruction of superconductivity by electrical current, *Nature* 138 (1936), no. 3497, 804.

50. L.V. Shubnikov, N.E. Alexeyevsky, Krivaja perekhoda pri razrushenii sverkh-provodimosti elektricheskim tokom (Transition curve at the destruction of superconductivity by an electric current), *Zh. Exper. Teor. Fiz.* 6 (1936) 1200–1201.

51. F. Houtermans, V. Fomin, A. Leipunsky, L. Shubnikov, Production of heavy hydrogen nuclei from protons and neutrons, *Visti AN URSR* 4 (1936) 37.

**1937**

52. L.V. Shubnikov, V.I. Khotkevich, Ju.D. Shepelev, Ju.N. Rjabinin, Magnitnye svojstva sverkhprovodjashchikh metallov i splavov (Magnetic properties of superconducting metals and alloys), *Zh. Exper. Teor. Fiz.* 7 (1937) 221–237.
53. B.G. Lasarew, L.W. Schubnikow, Das magnetische moment des protons, *Phys. Z. Sow.* 11 (1937) 445–457.
54. L. Schubnikow and I. Nakhutin, Electrical conductivity of a superconducting sphere in the intermediate state, *Nature* 139, no. 3518 (1937) 589–590.
55. L.V. Shubnikov, I.E. Nakhutin, Electrical conductivity of a superconducting sphere in the intermediate state, *Zh. Exper. Teor. Fiz. (USSR)* 7 (1937) 566.
56. L.W. Schubnikow, S.S. Schalyt, Eigenschaften einiger paramagnetischer Salze, *Phys. Z. Sow.* 11 (1937) 566–570.

**1938**

57. V.I. Khotkevich, K voprosu o kriticheskikh znachenijakh polja i toka dlja sverkh-provodjashchego olova (On the question of the critical values of field and current for superconducting tin), *Zh. Exper. Teor. Fiz.* 8 (1938) 515–517.
58. S.S. Shalyt, Magnitnie svojstva nekotorykh paramagnetnykh solej (Magnetic proper-ties of some paramagnetic salts), *Zh. Exper. Teor. Fiz.* 8 (1938) 518–530.
59. I.E. Nakhutin, Sverkhprovodimost' v promezhutochnom sostjanii (Superconductivity in the intermediate state), *Zh. Exper. Teor. Fiz.* 8 (1938) 713–716.
60. A.K. Kikoin, Teploprovodnost' tverdogo gelija (Thermal conductivity of solid helium), *Zh. Exper. Teor. Fiz.* 7 (1938) 840–843.
61. N.E. Alekseevsky, Zavisimost' kriticheskogo toka ot vneshnego polja v sverkh-provodjashchikh splavakh (Dependence of the critical current on the external field in superconducting alloys), *Zh. Exper. Teor. Fiz.* 8 (1938) 1098–1103.

**1939**

62. O.N. Trapeznikowa and G.A. Miljutin, Specific heat of methane under pressure, *Nature* 144 (1939) no. 3649, 632.

# Timeline of Shubnikov's Life

- 29 September 1901 birth in St. Petersburg
- September 1918 entered Petrograd University
- January 1919 Laboratory assistant at the State Optical Institute
- Early autumn 1921 sailing trip on Lake Ladoga, arrival in Finland
- 1922–1923 in Berlin in Germany
- Summer 1923 return to Petrograd
- Student at the Petrograd Polytechnic Institute, combined with work at LFTI
- 1924–1926 publication of papers with Obreimov

Development of the Shubnikov–Obreimov method of crystal growth

- November 1926 arrival in Leiden

Preparation of extremely pure bismuth crystals

Discovery of the Shubnikov–de Haas effect

- Summer 1930 return to Leningrad
- Autumn 1931 (after being idle for almost a year) appointed head of the cryogenic lab at UFTI
- 1935 conflict at UFTI about applied and fundamental physics research, resulting in the arrest of Korets
- 1936 conflict at Kharkov University in connection with the threat of Landau's dismissal
- 1936 Discovery of type-II superconductivity
- 26 December 1936 Shubnikov's resignation letter to the university
- 6 August 1937 arrest of Shubnikov
- 10 November 1937 executed by shooting
- 1956 Rehabilitation of Shubnikov

Details about Shubnikov's death become known

- 2001 L.V. Shubnikov Prize for outstanding research in experimental physics established by the Ukrainian Academy of Sciences[70]
- 2001 Shubnikov centenary conference in Kharkov

---

[70]This prize should be distinguished from the A.V. Shubnikov Prize established by the Russian Academy of Sciences in honour of Lev Shubnikov's uncle, the crystallographer Aleksej Shubnikov.

© Springer International Publishing AG 2018
L. J. Reinders, *The Life, Science and Times of Lev Vasilevich Shubnikov*,
Springer Biographies, https://doi.org/10.1007/978-3-319-72098-2

# Index

© Springer International Publishing AG 2018
L. J. Reinders, *The Life, Science and Times of Lev Vasilevich Shubnikov*,
Springer Biographies, https://doi.org/10.1007/978-3-319-72098-2

Printed in the United States
By Bookmasters